CIRCUIT BREAKERS

A TECHNICIAN'S GUIDE TO LOW- AND MEDIUM-VOLTAGE CIRCUIT BREAKERS

atp
AMERICAN TECHNICAL PUBLISHERS
Orland Park, Illinois 60467-5756

James R. White

American Technical Publishers, Inc., Editorial Staff

Editor in Chief:
Jonathan F. Gosse
Vice President—Production:
Peter A. Zurlis
Digital Media Manager:
Carl R. Hansen
Technical Editor:
James T. Gresens
Copy Editor:
Talia J. Lambarki
Cover Design:
Jennifer M. Hines
Illustration/Layout:
Joshua P. Hugo
Nicholas W. Basham
Thomas E. Zabinski
Digital Resources:
Robert E. Stickley
Daniel Kundrat
Cory S. Butler

1 2 3 4 5 6 7 8 9 – 15 – 9 8 7 6 5 4 3 2 1

Printed in the United States of America

ISBN 978-0-8269-1970-0

This book is printed on recycled paper.

ABOUT THE AUTHOR

James R. (Jim) White has been the training director of Shermco Industries, Inc., in Irving, Texas, since 2001. Jim is certified by the National Fire Protection Association (NFPA) as a certified electrical safety compliance professional (CESCP). He is the principal member for Shermco Industries on the NFPA Technical Committee "*Recommended Practice for Electrical Equipment Maintenance*" (NFPA 70B). Jim represents the InterNational Electrical Testing Association (NETA) as an alternate member of the NFPA Technical Committee, *Standard for Electrical Safety in the Workplace*, (NFPA 70E), is NETA's principle representative on the NEC Code Making Panel CMP-13, and represents NETA on the ASTM F18 Committee, *Electrical Protective Equipment For Workers*. Jim is an IEEE senior member and, in 2011, received the IEEE/PCIC Electrical Safety Excellence Award. In 2013, Jim received NETA's Outstanding Achievement Award. Jim is a past chairman (2008) of the IEEE Electrical Safety Workshop and is the author of two books available through American Technical Publishers titled *Significant Changes to NFPA 70E–2012 Edition* and *Electrical Safety: A Practical Guide to OSHA and NFPA 70E*.

Jim was part of the task group that developed the test bank for the NFPA's CESCP. He also developed and taught training programs, such as Basic and Advanced Protective Relay Maintenance and Testing, Low- and Medium-Voltage Circuit Breaker Maintenance and Testing, Transformer Testing and Maintenance, Electrical Safety for Qualified Persons, Electrical Safety for Managers, Electrical Safety for Unqualified Persons, and Understanding the NFPA 70E.

Jim presents 8 to 10 papers each year at technical and engineering conferences, such as IEEE Electrical Safety Workshop, IEEE Petroleum & Chemical Industry Committee (PCIC), IEEE Pulp & Paper Industry Committee (PPIC), NETA PowerTest, Association for Iron & Steel Technologies, and OSHA's VPPPA conferences. He co-teaches the Electrical Safety-Related Work Practices training program at the IBEW/NJATC National Training Institute each year. Jim's articles have appeared in magazines, including *EC&M*, *Electricity Today*, *Occupational Safety & Health*, *Windpower Magazine*, *NETAWorld Journal*, and *IEEE IAS Transactions*.

ACKNOWLEDGMENTS

The author and publisher would like to thank the following companies, organizations, and individuals for providing images.

ABB Inc.
AEMC® Instruments
Circuit Breaker Analyzer Inc./Group CBS
Cooper Bussman
Cutler-Hammer
DuPont Company
EA Technology Ltd.
Fluke Corporation
General Electric Company
Group CBS
Megger Group Limited
Shermco Industries
Vacuum Interrupters, Inc./Group CBS
Vanguard Instruments Company, Inc.

Technical Review provided by:

David McDonald
Automation Systems Estimator
Integrity Integration Resources

James H. Miller
National Breaker Director
Shermco Industries

CONTENTS

DIGITAL RESOURCES

- Quick Quizzes®
- Illustrated Glossary
- Flash Cards
- Case Studies
- Chapter Reviews
- Media Library
- ATPeResources.com

INTRODUCTION

Circuit Breakers: A Technician's Guide to Low- and Medium-Voltage Circuit Breakers is a new text/workbook that covers low- and medium-voltage circuit breakers used in commercial, industrial, and utility applications and provides comprehensive information on circuit breaker construction, operation, troubleshooting, and maintenance. This well-illustrated book begins with an introduction to circuit breaker construction and maintenance and then presents information on the insertion and removal (racking) of circuit breakers in/from their cubicles and other tasks required by technicians who work with these devices. Molded-case, insulated-case, vacuum, gas, and air-blast circuit breaker designs are covered as well as safety and protection from arc-flash hazards.

Each type of circuit breaker is introduced with descriptions, operational procedures, common applications, component examples, and operating characteristics. The most commonly used circuit breakers are depicted throughout the book.

Chapter objectives at the beginning of each chapter provide learning goals related to the topics introduced. Case studies present circuit breaker applications and related cases and events from the field. Detailed illustrations depict components, devices, systems, and operations. Chapter summaries at the end of each chapter provide a brief synopsis of the material covered. Chapter reviews at the end of each chapter provide a variety of assessment opportunities.

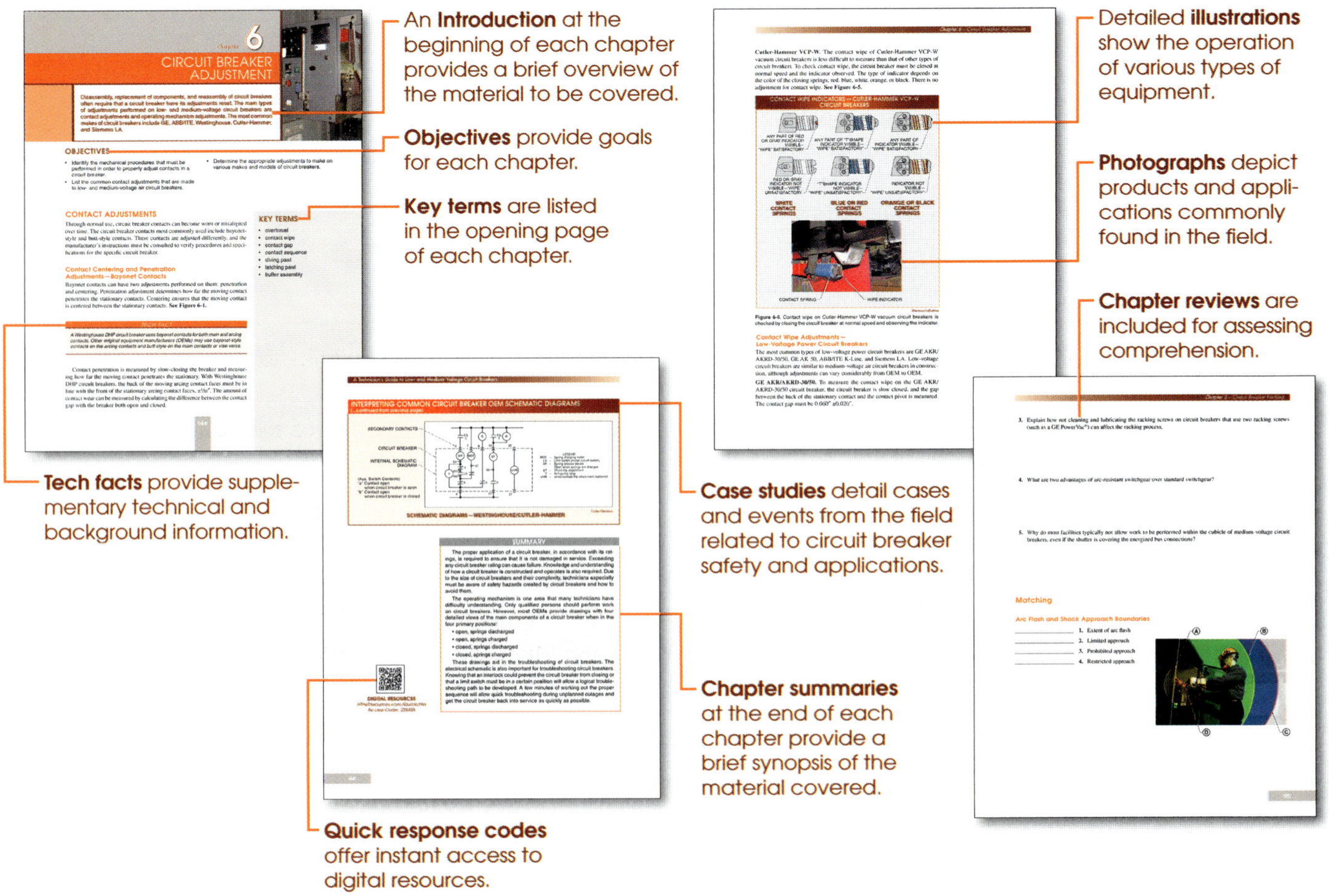

DIGITAL RESOURCES

Circuit Breakers: A Technician's Guide to Low- and Medium-Voltage Circuit Breakers also includes access to digital resources that reinforce and enhance learning. These resources can be accessed by either:

- keying ATPeResources.com/QuickLinks into a webbrowser and entering the QuickLink™ code: **223436**
- using a Quick Response (QR) reader app to scan the QR Code on a mobile device

The Digital Resources include the following:

- **Quick Quizzes®** that provide interactive questions for each chapter with embedded links to highlighted content within the textbook and to the Illustrated Glossary
- **Illustrated Glossary** that provides a helpful reference to commonly used terms. Selected terms are linked to interactive illustrations or media clips
- **Flash Cards** that provide a self-study/review tool for definitions and/or images found in the textbook
- **Case Studies** that consist of examples of actual field situations and cause, effect, and corrective actions
- **Chapter Reviews** that correlate to chapter objectives and serve as a knowledge check of chapter material
- **Media Library** that consists of animated illustrations that reinforce and expand upon textbook content
- **ATPeResources.com**, which provides access to online reference materials for continued learning

PREFACE

Some of the most critical safety-related pieces of equipment in any electrical power system are the overcurrent protective devices (OCPDs). Unfortunately, these tend to be some of the least-maintained or least-considered devices. Many companies and maintenance supervisors tend to think of them as being maintenance-free, much like a light switch. Flip it, and the lights come on; flip it in the opposite direction, and the lights go out. What could be easier? The real issue is that these devices are electromechanical, which means they are in an electrical circuit and performing an electrical task, but they are also primarily mechanical devices.

Circuit breakers require regular maintenance, testing, and inspections in order to provide any type of electrical power system protection. Some circuit breakers and other OCPDs will sit in service for 5, 10, even 20 years and never have true maintenance performed on them. Someone will blow out the dirt, wipe them down, and maybe do an overpotential test on them but give absolutely no thought that these devices are lubricated by the manufacturer when built and require a change in lubrication while they are in service. Electromechanical devices require mechanical maintenance, plus electrical testing. Circuit breakers are relatively simple devices, but they affect every part of an electrical system when they malfunction.

As with all other knowledge and skills learned in life, there are a number of people who spent considerable time and resources to teach me about electrical power systems, circuit breakers, transformers, protective relays, etc. When I first began in utility work, I worked with a group of "old-school" electrical types. I had been in construction and did some facilities work up to that point, but I did not have the depth of knowledge I needed to stay alive working with medium- and high-voltage systems. Lloyd Thompson and his brother, Frank, were two who taught me a lot. I'm sure Frank is long-retired. Lloyd was killed a couple of years after I left the power plant in an electrical accident on a network protector. Phil Welty was the electrical lead, and Andy Klassett was the foreman. Both of them were good guys who knew their electrical work well.

I would be remiss if I didn't mention Tom Reichle, one of the few ex-Army nuclear technicians who ever existed. Tom and I worked at the old Multi-Amp Institute, and he was one of the most knowledgeable circuit breaker instructors I have ever met, then and now. If he didn't know something off the top of his head, he'd have a piece of paper squirreled away somewhere that had the answer on it. Jim Miller at the Shermco Industries Circuit Breaker Shop is another resource as well as Kevin Carter and Shaun Nelmes. Kevin is the resident ITE/ABB low-voltage power circuit breaker guru, while Shaun mostly handles medium-voltage circuit breakers. When we conduct our circuit breaker classes, they always have time to talk to the students, sharing their real-time tips and helpful hints and passing on what they have gained through the years.

Last, I'd like to recognize my wife of 38 years, Vickie. She has allowed me the freedom to take on projects like this one. Without her support, this book may never have been written. She has also blessed me with two great sons, Jason and Justin, both of whom now work for Shermco's engineering field services division.

James R White

chapter 1

CIRCUIT BREAKER CONSTRUCTION AND OPERATION

While circuit breakers used in residential panelboards are small enough to be held in the hand, circuit breakers used in commercial and industrial locations and with utilities are large and can weigh thousands of pounds. Circuit breakers used in commercial, industrial, and utility applications are rated in much the same way as the smaller panelboard circuit breakers but may have additional ratings, such as megavolt ampere or short-time. Circuit breakers are classified by their operating voltage (low or medium), construction (molded case or insulated case), or method of arc interruption (air, air-magnetic, oil, vacuum, etc.). When performing circuit breaker installation, maintenance, and troubleshooting, an understanding of specific manufacturer instructions is required.

OBJECTIVES

- List and define the types of circuit breaker ratings.
- Identify low- and medium-voltage circuit breaker classifications.
- Identify and describe the primary and secondary components of all circuit breaker types.
- Define the purpose of the different types of circuit breaker contact assemblies.
- State the differences in arc chute construction between low- and medium-voltage air circuit breakers.
- List operating mechanism positions.
- List the components of a vacuum interrupter.
- Identify the components in a typical circuit breaker schematic diagram.

KEY TERMS

- circuit breaker
- overcurrent protective device (OCPD)
- nameplate
- medium-voltage power circuit breaker
- molded-case circuit breaker (MCCB)
- insulated-case circuit breaker (ICCB)
- arc
- arc chute
- thermal-magnetic overcurrent protective device
- arcing contact
- backboard assembly
- undervoltage overcurrent protective device
- interphase barrier
- schematic diagram

CIRCUIT BREAKER RATINGS

A *circuit breaker* is an electrical device that opens and closes a circuit by nonautomatic means and automatically opens a circuit when a predetermined current overload is reached, without damage to itself. Circuit breakers provide protection to circuits, and their function is similar to fuses.

A *fuse* is an electrical overcurrent protective device with a fusible portion that is heated and broken by the passage of excessive current. The difference between circuit breakers and fuses is that circuit breakers can be reset and used repeatedly while fuses must be replaced once they have been opened (blown) from an overcurrent condition. An *overcurrent protective device (OCPD)* is a circuit breaker, fuse, or switch with an element that disconnects or discontinues current flow when the amount of current exceeds the design load. An OCPD is also known as a trip device.

Circuit breaker original equipment manufacturers (OEMs) design and build products with various ratings that must not be exceeded. Exceeding these ratings can cause catastrophic failure with resulting damage to nearby equipment as well as injury or death.

Circuit breaker ratings are determined by the National Electrical Manufacturer's Association (NEMA) and the Institute of Electrical and Electronics Engineers. NEMA is a trade organization that follows ANSI guidelines for the development of its standards. Usually, all members that develop and vote on a NEMA standard are OEMs. The *Institute of Electrical and Electronics Engineers (IEEE)* is a professional association that is dedicated to advancing technological innovation and excellence of electrical and electronic equipment.

As defined in IEEE C37.100, a circuit breaker is *"A mechanical switching device, capable of making, carrying, and breaking currents under normal circuit conditions and also, making and carrying for a specified time and breaking currents under specified abnormal circuit conditions such as those of short circuit."*

According to Article 100 of the NEC®, a circuit breaker is *"A device designed to open and close a circuit by nonautomatic means, and to open the circuit automatically on a predetermined overcurrent without damage to itself when properly applied within its rating."* IEEE Standard C37.16 covers circuit breaker ratings, application, and operating conditions for low-voltage power circuit breakers. Circuit breaker ratings, application, and operating conditions for high-voltage power circuit breakers are contained in IEEE Standard C37.04.

It should be noted that the NEC® definition states that a circuit breaker will open a circuit *"without damage to itself when properly applied within its rating."* Consequently, a circuit breaker could fail if it is applied outside of its rating. Therefore, it is important to ensure that circuit breakers are properly applied within their ratings as specified on the nameplate located on the circuit breaker. A *nameplate* is a metal plate attached to a device, such as a circuit breaker, that lists its technical specifications. **See Figure 1-1.** The circuit breaker ratings most commonly referred to are voltage, frequency, frame, continuous current, short-time current, short-circuit current, fuse, control voltage, and megavolt ampere (MVA) ratings.

CIRCUIT BREAKER NAMEPLATES

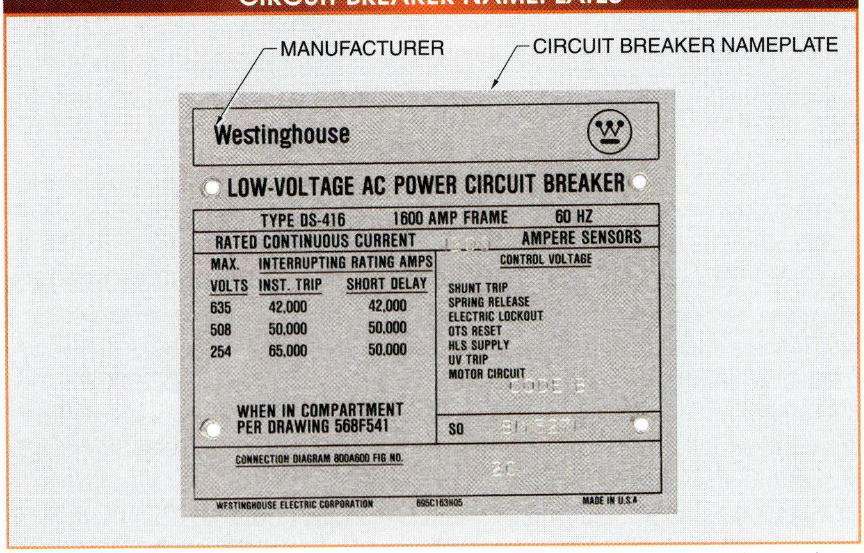

Shermco Industries

Figure 1-1. It is important to ensure that circuit breakers are properly applied within their ratings as specified on the circuit breaker nameplate.

Voltage Rating

According to IEEE Standard C37.13, *"The rated maximum voltage of a circuit breaker is the highest rms voltage, three-phase or single-phase, at which it is designed to perform. The circuit breaker shall be rated at one or more of the following maximum voltages: 635 V, 508 V, or 254 V. For fused circuit breakers, the 635 V rated maximum voltage becomes 600 V to match the fuse rating."* The voltage rating of a circuit breaker is the maximum continuous voltage that can be applied to the circuit breaker without causing a dielectric failure of its insulation. Often a circuit breaker has more than one voltage listed on its nameplate. The actual voltage rating of the circuit breaker is the highest voltage listed on the nameplate.

Frequency Rating

According to IEEE Standard C37.13, *"The rated frequency of a circuit breaker is the frequency at which it is designed to perform. The standard frequency is 60 Hz. Application at other frequencies should receive special consideration".* Circuit breakers are typically rated at 50 Hz or 60 Hz and are sometimes rated for use in direct current (DC) systems. One special consideration for using circuit breakers at other frequencies is that the continuous current rating of 60 Hz circuit breakers applied to DC systems is often decreased, sometimes by as much as 20%. This is due to DC being more difficult to interrupt than alternating current (AC) during short-circuit conditions.

> **TECH TIP**
>
> *Maximum DC Voltage ratings are often less than the AC ratings. Circuit breaker poles in series are usually required in order to obtain the published DC ratings.*

Frame Rating

The *frame rating* is the continuous current rating of all current-carrying parts of a low-voltage circuit breaker, excluding its OCPD. This includes the primary disconnects or stabs, contact assemblies, and any other parts that carry current. The OCPD, also known as a trip unit, is not included because it may have a rating lower than the frame rating of the circuit breaker. When the rating of the OCPD is less than the frame rating, it becomes the continuous current rating of the circuit breaker.

For example, many electrical drawings show low-voltage circuit breakers rated as "1600 A Frame" and "800 A Trip." This type of rating indicates that the current-carrying components of a circuit breaker can carry up to 1600 A of current continuously but that the OCPD will operate whenever the current through it exceeds 800 A.

Continuous Current Rating

According to IEEE Standard C37.13, *"The rated continuous current of a circuit breaker is the designated limit of rms current at rated frequency that it shall be required to carry continuously without exceeding the temperature limitations designated in Section 7. The preferred continuous current ratings of the various frame sizes are listed in ANSI C37.16. The rated continuous current of a circuit breaker equipped with direct-acting trip devices or fuses of a lower rating than the frame size of the circuit breaker is determined by the rating of those devices."*

Another way to define continuous current is that it is the amount of current a device can carry without exceeding its rated temperature rise over a set period of time. As current flows through a conductor, it creates heat. This is often referred to as copper loss (I^2R).

As current flow through a conductor increases, the heat generated by copper loss also increases. At some point, the heat created by current flow can be enough to cause thermal damage to the surrounding insulation.

Most electrical devices have a rated temperature rise of 40°C (104°F) that is based on the thermal characteristics of the device insulation. The methodology used in evaluating the electrical insulation thermal characteristics can be referenced in IEEE Standard 1, Section 7, *IEEE Recommended Practice—General Principles for Temperature Limits in the Rating of Electrical Equipment and for the Evaluation of Electrical Insulation.*

The *rated temperature rise* is the temperature rise of an electrical device or component above ambient, or surrounding air, temperature. For example, if the temperature of a device is 65°C (149°F) and the ambient temperature is 30°C (86°F), the temperature rise would be 35°C. Using this basis excludes the effects of ambient temperature and focuses on the actual temperature of the device or component.

The second sentence in this definition refers to the "preferred continuous current ratings". IEEE Standard C37.16, *Low-Voltage Power Circuit Breakers and AC Power Circuit Protectors — Preferred Ratings, Related Requirements, and Application Recommendations,* provides recommended combinations of circuit breaker frame size, OCPD ratings, and voltage ratings.

The third sentence of the standard is true for all circuit breakers, not just direct-acting OCPDs or fuses. *"The rated continuous current of a circuit breaker equipped with direct-acting trip devices or fuses of a lower rating than the frame size of the circuit breaker is determined by the rating of those devices."* A direct-acting OCPD is also referred to by the IEEE as an electromechanical or series OCPD.

Most series OCPDs were of an oil dashpot design, although Westinghouse manufactured an air dashpot device. These direct-acting OCPDs were connected in series with the load current. When they operated, they physically caused the circuit breaker to open by operating the trip bar. As with circuit breakers, series OCPD ratings can be identified by their nameplates. **See Figure 1-2.**

OVERCURRENT PROTECTIVE DEVICES (OCPDs)

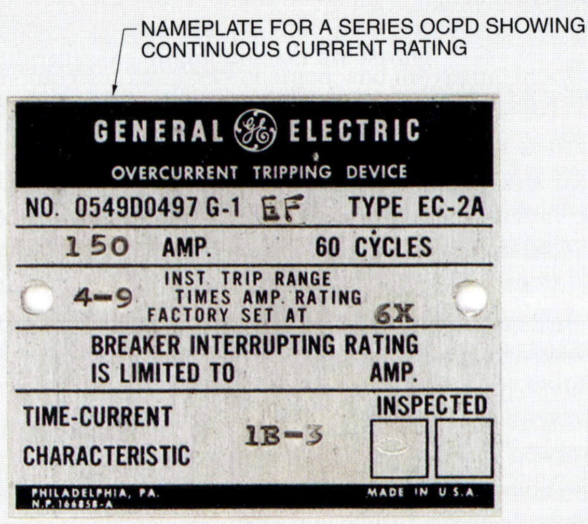

Shermco Industries

Figure 1-2. An OCPD (trip device) can be identified by its nameplate.

The actual continuous current rating of a low-voltage circuit breaker is always determined by the rating of its OCPD. The continuous current rating cannot exceed the frame rating of the circuit breaker but can be equal to it for a low-voltage power circuit breaker. For example, a Westinghouse/Cutler-Hammer DS circuit breaker may have a 2000 A frame rating while the OCPD may only have a rating of 1200 A. A molded-case circuit breaker can have a continuous current rating that is only 80% of its frame rating. Molded-case circuit breakers are typically smaller than other types of circuit breakers and have enclosed cases. This construction limits the amount of airflow through them.

The DS-style circuit breaker used in this example has a solid-state OCPD and uses current sensors to set the continuous current rating. This allows one circuit breaker to have several continuous current ratings by setting the taps on the current sensor to the appropriate value. Current sensors can be single-tap or multiple tap, depending on how the circuit breaker is ordered.

Note: An ITE K1600 circuit breaker with a 1600 A frame rating could have either an electronic OCPD or a series OCPD (OD-3, OD-4, or OD-5). With a series OCPD, the continuous current rating would be determined by the coil rating of the OCPD. For the example 1600 A frame ITE circuit breaker, ITE listed coil ratings from 300 A to 1600 A. Unlike the electronic OCPD, which may have several continuous current ratings by using a multitap sensor, the series OCPD would only have the one rating based on its coil rating.

IEEE Standard C37.06.5.3.1 provides the following conditions on which continuous current ratings are based:

- Circuit breakers are used under usual service conditions.
- The total temperature limits of the materials used for circuit breaker parts shall be considered. A temperature rise reference is given to permit testing at reduced ambient temperature.
- Circuit breakers designed for installation in enclosures shall meet these ratings when in their proper enclosure and based on 40°C ambient temperature outside the enclosure.
- Outdoor circuit breakers and indoor circuit breakers without enclosures shall have ratings based on 40°C ambient temperature.

Most series-type OCPDs have a range of minimum pickup settings from 80% to 160% of the continuous current rating. Newer solid-state and digital OCPDs can have minimum pickup settings that range from 50% to 120%. Settings above 100% of the continuous current rating do not increase the amount of current the device can safely carry but are used to compensate for transient overcurrent conditions. For example, an electric motor that has an extra-long startup time could cause a circuit breaker set at 100% to false trip. By increasing the minimum pickup to 140%, the motor would be able to start as designed. The primary problem with this easy fix is that the OCPD acts as if it has a higher continuous current rating than it actually does.

SAFETY TIP

Never set series-type OCPDs above 100% of their maximum continuous current rating without having an engineer review effects on the electrical power systems.

Short-Time Current Ratings

The short-time current rating of a circuit breaker is the maximum current it can safely carry for two 0.5 sec periods, with a 15 sec rest period between them at zero current. According to IEEE Standard C37.13, *"For an unfused circuit breaker, the rated short-time current is the designated limit of available (prospective) current at which it shall be required to perform its short-time current duty cycle (two periods of ½ s current flow, separated by a 15 s interval of zero current) at rated maximum voltage under the prescribed test conditions. This current is expressed as the rms symmetrical value of current measured from the available current wave envelope at a time ½ cycle after short-circuits initiation."*

Note: Current-limiting fuses do not have a rated short-time current. Only the circuit breaker element of the fused circuit breaker assembly can have such a rating.

Shermco Industries

Electrical equipment is provided with overcurrent protection to protect it from high currents caused by overloads and short circuits.

Short-Circuit Current Ratings

According to IEEE Standard C37, "*The rated short-circuit current of an unfused circuit breaker is the designated limit of available (prospective) current at which it shall be required to perform its short-circuit current duty cycle (0 – 15 s – CO) at rated maximum voltage under the prescribed test conditions. This current is expressed as the rms symmetrical value of current measured from the available current wave envelope at a time ½ cycle after short-circuit initiation. Unfused circuit breakers shall be capable of performing the short-circuit current duty cycle with all degrees of current asymmetry produced by three-phase or single-phase circuits having a short-circuit power factor of 15% or greater (X/R ratio of 6.6 or less).*"

The *short-circuit current rating,* also known as the interrupting rating, is the maximum short-circuit current a circuit breaker can safely clear without external damage. External damage is specified since there will probably be some internal damage, such as erosion of the arcing contacts, carbon and other arcing by-products scattered inside the arc chutes, and thermal damage to the insulation system, due to the extremely high temperatures of the arc interruption process. However, there will be no external damage, and the circuit breaker should still be serviceable. The circuit breaker should be inspected after an arc interruption.

According to IEEE Standard C37.13, *"Unfused circuit breakers with direct-acting instantaneous phase trip elements shall have short-circuit current ratings at each rated maximum voltage."* It is common for OEMs to state the short-circuit current ratings of their OCPDs at standard system voltage levels. **See Figure 1-3.** As the system voltage increases and the arc becomes more persistent and difficult to extinguish, the rating often has to be decreased.

SHORT-CIRCUIT CURRENT (INTERRUPTING) RATINGS

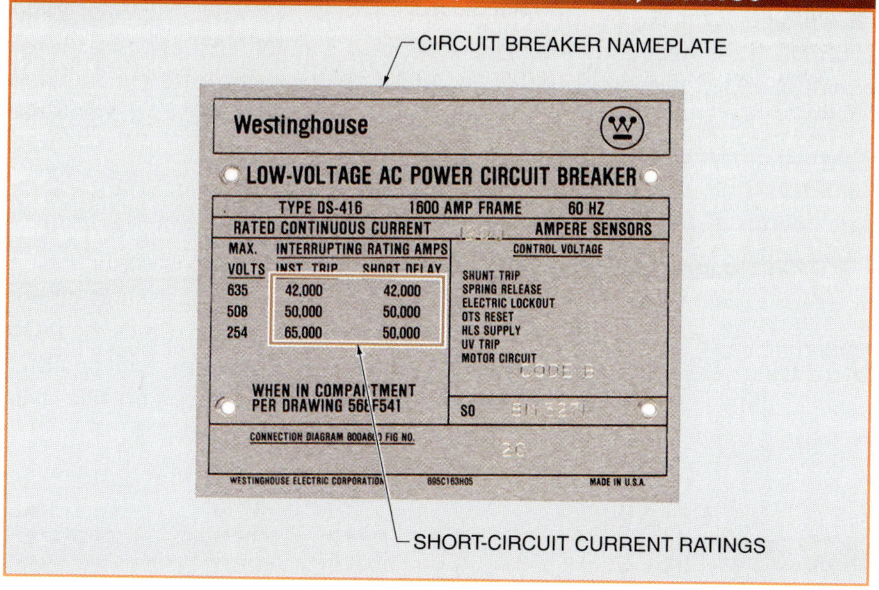

Shermco Industries

Figure 1-3. The short-circuit current rating is the maximum short-circuit current a circuit breaker can safely clear without being damaged externally. It is usually listed on the circuit breaker nameplate.

Fuse Ratings

Certain types of circuit breakers are protected by current-limiting fuses. A circuit breaker that uses current-limiting fuses will not have an interrupting rating but will use the interrupting rating of the fuse. By standard, current-limiting fuses must operate in half a cycle or less, although most fuse OEMs state that their fuses typically operate in a quarter of a cycle.

According to IEEE Standard C37.13, *"The rated short-circuit current of a fused circuit breaker is the designated limit of available (prospective) current at which it shall be required to perform its short-circuit current duty cycle at rated maximum voltage under the prescribed test conditions. The short-circuit current duty cycle consists of an O (OPEN) followed by a CO (CLOSE – OPEN) operation. The time between the O and CO operations is the time necessary to replace fuses and to reset the open-fuse trip device. This current is expressed as the rms symmetrical value of current measured from the available current wave envelope at a time ½ cycle after short-circuit initiation. Fused circuit breakers shall be capable of performing the short-circuit current duty cycle with all degrees of current asymmetry produced by three-phase or single-phase circuits having a short-circuit power factor of 20% or greater (X/R ratio of 4.9 or less)."*

Control Voltage Ratings

According to IEEE Standard C37.13, *"The rated control voltage is the voltage at which the mechanism of the circuit breaker is designed to operate when measured at the control power terminals of the operating mechanism with the highest operating current flowing. Rated control voltages and their ranges for low-voltage power circuit breakers are listed in Table 23 of ANSI C37.16."*

Coils, latches, motors, and other electrical components used to control a circuit breaker use lower voltage for the control circuit operation, most often DC voltage. Typical control voltages are 24 VDC, 48 VDC, 125 VDC, and 250 VDC. Less frequently, 120 VAC, 240 VAC, and 460 VAC are used. Some circuit breakers may have both AC and DC voltages of different nominal ratings. Control power will be brought into the circuit breaker through the secondary disconnects.

Megavolt Ampere Ratings

The short-circuit capability of oil, gas, and some older models of air circuit breakers is rated in megavolt amperes rather than or in addition to short-circuit current. A *megavolt ampere (MVA) rating* is the rated voltage times the rated interrupting current of a circuit breaker. Common examples are circuit breakers rated 250 MVA, 500 MVA, and 1500 A. MVA ratings for circuit breakers can be located in OEM manuals or on the nameplates. **See Figure 1-4.**

MEGAVOLT AMPERE (MVA) RATINGS

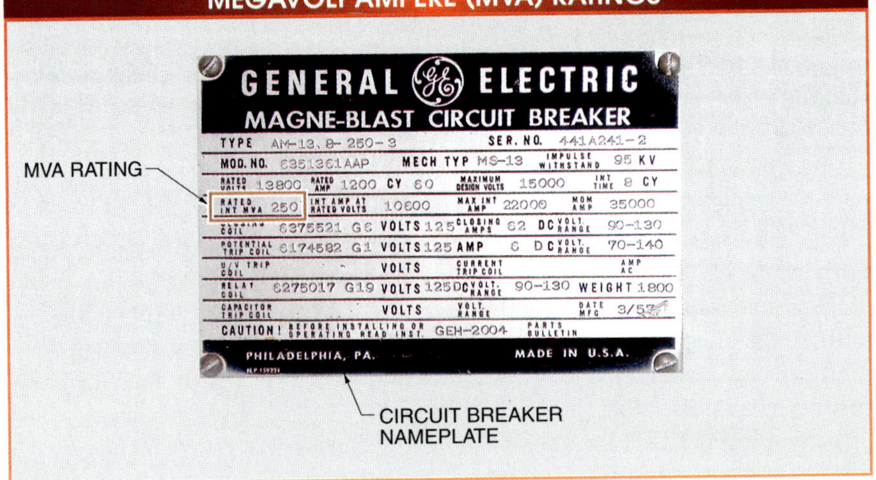

Shermco Industries

Figure 1-4. The short-circuit capability of some types of circuit breakers are rated in megavolt amperes rather than in short-circuit current.

CIRCUIT BREAKER CLASSIFICATIONS

Low-voltage circuit breakers are typically used in applications such as lighting circuits, motor control centers, feeder circuits and any application in which electrical power needs to be turned on and off and overcurrent protection is required. Medium-voltage circuit breakers are used in many of the same type of applications as low-voltage circuit breakers but rely on protective relays rather than built-in OCPDs to provide a trip signal.

Low-Voltage Circuit Breakers

There are many methods of identifying circuit breakers. Low-voltage circuit breakers are typically classified by their configuration. Types of low-voltage circuit breakers include molded-case circuit breakers, insulated-case circuit breakers, and low-voltage power circuit breakers. Common voltage levels for low-voltage circuit breakers include 120/208 V and 277/480 V, and the voltage can range up to 635 V.

Molded-Case Circuit Breakers. With molded-case circuit breakers (MCCBs), the case supports the components and is also part of the insulating structure. MCCBs typically have thermal-magnetic OCPDs built into their cases. According to IEEE standard C37.100, *"A molded case circuit breaker is one that is assembled as an integral unit in a supporting and enclosing housing of insulating material."* The most common types of MCCBs include sealed-case, interchangeable trip, and molded-case switch circuit breakers.

A sealed-case circuit breaker is a type of MCCB that cannot be opened for maintenance or renewal because its UL listing will be nullified. MCCBs that have tar over the case screws, riveted covers, or paper seals either on the joint between the two case sections or over the case screws are considered sealed-case MCCBs. **See Figure 1-5.**

SEALED-CASE CIRCUIT BREAKERS

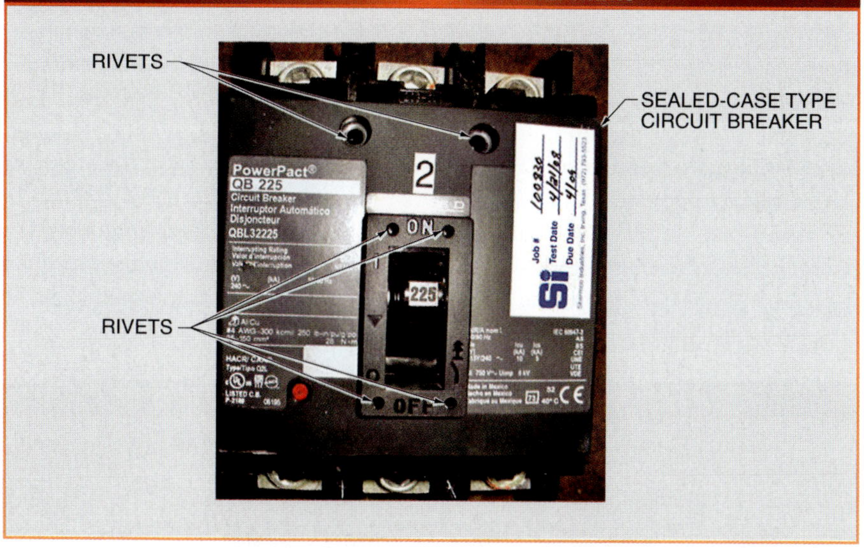

Shermco Industries

Figure 1-5. MCCBs that have riveted covers are considered sealed-case type.

With interchangeable trip circuit breakers, the overcurrent unit can be changed as needed. The OCPD rating cannot exceed 80% of the circuit breaker frame rating. Changes to the continuous current rating of electronic interchangeable trip units are made by changing the rating plug. If the interchangeable trip unit is electromechanical, the entire trip unit has to be changed.

Molded-case switches are externally identical to thermal-magnetic MCCBs but do not have built-in OCPDs. The only current rating is the frame rating. They are often used with motor control centers in which protection is provided by some other device.

Insulated-Case Circuit Breakers. Insulated-case circuit breakers (ICCBs) are similar in appearance to MCCBs but have higher short-circuit current ratings and usually have electronic OCPDs rather than thermal-magnetic trip units. ICCBs can have a bolt-in or draw-out design and are often used in place of low-voltage power circuit breakers. ICCBs typically offer high short-circuit current ratings, have electronic OCPDs, and cost less than low-voltage power circuit breakers of comparable ratings. **See Figure 1-6.**

INSULATED-CASE CIRCUIT BREAKERS (ICCBs)

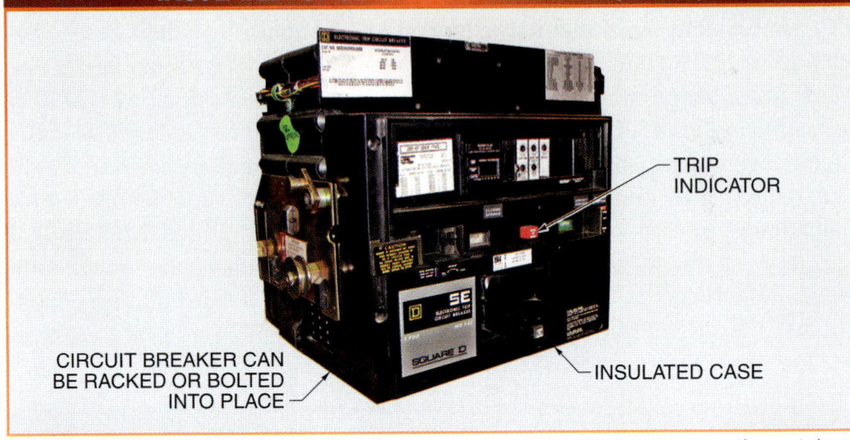

Shermco Industries

Figure 1-6. ICCBs are similar in appearance to MCCBs but have extremely higher short-circuit current ratings and usually have electronic OCPDs rather than thermal-magnetic trip units.

Low-Voltage Power Circuit Breakers. Low-voltage power circuit breakers (LVPCBs) are defined by IEEE C37.100 as "*for use on circuits rated 1000 volts AC and below, or 3000 volts DC and below, but not including molded-case circuit breakers.*" An LVPCB is also referred to as an air-frame breaker because the frame of the circuit breaker is only used to support the components and is not part of the insulation system. **See Figure 1-7.**

LOW-VOLTAGE POWER CIRCUIT BREAKERS (LVPCBs)

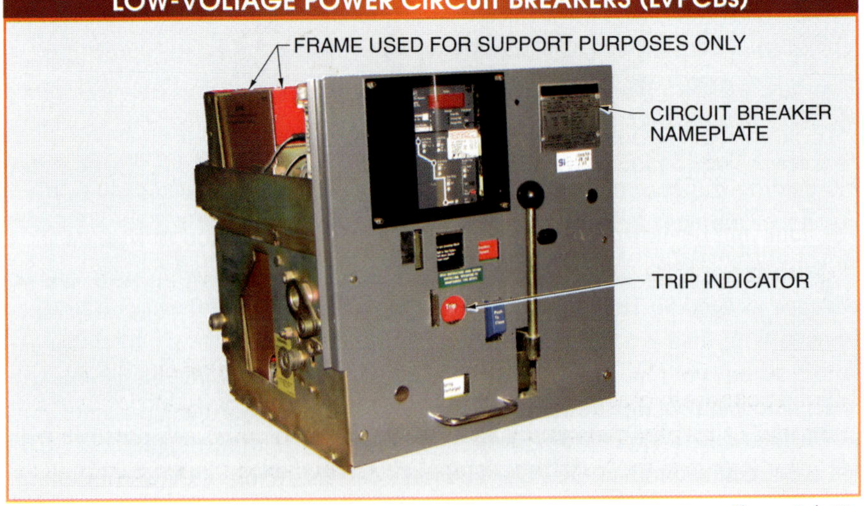

Shermco Industries

Figure 1-7. LVPCBs are used on circuits rated 1000 VAC (3000 VDC) and below.

Medium-Voltage Power Circuit Breakers

A *medium-voltage power circuit breaker (MVPCB)* is a circuit breaker that operates with voltages ranging from 2300 V to 69,000 V. The most common method of classifying these circuit breakers is by using their arc-quenching medium. MVPCBs are classified as air-magnetic, vacuum, oil, air-blast, and gas circuit breakers. **See Figure 1-8.** *Note:* Most metal-clad circuit breakers, which are a type of medium-voltage circuit breaker, operate at 4160 V and 13,200 V/13,800 V, although vacuum circuit breakers can have voltage ratings up to 38,000 V.

MEDIUM-VOLTAGE POWER CIRCUIT BREAKERS (MVPCBs)

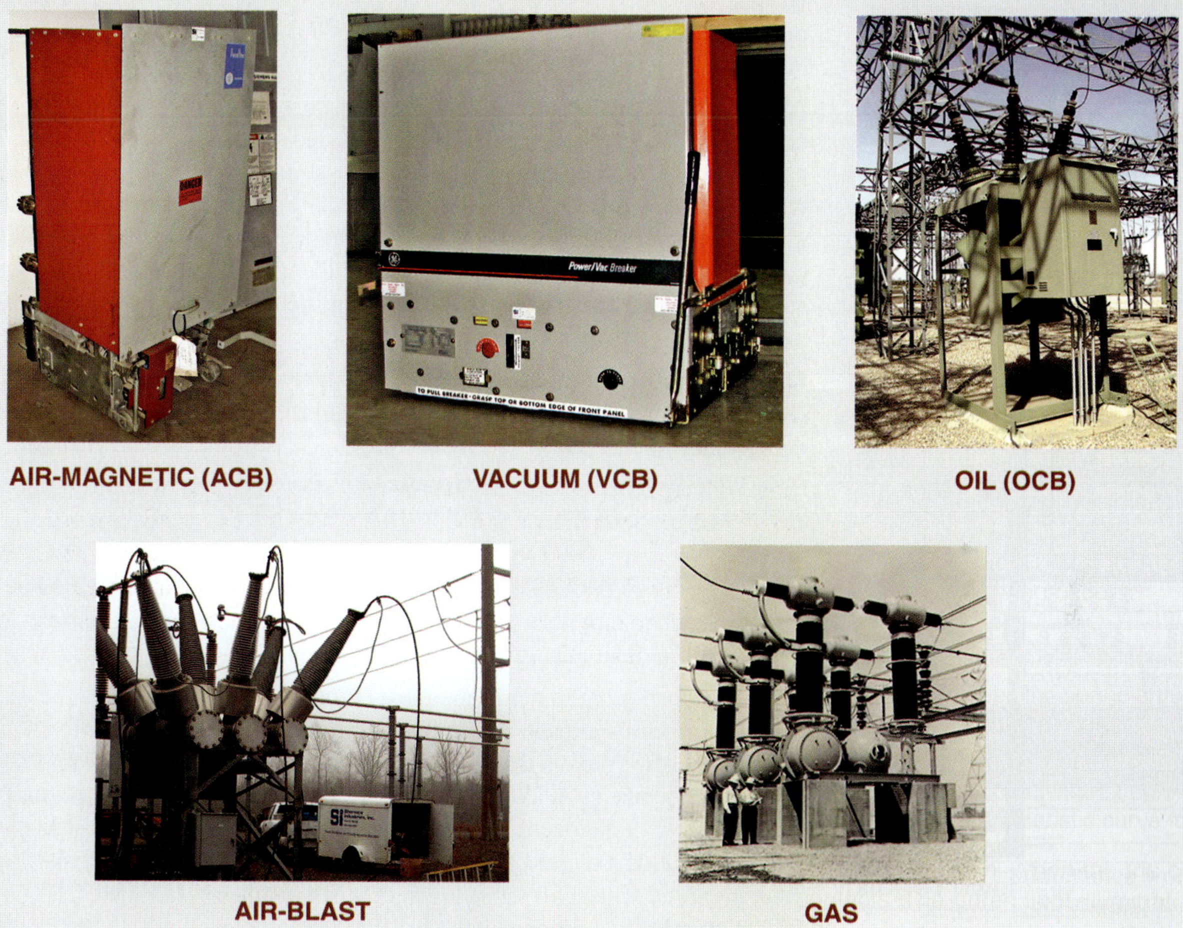

Shermco Industries

Figure 1-8. MVPCBs are classified as air-magnetic, vacuum, oil, air-blast, and gas circuit breakers.

Air-Magnetic Circuit Breakers. Air-magnetic circuit breakers (ACBs) are the most common type of MVPCB. With ACBs, arc interruption takes place in air and is usually assisted by arc chutes, blowout coils, and puffer tubes. The practical limit for ACBs using arc chutes (metal-clad switchgear) is about 15 kV. Due to complex designs, maintenance requirements, and size, installed ACBs are being replaced by vacuum circuit breakers.

Vacuum Circuit Breakers. Vacuum breakers (VCBs) are used to replace air-magnetic breakers, mostly because they are smaller, lighter, and require a lot less maintenance. VCBs use self-contained bottles with an internal vacuum to interrupt arcing. Since the vacuum in the bottles is not a perfect vacuum, a small amount of arcing occurs. The vacuum vaporizes a small amount of contact material, but most of the vaporized metal reforms onto the contacts. The practical limit for single-bottle vacuum circuit breaker technology is 15 kV, but by installing two bottles in series, voltages up to 38 kV can be interrupted.

Oil Circuit Breakers. Oil circuit breakers (OCBs) use mineral oil in large tanks to interrupt arcs. Arc interrupters force the oil through the arcs. At high voltages (230 kV), oil blast interrupters force oil through the arcs by using internal pumps and pistons. However, OCBs are not efficient at these voltages because the quantity of oil needed increases tremendously as the voltage rating increases. As with ACBs, OCBs are being phased out of service in favor of more efficient types of circuit breakers.

Air-Blast Circuit Breakers. Air-blast circuit breakers are one of the earliest types of circuit breakers used with 345 kV and 550 kV systems. With air-blast circuit breakers, dry, high-pressure air (about 240 psi) is forced across the arc when the contacts open. The high-pressure air extinguishes the arc. Moisture is always an issue with this type of circuit breaker, even more than with OCBs or gas circuit breakers. Leaks associated with the use of high-pressure air systems are also problematic. These types of circuit breakers have mostly been phased out by the utilities that used them.

Gas Circuit Breakers. Gas circuit breakers, which use sulfur hexafluoride (SF_6) gas, are rapidly replacing oil and air-blast circuit breakers. SF_6 is a high-density gas that has high dielectric strength. Gas circuit breakers use SF_6 gas to interrupt arcs. The dielectric strength of the gas increases as pressure increases, and the gas is used in circuit breakers from 15 kV to 765 kV.

MOLDED- AND INSULATED-CASE CIRCUIT BREAKERS

A *molded-case circuit breaker (MCCB)* is a circuit breaker that uses a molded case to house and support its current-carrying components as well as to be a part of the insulation system. The most common type of MCCB is the thermal-magnetic general-purpose circuit breaker. **See Figure 1-9.**

MCCBs often have a thermal overcurrent trip element to provide protection against overloads, such as what is caused when a coupling is misaligned on an electric motor or an electrical device draws too much current. An instantaneous overcurrent element is also provided to protect against short circuits, such as what is caused when two wires touch or when insulation fails. MCCBs have the following primary components:

- frame or case
- contact assemblies
- arc chutes
- OCPDs
- operating mechanism
- terminal connections

An *insulated-case circuit breaker (ICCB)* is a circuit breaker that is similar in construction to an MCCB but typically uses an electronic or digital OCPD and has much higher interrupting ratings.

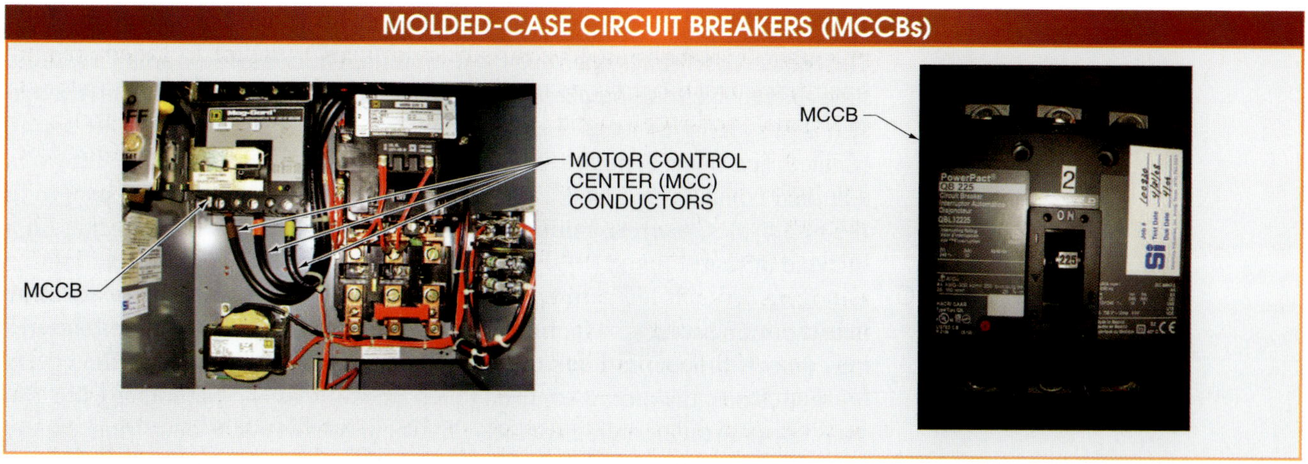

Shermco Industries

Figure 1-9. MCCBs are typically used in residential, commercial, and industrial applications to protect lighting panels and conductors.

MCCB Frames and Cases

The frame on an MCCB is an enclosed unit that surrounds and supports the other components while providing insulation. Sealed-case circuit breakers cannot be opened or serviced, except for testing, inspection, and cleaning of the outside. Sealed-case circuit breakers can be identified in several ways: by a tarlike substance, riveted cases, or paper seals over the case screws. Opening sealed-case circuit breakers invalidates their UL listing, which can cause issues if there is a fire or other accident involving those circuit breakers.

Large-frame MCCBs and most ICCBs have replaceable OCPDs and are also known as interchangeable trip circuit breakers. Interchangeable trip circuit breakers have cases that can be opened to be serviced and maintained. OEMs can provide certain parts so that they can be renewed. The OCPD can be replaced with an element that can be sized up to the frame rating of ICCBs and up to 80% of the frame rating of MCCBs. **See Figure 1-10.**

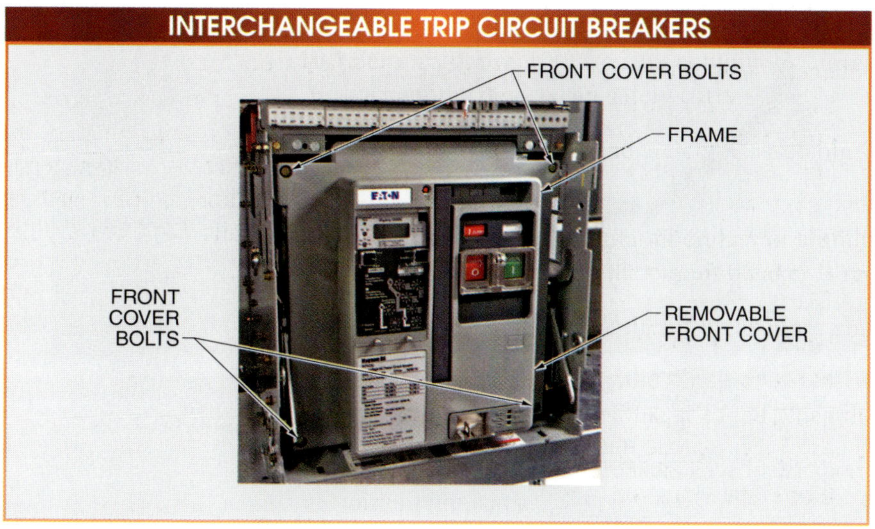

Shermco Industries

Figure 1-10. Interchangeable trip circuit breakers are MCCBs with cases that can be opened for servicing and maintenance.

MCCB Contact Assemblies

The contact assemblies open and close circuits. The contacts on small MCCBs, such as branch circuit breakers used in panelboards, carry the load current and also act as arcing contacts. Large-frame circuit breakers have separate arcing contacts and main contacts. One advantage that MCCBs and ICCBs have is that their contacts are small, light, and can interrupt an arc quickly, such as in 1½ to 2 cycles. Current-limiting versions can clear a fault even quicker, in a ½ cycle or less.

Arcing contacts aid in interrupting arcs and are composed of a harder alloy than the main contacts, which are designed to carry only load current. The arcing contacts (upper) extend ahead of the main contacts (lower). As the circuit breaker closes, the arcing contacts touch (make) first. Therefore, any arc that occurs does so on the arcing contacts. The main contacts then touch immediately after the arcing contacts touch. The main contacts are primarily composed of silver and are softer than the arcing contacts, which means that they will erode quickly if the arcing contacts are misadjusted or worn.

New generation current-limiting circuit breakers differ from standard MCCBs and ICCBs primarily by their contact structures. Standard MCCBs and ICCBs use single pivot-point mechanisms for the contacts, while current-limiting circuit breakers often use dual pivot mechanisms. **See Figure 1-11.** The magnetic fields around each of the contacts repel and rapidly force the contacts apart. As the short-circuit current flowing through them increases, the magnetic fields become stronger, and the contacts open faster.

STANDARD vs CURRENT-LIMITING CIRCUIT BREAKERS

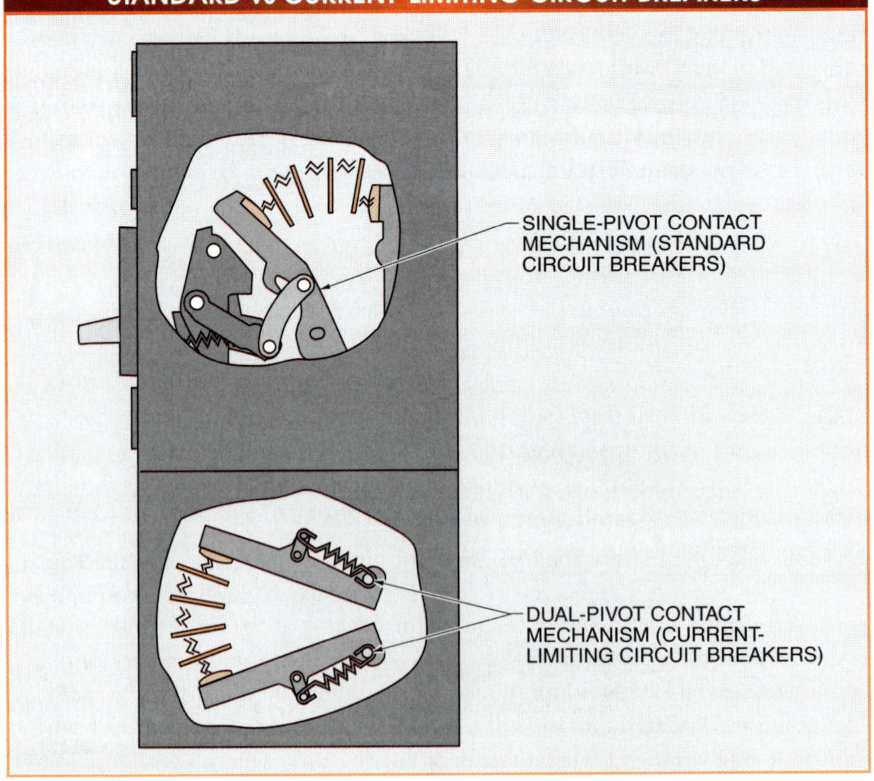

Figure 1-11. Standard MCCBs and ICCBs use single pivot-point mechanisms for the contacts, while current-limiting circuit breakers often use dual pivot mechanisms.

For these circuit breakers (and current-limiting fuses) to be current-limiting, the short-circuit current must be of a value high enough to cause it to be in its current-limiting region. If the short-circuit current is below this value, it responds as a standard circuit breaker.

CURRENT-LIMITING FUSES

When the available short-circuit current in a circuit exceeds the interrupting rating of the circuit breaker, current-limiting fuses (or circuit breakers) must be used. Until the late 1990s, technology was not available to manufacture current-limiting circuit breakers. For MCCBs, the use of current-limiting fuses was required, typically in the form of a fuse pack connected in series with the load current. The trip bar in the fuse pack was used to trip the circuit breaker when one of the fuses operated. The fuse or fuse pack needed to be replaced in order to reset the circuit breaker.

Shermco Industries

MCCB Arc Chutes

An *arc* is a sustained discharge of electricity across a gap in a circuit or between electrodes, usually accompanied by the electrodes (contacts) being vaporized and/or melted by the extreme heat of the arc. An *arc chute,* also known as arc extinguisher, is a structure that contains arc dividers. As the contacts part, the arc is drawn between the arcing contacts. The arc rises (due to its extreme temperature) and, as it does so, is stretched by the arc dividers. This cools the arc so it can be extinguished. MCCBs and ICCBs use arc chutes to stretch arcs, cool them down, and extinguish them, all in 1½ to 2 cycles. **See Figure 1-12.**

MCCB ARC CHUTES

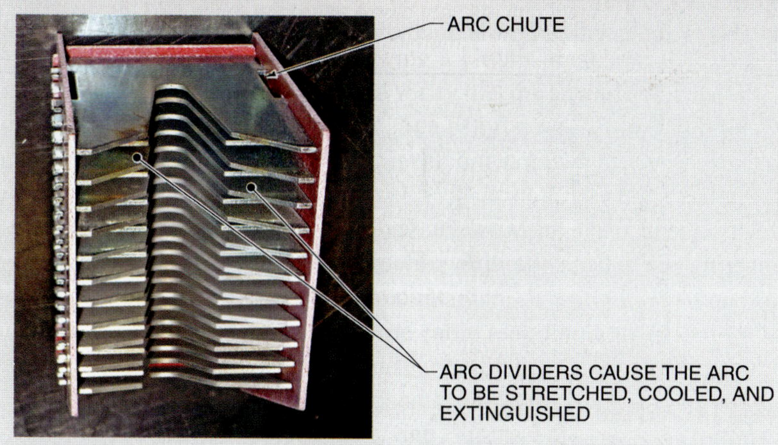

Shermco Industries

Figure 1-12. An MCCB and ICCB use an arc chute to stretch an arc, cool it down, and extinguish it, all in 1½ to 2 cycles.

OCPDs

Small-frame MCCBs typically use thermal-magnetic OCPDs. A *thermal-magnetic OCPD* is an OCPD that reacts to the heat created by the copper loss (I^2R) when current passes through a conductor. Copper loss is caused by the resistance of the conductor to current passing through it. This loss is expressed as heat. The higher the current flow through a conductor, the more heat is created.

A thermal-magnetic OCPD uses a bimetallic strip placed in the current path. The bimetallic strip is made of two metals that have different rates of expansion when heated. The bimetallic strip is constructed so that the metal having the higher rate of expansion forces the bimetallic strip to deflect, or bend, and release the trip latch. This occurs when the circuit breaker senses an overcurrent condition that lasts a predetermined amount of time.

A thermal-magnetic OCPD provides protection against overcurrent and short circuits in MCCBs. A thermal-magnetic OCPD is also known as a general-purpose trip unit. Other names for a thermal-magnetic OCPD are trip device and trip unit and are often used interchangeably.

On large-frame MCCBs, an electronic OCPD is typically used. Information pertaining to a specific OCPD can be found on the OEM nameplate affixed to the unit. **See Figure 1-13.**

THERMAL-MAGNETIC OCPDs

Shermco Industries

Figure 1-13. A thermal-magnetic OCPD provides protection against overcurrents and short circuits in MCCBs and is sometimes referred to as a general-purpose trip unit.

An MCCB can only have an OCPD with a continuous current rating of 80% of the frame rating, while an ICCB can have an OCPD with a continuous current rating equal to the frame rating. This is because a thermal-magnetic OCPD has a very broad time-current characteristic curve, which means that OEMs have to make an extra allowance for the circuit breaker to trip without damaging itself from the heat generated by the excess current flow. ICCBs electronic OCPDs that are more accurate, and therefore reduce the possibility of thermally damaging the circuit breaker frame.

MCCB Operating Mechanisms

The operating mechanism of an MCCB opens and closes the contact assemblies and has three positions: open, closed, and trip. Branch circuit breakers of the type used for panelboards and lighting panels are of a fairly simple design. **See Figure 1-14.**

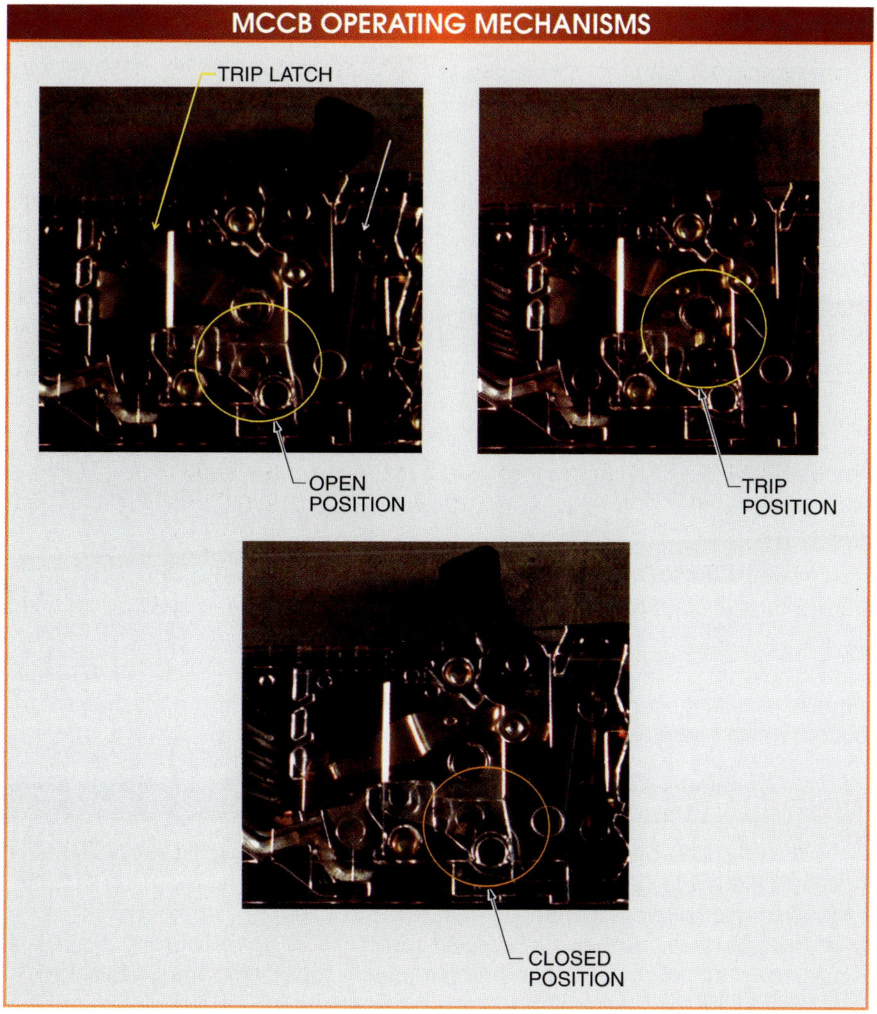

Shermco Industries

Figure 1-14. The operating mechanism of a molded-case circuit breaker opens and closes the contact assemblies. It has three positions: open, trip, and closed.

With the contacts closed, the trip latch is in the latched position (yellow circle). As the contacts are opened and closed, the trip latch position does not move. This type of trip latch is one of the major issues with MCCBs in that it, and other parts of the operating mechanism, is lubricated at the factory. Current flow through the contacts creates heat, which dries out the lubricant over time. As the factory-applied lubricant dries, it thickens and slows the circuit breaker performance. As it continues to dry, it begins to flake off, and metal-to-metal wear occurs. This metal-to-metal wear and the corrosion that can occur on the trip latch can easily cause the circuit breaker to fail to open as required.

The only time the trip latch changes position is when the circuit breaker is tripped. Note how the trip latch is stationary in the open and closed positions but is different in the trip position. Trip latch malfunction is one of the primary causes of MCCBs failing to operate in accordance with the OEM specifications.

Modern large-frame MCCBs often include red mechanical trip buttons. The trip button operates the trip latch directly. The trip latch (yellow arrow in Figure 1-14) does not move when the circuit breaker is toggled from the open to the closed position. It does move, however, when the circuit breaker is tripped. **See Figure 1-15.**

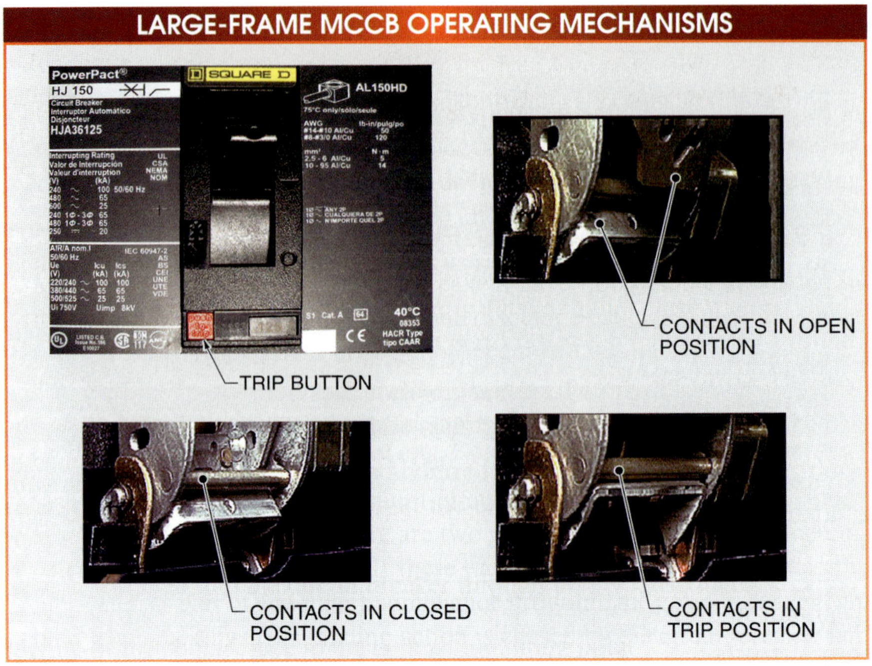

Shermco Industries

Figure 1-15. A modern, large-frame MCCB often includes a red mechanical trip button, which operates the trip latch directly (mechanically).

CIRCUIT BREAKER AGING AND TESTING

A study done by the Nuclear Regulatory Commission (NRC) in NUREG/CR-5762, Wyle 60101, *Comprehensive Aging Assessment of Circuit Breakers and Relays,* (written in March 1992) covers the subject of failures in MCCBs that had been in service three to five years with no maintenance. In this report, various problems were found with the 11 breakers surveyed. Of the 11 circuit breakers, 5 had long-time delay defects and 4 had instantaneous trip problems. Some circuit breakers had multiple-pole failures and some had both instantaneous and long-time delay problems. Although the survey was limited in number, it is typical of problems seen in the field during testing.

The NRC recommended primary injection testing of the circuit breakers every three years and, if they could not be tested, operating the "Push-to-Test" or "Twist-to-Test" mechanism every year. If a circuit breaker does not have such test features, the NRC recommended operating the toggle (handle) several times rapidly twice a year to help maintain functionality.

TECH FACT

Draw-out construction makes maintenance and testing simpler, less expensive, and safer. The circuit breaker is mounted on a draw-out cradle, which slides in and out of its enclosure. In order for the circuit breaker to connect to the power bus, primary disconnects are added to the circuit breaker and the power bus so they can mate properly. These parts add to the required maintenance and complexity but pay off as the equipment begins to age. With a draw-out circuit breaker, the circuit breaker is racked out of its enclosure by way of a screw-type mechanism. The circuit it is connected to is de-energized, and maintenance and testing are performed without interruption to other circuits.

MCCB Terminal Connections

The safe installation of MCCBs and ICCBs depends on proper terminations. If the terminations are not completed properly, they can start fires and damage equipment. Many large ICCBs are either bolted directly to the bus or are of a draw-out construction. **See Figure 1-16.** Problems with these types of connections are fairly infrequent.

MCCBs are often connected using stranded-conductor cable or wire, which can cause problems because they have a tendency to loosen over time due to heat cycling. With the terminal connections of standard three-phase thermal-magnetic industrial circuit breakers, the conductors are fitted into the terminal lugs and torqued to specification. The terminal lug can only be used for a specified range of wire sizes and wire types. If the conductor is too small, it will not have the surface area within the lug to carry the expected amount of current.

Figure 1-16. Many large ICCBs are either bolted directly to the bus or are of a draw-out design.

When a small conductor is connected to a terminal lug that should be used for a much larger conductor, there is only point-to-point contact between the conductor and the terminal lug. This type of connection causes overheating at the connection and, if not corrected, will cause the conductor to become annealed. **See Figure 1-17.**

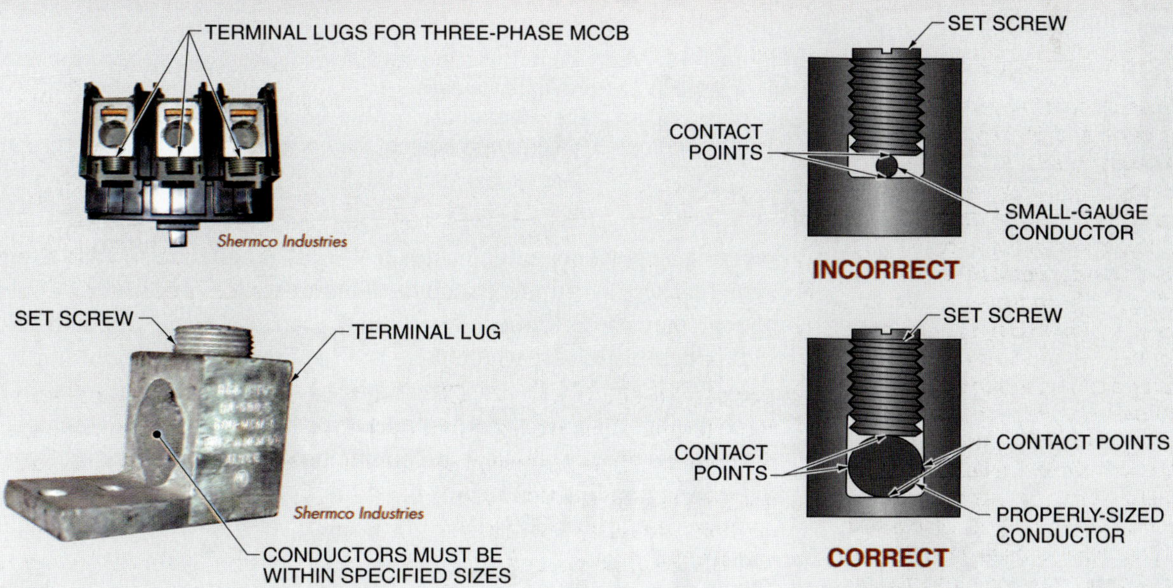

Figure 1-17. Terminal lugs can only be used for a specified range of conductor sizes. If the conductor is too small, it will not have the surface area within the lug to carry the expected amount of current.

When a conductor becomes annealed, it does not carry the proper amount of current due to increased impedance. The increased impedance causes further heating, which then causes the conductor to have high impedance. Often the insulation around an annealed conductor becomes completely burned off due to the heat being generated. An annealed conductor must be replaced or the annealed portion must be cut off and a new piece of conductor must be spliced in.

The other issue concerning terminal lugs is improper torquing. If the cable loosens inside the terminal lug, the connection will heat due to increased impedance. This additional heating can also cause the conductor to become annealed. Often when the terminal lug set screw loosens, arcing occurs inside the set screw threads. Typically, this cannot be seen from the outside, so the technician may retighten the set screw and believe the problem is solved. However, the arcing inside the threads typically prevents the set screw from tightening any farther than where the arcing took place. Regardless of how much force is applied to the set screw, it is never completely tightened against the conductor and overheating continues. **See Figure 1-18.**

IMPROPER TORQUING

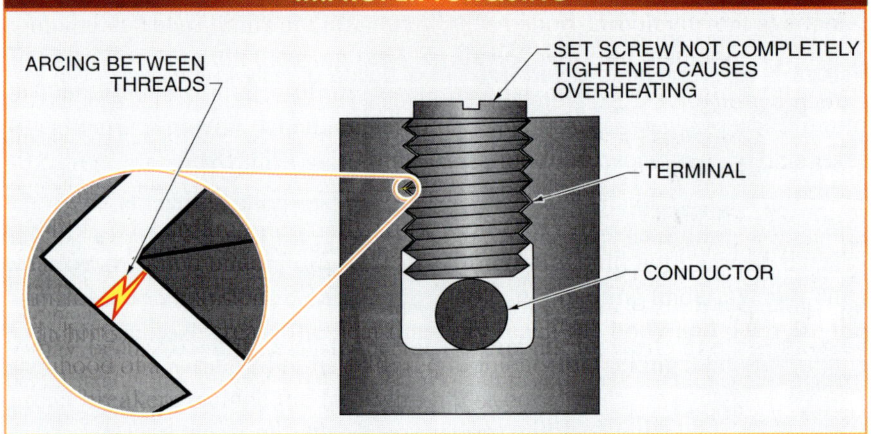

Figure 1-18. When the terminal lug set screw loosens, arcing occurs inside the set screw threads and prevents the set screw from tightening any farther than where the arcing has taken place. Regardless of how much force is applied to the set screw, it is never completely tightened against the conductor and overheating continues.

AIR-FRAME CIRCUIT BREAKERS

Low-voltage power circuit breakers (LVPCB) are often referred to as air-frame circuit breakers. Air-frame circuit breakers are typically of the draw-out design, although they can be bolted directly to the bus. Air-frame circuit breakers are used for applications in which there is a requirement for a more sturdy design than MCCBs and ICCBs offer. **See Figure 1-19.** In addition, the circuit breaker can be used as an across-the-line starter for large low-voltage motors.

Air-frame circuit breakers are completely renewable. That is, they can be rebuilt. Any part that is damaged or compromised can be repaired or replaced. Air-frame circuit breakers have six primary components, which include the following:

- disconnects (stabs)
- contact assemblies
- arc chutes

- operating mechanisms
- OCPDs
- frame and backboard assemblies

AIR-FRAME CIRCUIT BREAKERS

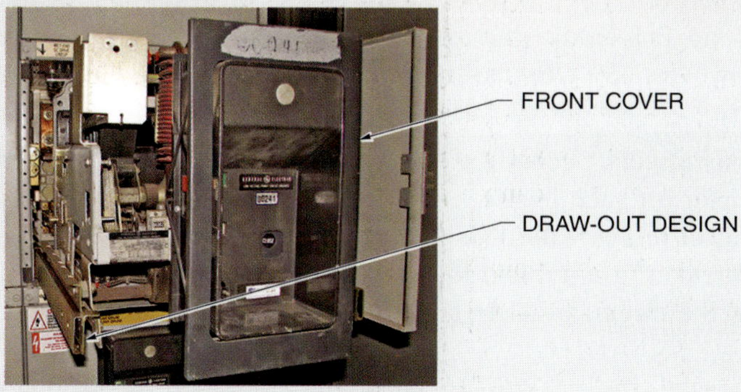

Shermco Industries

Figure 1-19. An LVPCB is often referred to as an air-frame circuit breaker and is typically of the draw-out design, although it can be bolted directly to a bus.

Air-Frame Circuit Breaker Disconnects

Circuit breaker disconnects (stabs) are current-carrying components that connect the circuit breaker to the bus or other connections. Disconnects in a circuit breaker include main, secondary, and ground disconnects.

Main Disconnects. Main disconnects connect a circuit breaker to the primary bus and allows load current to flow through the circuit breaker. Air-frame circuit breakers typically use either vertical or horizontal rows of spring-loaded fingers to make contact with the busbar.

Air-frame circuit breakers with high frame ratings (3000 A to 4000 A) often have round (tulip-shaped) main disconnects. The round shape provides more surface area contact and can carry higher load currents than flat bus disconnects. Other types of air-frame circuit breakers with large continuous current ratings have a double row of primary disconnects rather than round disconnects. **See Figure 1-20.**

CIRCUIT BREAKERS—MAIN DISCONNECTS

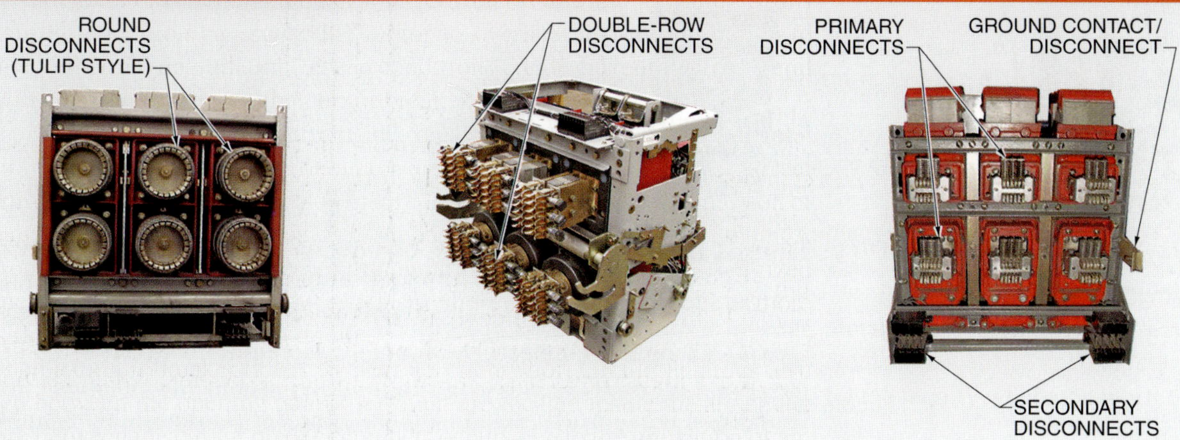

Shermco Industries

Figure 1-20. LVPCBs with high frame ratings (3000 A to 4000 A) often have round main disconnects, while those with large continuous current ratings have a double row of primary disconnects.

If current-limiting fuses are installed on an LVPCB, the line-side stabs will sometimes be extended farther back from the circuit breaker backboard than the load-side stabs. On some circuit breakers that use current-limiting fuses, the load-side stabs have extenders installed to make the stab distances from the backboard uniform. The specific arrangement varies from OEM to OEM. It is typical for LVPCBs to have a draw-out construction in which the circuit breaker sits in a cradle that can be moved in and out of the enclosure or cubicle. However, LVPCBs can be bolted directly to the bus as a cost-cutting measure.

Secondary Disconnects. Secondary disconnects are used to bring power into the circuit breaker for such features as the spring-charging motor, indicator lights, and control relays within the circuit breaker. For a low-voltage air-frame circuit breaker, it is typical to connect and disconnect the secondary disconnects as the circuit breaker is racked in and out of its cubicle. **See Figure 1-21.**

SECONDARY DISCONNECTS

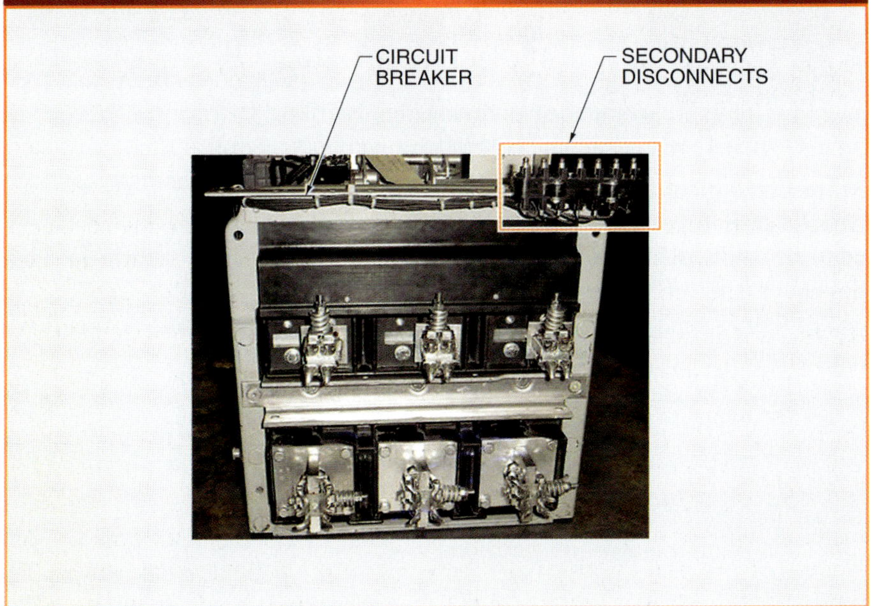

Shermco Industries

Figure 1-21. Secondary disconnects are used to bring power from the control circuits into the circuit breakers.

Ground Disconnects. The frame of a circuit breaker must be connected to the ground bus so that a short circuit or fault will be taken to ground immediately and the protective devices will operate as quickly as possible. The ground disconnect connects the frame of the circuit breaker to the ground bus. **See Figure 1-22.** The connection to the ground disconnect is the first connection made when racking a circuit breaker into the cubicle and is the last connection broken when racking it out of the cubicle. This ensures that the frame of the circuit breaker is grounded whenever there is the possibility of the frame being energized.

GROUND DISCONNECTS

Shermco Industries

Figure 1-22. The ground disconnect is used to connect the frame of the circuit breaker to the ground bus.

Air-Frame Circuit Breaker Contact Assemblies

Contact assemblies for air-frame circuit breakers include arcing contacts, main contacts, and auxiliary contacts. Contact assemblies can have a butt or bayonet design, depending on the circuit breaker manufacturer's requirements.

Arcing Contacts. An *arcing contact* is the contact of a switching device on which an arc is drawn after the main contacts have parted. Arcing contacts are designed to prevent the main contacts from being damaged as the circuit breaker opens and closes. The arcing contacts on low-voltage circuit breakers are often made of a silver and zinc alloy.

Whenever a circuit breaker opens its arcing contacts, some of the contact material melts and some of it vaporizes from the face of the contacts. The amount of material that is eroded depends on the current value being interrupted. This molten and/or vaporized contact material is primarily ejected up into the arc chutes where it can collect. Cleaning this metal spatter is part of a normal preventive maintenance (PM) program.

When the circuit breaker opens, the main contacts part first and transfer the current to the arcing contacts. The arcing contacts then part and draw the arc across them. When the circuit breaker closes, the arcing contacts make first, which again draws the arc across them. This prevents the main contacts from carrying the arc and preserves them from damage. This also means that the arcing contacts wear/erode long before the main contacts. If the main contacts show any signs of pitting or damage from arcing, the arcing contacts may be misaligned or worn out. **See Figure 1-23.**

Arcing contacts often have an arcing horn on the top of the contact structure. The arcing horn aids in transferring the arc from the arcing contact to the arc runner in the arc chute.

ARCING CONTACTS

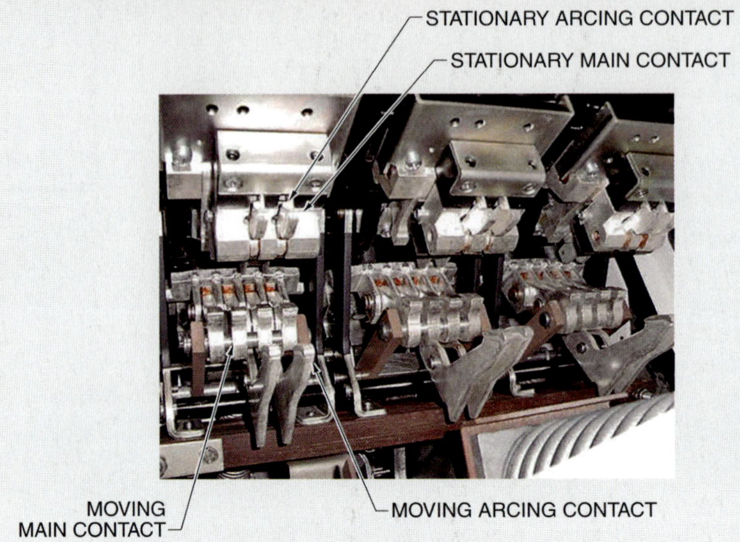

Figure 1-23. Arcing contacts are designed to prevent the main contacts from being damaged as the circuit breaker opens and closes.

Main Contacts. The main contacts in a circuit breaker are constructed of a softer alloy using less zinc and more silver than arcing contacts. They carry the load current and must have a low impedance to current flow. Main contacts are larger than arcing contacts, which decreases their impedance.

Some circuit breakers also have intermediate contacts. The intermediate contacts are only used on the largest frame circuit breakers such as the GE AK-50, AK-75, and AK-100. Intermediate contacts act as a second set of arcing contacts. When the main contacts open, the load is transferred to the intermediate contacts and then to the arcing contacts. Intermediate contacts do not actually make or break the load current. Rather, they transfer the load current from the main contacts to the arcing contacts when opening and from the arcing contacts to the main contacts when closing. Since the intermediate contacts extend just a bit farther than the main contacts, they will make before and break just after the main contacts do.

Arcing contacts prevent the main contacts from eroding by drawing the arc across themselves. If the main contacts carried the arc, they would rapidly burn off. This would cause the circuit breaker to restrike, which means it cannot interrupt the arc. A restrike often results in the total destruction of the circuit breaker and switchgear. **See Figure 1-24.**

Auxiliary Contacts. Auxiliary contacts supply control power for electrical functions within circuit breakers, such as switching the spring-charging motor on and off, controlling indicator lights, and controlling open/close circuits. With LVPCBs, auxiliary contacts are mounted directly onto the frame of the circuit breakers.

CIRCUIT BREAKER RESTRIKES

Shermco Industries

Figure 1-24. If main contacts are exposed to arcing, they will rapidly burn off and cause the circuit breakers to restrike, or not interrupt the arcs.

TECH FACT

Auxiliary contacts are designated on drawings as 52a (for normally open contacts) and 52b (for normally closed contacts). The 52a contacts are in the same position as the main contacts of the circuit breaker, while 52b contacts are opposite. When the main contacts are open, 52a contacts are open and 52b contacts are closed.

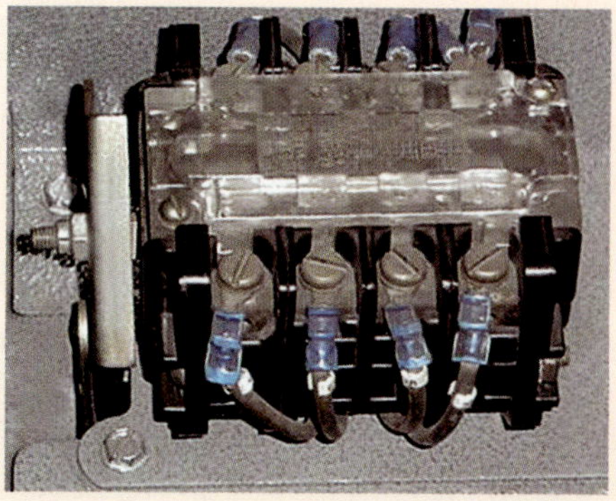

Shermco Industries

Auxiliary contacts supply control power for electrical functions within circuit breakers, such as switching the spring-charging motor on and off, controlling indicator lights, and controlling open/close circuits.

Low- and Medium-Voltage Circuit Breaker Contact Design. Both arcing and main contacts can be butt- or bayonet-style contacts. Butt-style contacts usually have one flat and one contoured contact face. Some circuit breakers have a combination of butt and bayonet contacts. **See Figure 1-25.**

The purpose of the contour is to make a wiping motion when the contacts make. The moving contact hits the stationary contact then makes a wiping motion as it slides along the face of the stationary contact. This wiping motion prevents the contacts from welding together and also cleans the contacts.

In addition, the stationary contacts are usually spring-loaded so that when the moving contact assembly hits the stationary contact assembly, it is pushed backward. This promotes the wiping action and also prevents contact bounce, which would occur if the moving contacts were immobile. When the contacts close, the combination of the compression of the stationary contact and the contoured surface causes a wiping motion against the flat-surface contact without contact bounce.

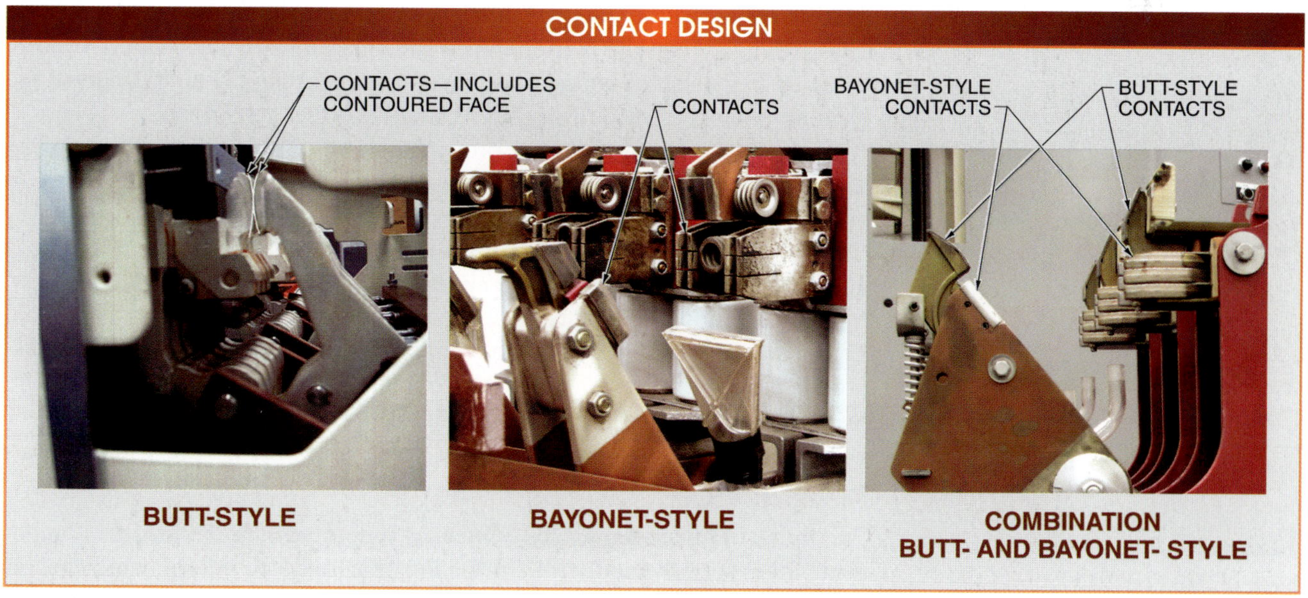

Shermco Industries

Figure 1-25. Circuit breaker contact designs can have a butt, bayonet, or combination butt and bayonet style.

Bayonet-style contacts are found on medium-voltage air circuit breakers and late generation LVPCBs. The current-carrying portions of the contacts are on the sides rather than the faces of the contacts. As with butt-style contacts, it is important not to change the shape of the contact faces, as this could cause the circuit breaker to malfunction.

Air-Frame Circuit Breaker Arc Chutes

Arc chutes cool down the electrical arc and extinguish it in four cycles (LVP-CBs) to six cycles (medium-voltage air circuit breakers). Because arcs are harder to extinguish in medium-voltage circuit breakers than in low-voltage circuit breakers, extra cycles are required to extinguish arcs.

Air-Frame Circuit Breaker Operating Mechanisms

Modern operating mechanisms are quick make, quick break, which means that the speed of operation is independent of the speed of the control handle. Modern circuit breaker operating mechanisms are called stored energy mechanisms, since there are both opening springs and closing springs.

SAFETY TIP

One set of springs in circuit breaker operating mechanisms usually has tension placed on it. For this reason, extreme care must be used when working on or near circuit breaker contacts. They have heavy moving contact assemblies and powerful springs that can cause injury. Even when opening, these devices have such force that they could break bones and lacerate skin. Allow plenty of clearance between any body parts and the moving parts of a circuit breaker. Always wear proper personal protective equipment (PPE) when working with circuit breakers.

One obsolete design includes a simple prop and roller operating mechanism that has an accordion-style linkage to connect the operating mechanism to the contact assemblies. When the circuit breakers close, the closing springs force the contacts closed. At the same time, the prop is lifted onto the top of the roller. The opening springs are now stretched and create an opening force. The prop, by being in the center of the roller, is now in a mechanical bind, since all the opening force is applied through the prop-and-roller centerlines.

To trip the circuit breaker, the prop is knocked off the roller by the trip bar. The prop falls off the roller, the accordion linkage collapses, and the circuit breaker contacts open.

This type of operating mechanism is used on older circuit breakers. It is inherently weak and requires the circuit breaker to be larger than necessary. Modern operating mechanisms are more compact, which also makes them stronger and less likely to have problems if maintained properly. Regular disassembly and lubrication are necessary for all circuit breakers in order to ensure that they will function properly. **See Figure 1-26.**

Air-Frame Circuit Breaker Frames

The frame of an air-frame circuit breaker supports all circuit breaker components. It is typically made from metal and of an open-air design, which means that it is not part of the insulation system in the same way the frame of an MCCB or ICCB would be.

OPERATING MECHANISMS

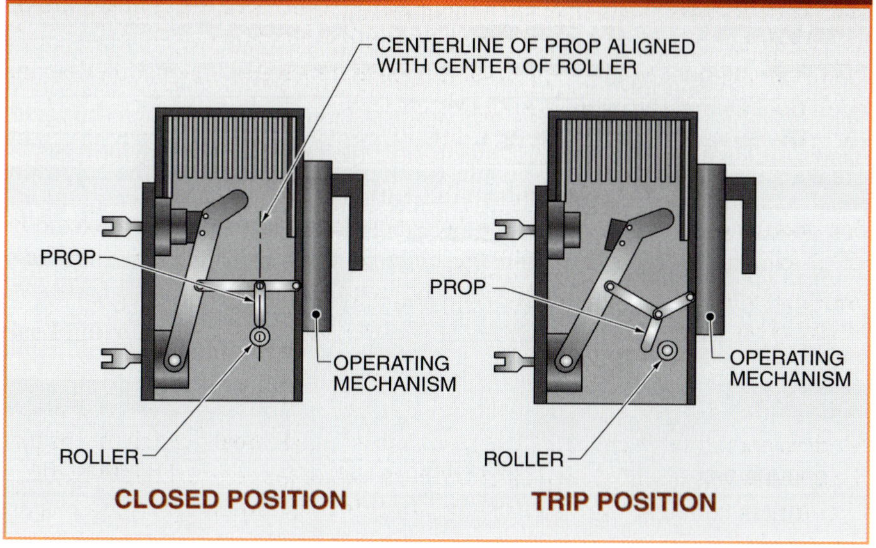

Figure 1-26. LVPCBs typically use prop-and-roller operating mechanisms.

A *backboard assembly* is the section of a circuit breaker that supports stationary contacts, the primary line, and load disconnects. The backboard assembly is mounted on the circuit breaker frame. Depending on the age of the circuit breaker, the backboard assembly may be composed of slate, polycarbonate, or fiberglass. **See Figure 1-27.**

FRAMES

Figure 1-27. A backboard assembly is mounted to the frame of an LVPCB. It supports the stationary contacts, primary line, and load disconnects.

MEDIUM-VOLTAGE AIR AND VACUUM CIRCUIT BREAKERS

Low- and medium-voltage air circuit breakers are similar in construction and operation. One of the primary differences between them is that the OCPDs on LVPCBs are mounted on the circuit breakers, while medium-voltage circuit breakers use protective relays, which are typically mounted on the enclosure door. Medium-voltage air circuit breakers have the following four major components:

- disconnects (stabs)
  - main stabs
  - secondary stabs
  - ground stabs
- contact assemblies
  - arcing contacts
  - main contacts
  - auxiliary mechanism-operated contacts (MOCs)
  - truck-operated contacts (TOCs)
- arc extinguishers (arc chutes)
- operating mechanisms (frames)

Vacuum circuit breakers do not use arcing contacts. Also, instead of large, heavy contact assemblies and arc chutes, they use vacuum bottles with smaller contacts to extinguish arcs.

Shermco Industries

The OCPD on medium-voltage circuit breakers is a protective relay, which is typically mounted in the enclosure door.

Main Disconnects

Most medium-voltage metal-clad circuit breakers have tulip-style (round) disconnects with springs around them to ensure good contact with the bus. However, certain OEMs sometimes place bus type, straight finger connections on small-frame circuit breakers. The main disconnects are mounted on the bushings, or pole units. Bushings are constructed of copper alloy, connect the primary disconnects to the stationary contact assembly, and are covered with insulation. Epoxy-coated paper, which can absorb moisture over time, is used as insulation for many medium-voltage air circuit breakers. Some manufacturers use porcelain. **See Figure 1-28.**

PRIMARY DISCONNECTS–MEDIUM-VOLTAGE CIRCUIT BREAKERS

Shermco Industries

Figure 1-28. Medium-voltage metal-clad circuit breakers have tulip-style (round) disconnects with springs around them to ensure good contact with the bus.

Secondary Disconnects

Secondary disconnects are often mounted on the contact block located on the lower right-hand corner of medium-voltage metal-clad circuit breakers (when facing the rear of the circuit breaker). Vertical lift circuit breakers, such as the GE MagneBlast, are an exception to this typical design. The secondary contact block of vertical lift circuit breakers is mounted on the top of the circuit breaker on the right-front corner (when facing the front of the circuit breaker). Many horizontal racking circuit breakers have a T-handle extension that must be operated in order to make the secondary disconnects and place the circuit breaker into the test position.

Unlike LVPCBs, medium-voltage metal-clad circuit breakers usually do not have a separate test position. They are racked to the disconnected position, and the secondary disconnect is extended until it makes contact with the stationary secondary disconnects located in the rear of the cubicle. **See Figure 1-29.**

Medium-Voltage Circuit Breaker Contact Assemblies

The function of the main and arcing contacts is the same in medium-voltage air and vacuum circuit breakers as in LVPCBs. Instead of zinc being used to harden the arcing contacts, tungsten is often used for medium-voltage air circuit breakers. The arc runners in the arc chute are connected to the line and load sides of the circuit breaker, respectively. This means that when the arc chute is installed on a medium-voltage circuit breaker, both it and the arc runners are energized. **See Figure 1-30.**

SECONDARY DISCONNECTS—MEDIUM-VOLTAGE CIRCUIT BREAKERS

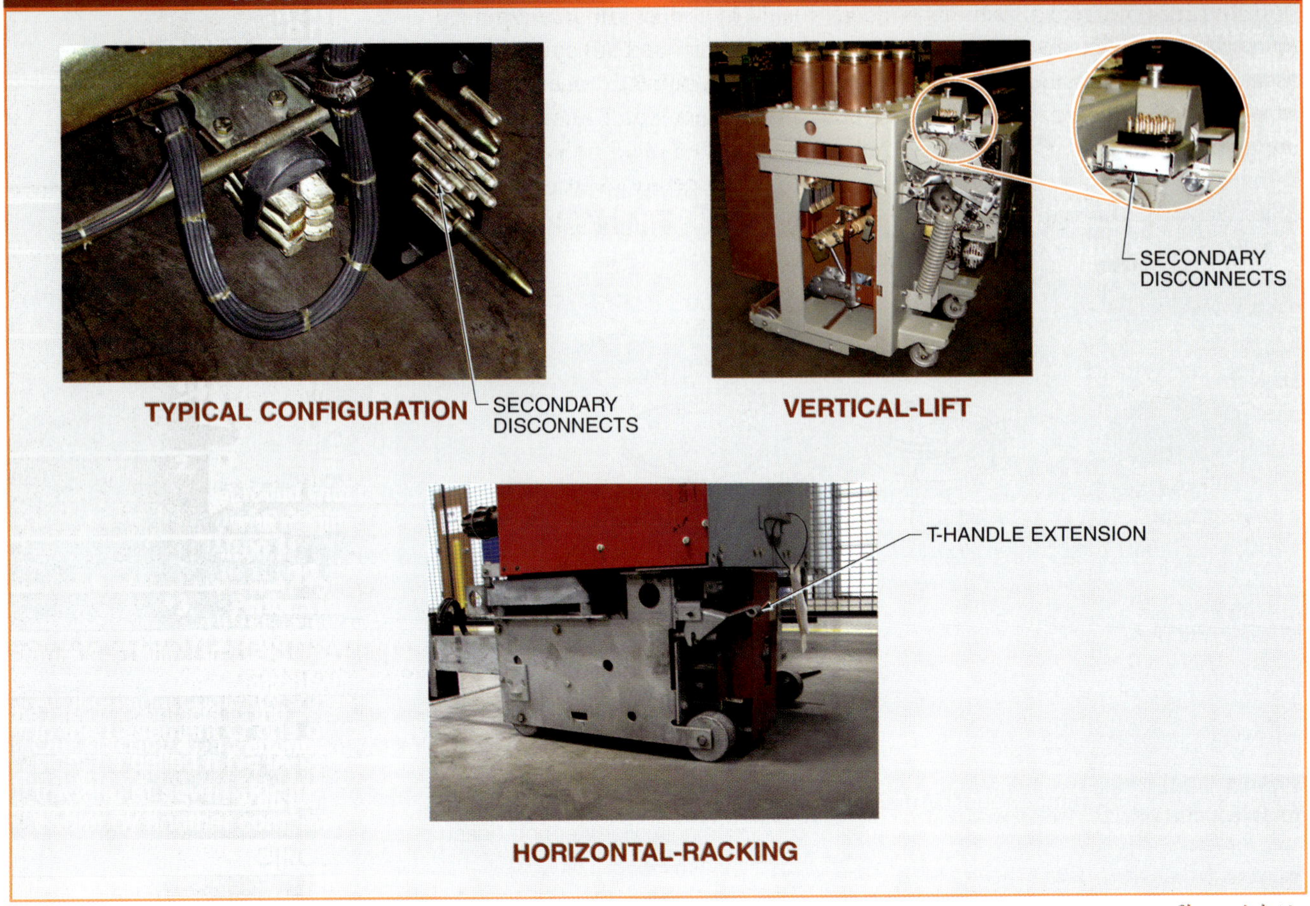

Shermco Industries

Figure 1-29. Depending on the design, secondary disconnects can be mounted near the top corner, or the bottom corner. On circuit breakers racked horizontally, a T-handle extension is used to connect the secondary contact block.

MEDIUM-VOLTAGE CIRCUIT BREAKER CONTACT ASSEMBLIES

Shermco Industries

Figure 1-30. Contact assemblies on medium-voltage circuit breakers include arc runners, arc runner connections, arcing contacts, and main contacts.

Auxiliary Contact Assemblies

The auxiliary contacts of medium-voltage metal-clad switchgear are typically mounted on the switchgear rather than the circuit breaker frame. They are known as mechanism-operated contact (MOC) and truck-operated contact (TOC) switches. MOCs are mechanically driven from the operating mechanism and are used for control and indicating circuits. **See Figure 1-31.** The TOC switch is used to disable the circuit breaker during racking operations. It is operated when the circuit breaker is racked in or out of its cell.

AUXILIARY CONTACT ASSEMBLIES

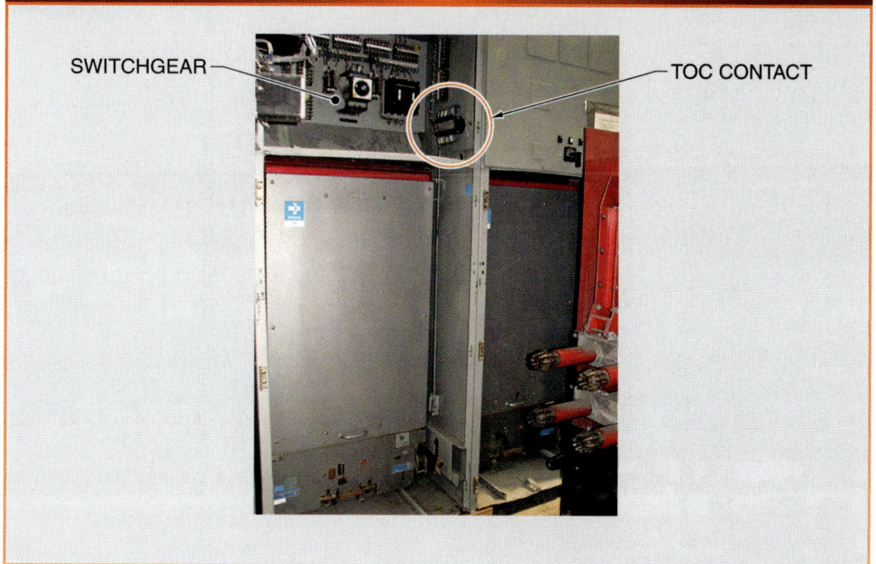

Group CBS

Figure 1-31. The auxiliary contacts (MOC and TOC) of medium-voltage metal-clad switchgear are typically mounted on the switchgear rather than the circuit breaker frame.

Arc Chutes

Another area in which low- and medium-voltage air circuit breakers are similar is the arc interruption process. The basic construction of low- and medium-voltage arc chutes is similar. However, the arc in a medium-voltage circuit breaker is more persistent, which means it is harder to extinguish. Medium-voltage circuit breakers add several aids, such as puffers, blowout coils, and pole pieces, to ensure that arcs are extinguished quickly and effectively.

Operating Mechanisms–Low- and Medium-Voltage Circuit Breakers

Modern operating mechanisms are quick make, quick break. This means that the speed of operation is independent of the speed of the control handle. Modern circuit breaker operating mechanisms are said to be stored energy mechanisms, since there are both opening springs and closing springs. One set of springs usually has tension on it. For this reason, extreme care should be taken when working on or near circuit breaker contacts. They have heavy moving contact assemblies and powerful springs.

If a technician were to place his hand between the moving and stationary contacts when it closed, he could be maimed. Even when opening, these devices have such force that they could break bones. Plenty of clearance should be allowed between any body part and the moving parts of a circuit breaker.

Closing springs accelerate the contacts to the closed position but do not hold the contacts closed. If only the closing springs were relied upon to keep contacts closed, over a period of time the springs would weaken, causing the contacts to bounce, vibrate, and burn. The contacts are held in the closed position by a prop-and-roller operating mechanism. The prop and roller puts the contact linkage into a mechanical bind, forcing the contacts to stay tightly closed.

LVPCBs and medium-voltage vacuum circuit breakers (such as the GE Power/Vac® ML-18) have advanced prop-and-roller operating mechanisms. **See Figure 1-32.** In Figure A, the closing springs are discharged and the circuit breaker is open.

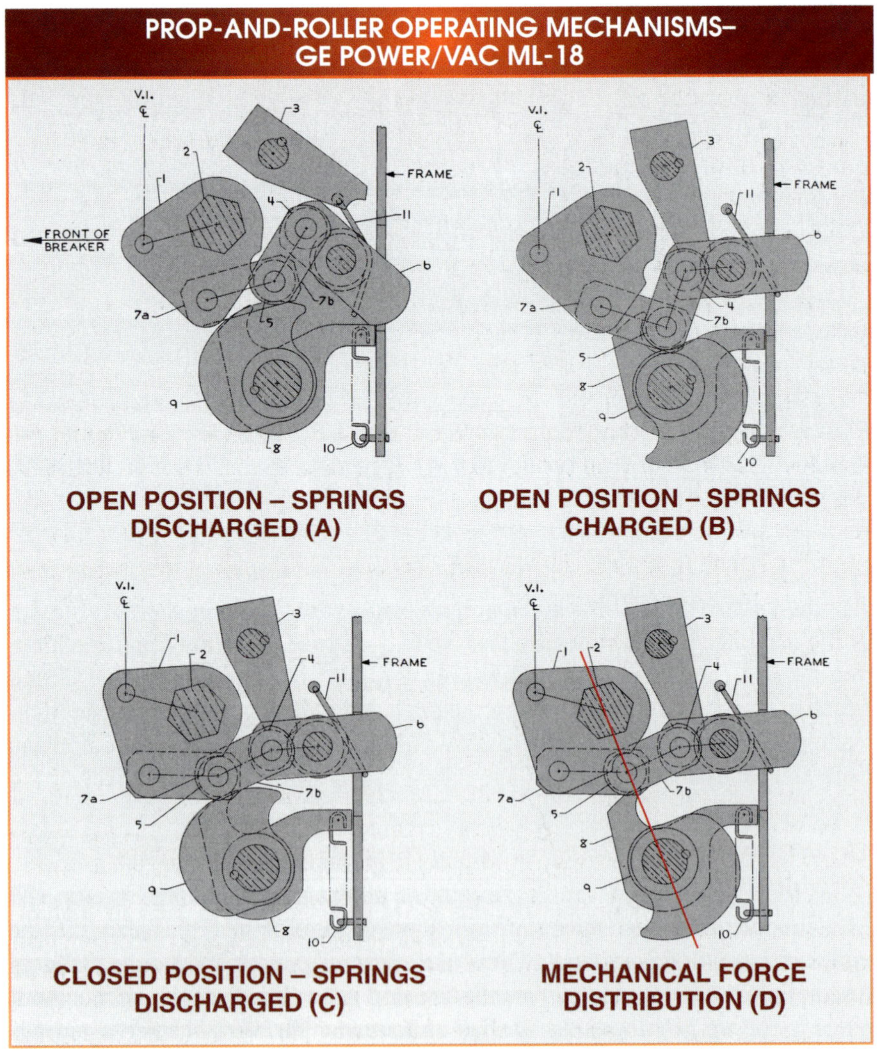

General Electric Company

Figure 1-32. A prop-and-roller operating mechanism is used by GE for its LVPCBs and medium-voltage vacuum circuit breakers (Power/Vac ML-18).

The closing toggles (7A and 7B) and the trip link (6) are collapsed, which allow the bellcrank (1) to rotate counterclockwise. This rotation forces the pushrod connected to the bellcrank down, opening the contacts in the vacuum bottle. A bellcrank (bellcrank lever) is used to change from one direction to another, usually at a 90° angle. It is almost always triangular in shape and is connected to a pivot pin on one end. The closing springs force the contacts closed, so the bellcrank is forced to rotate.

When electrical control power is applied to the circuit breaker, as in **1-32 (Figure B)**, the closing springs charge and the trip latch (3) drops against the trip roller (4). This forces the other closing toggle (7a) down also and into the crook of the prop (8). The trip link (6) also drops to a near-horizontal position. This sequence prepares the circuit breaker to close its contacts when the close button is pressed.

Pushing the close button discharges the closing springs, which forces the closing cam to rotate counterclockwise, as in **1-32 (Figure C).** When it does, both closing toggles and the trip link are pushed upward, and the bellcrank (1) rotates clockwise. The pushrod, connected to the vacuum bottle contacts, is forced upward, which closes the contacts. The contacts are now held in the closed position by the prop's interference with the closing roller (5). As long as the prop is in this position, the circuit breaker will stay closed.

A line through the center of the roller and pivot diagram illustrates the entire mechanical force of the mechanism concentrated through these points in **1-32 (Figure D).** To trip the circuit breaker, the trip latch (3) is rotated counterclockwise, which releases the closing toggles and the trip link. The bellcrank is forced counterclockwise, and the contacts open.

Usually, the closing springs recharge when the circuit breaker closes. Some manufacturers set up the sequence so that they must recharge the springs when the circuit breaker is tripped. Reviewing the internal schematic and how the various limit switches and interlocks are sequenced will indicate how the circuit breaker will operate.

Another example of a prop-and-roller mechanism is the Westinghouse/Cutler-Hammer DS 480 V circuit breaker operating mechanism. This is the same basic design that is used on their medium-voltage vacuum circuit breakers. The mechanism is shown with the contacts open and the springs discharged. Note the position of the main link (prop) as well as the closing cam. **See Figure 1-33 (Figure A).**

When the circuit breaker is inserted into the cubicle, the springs charge, which causes the closing cam to rotate approximately 90° (as shown in Figure B). The roller constraining link is rotated by the closing cam, which allows the trip latch to drop under the trip shaft. The edge of the trip latch is caught under the tip of the trip bar and held in place. The mechanism is held in the ready-to-close position by the closing springs applying force to the closing cam, which is prevented from rotating by the spring release latch.

When the close button is activated, the closing coil energizes, and the spring release latch allows the closing cam to rotate. (as shown in Figure C). As the contacts close, the springs discharge, and the main link is forced up by the closing cam rotation, which forces the contacts closed. All the force of the opening springs is now being applied to the centerline of the pivots of the bellcrank, main link, and closing cam, causing the contacts to be mechanically held closed. To trip the circuit breaker into the contact open position, the trip button is depressed, which rotates the trip shaft, allowing the trip latch to release. Since the roller constraining link is no longer being held by the trip shaft, the opening springs force the closing cam to rotate, allowing the contacts to open.

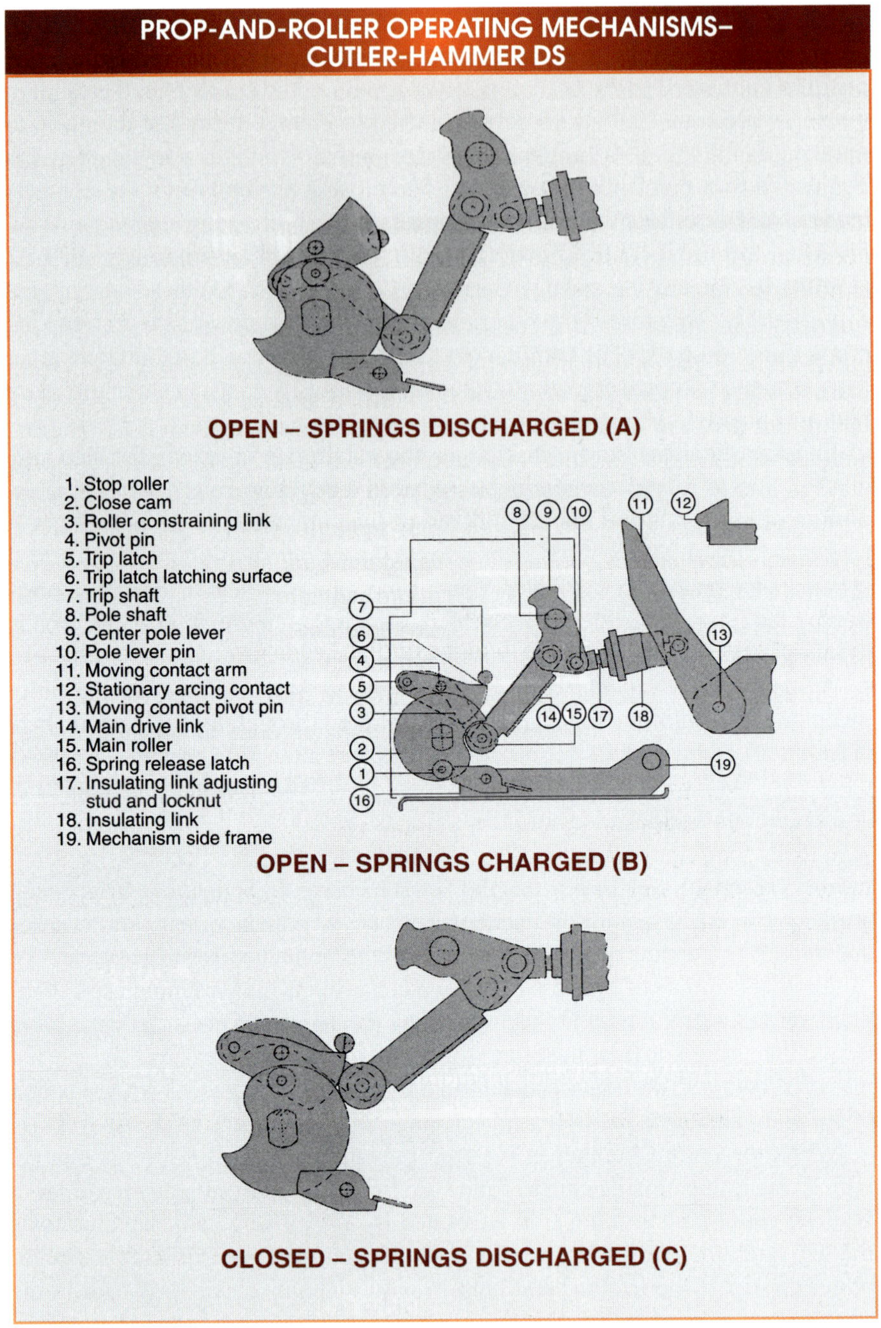

Cutler-Hammer

Figure 1-33. The operating mechanism of a Westinghouse/Cutler-Hammer DS 480 V circuit breaker has the same basic design as the one used on their medium-voltage vacuum circuit breakers.

A common type of LVPCB is the ABB/ITE K-Line 600 V circuit breaker. With this type of circuit breaker, the small bearing in the main roller (arrow) often seizes due to dirt or corrosion. This causes a trip-free condition or prevents the circuit breaker from tripping when the trip button is depressed. **See Figure 1-34.**

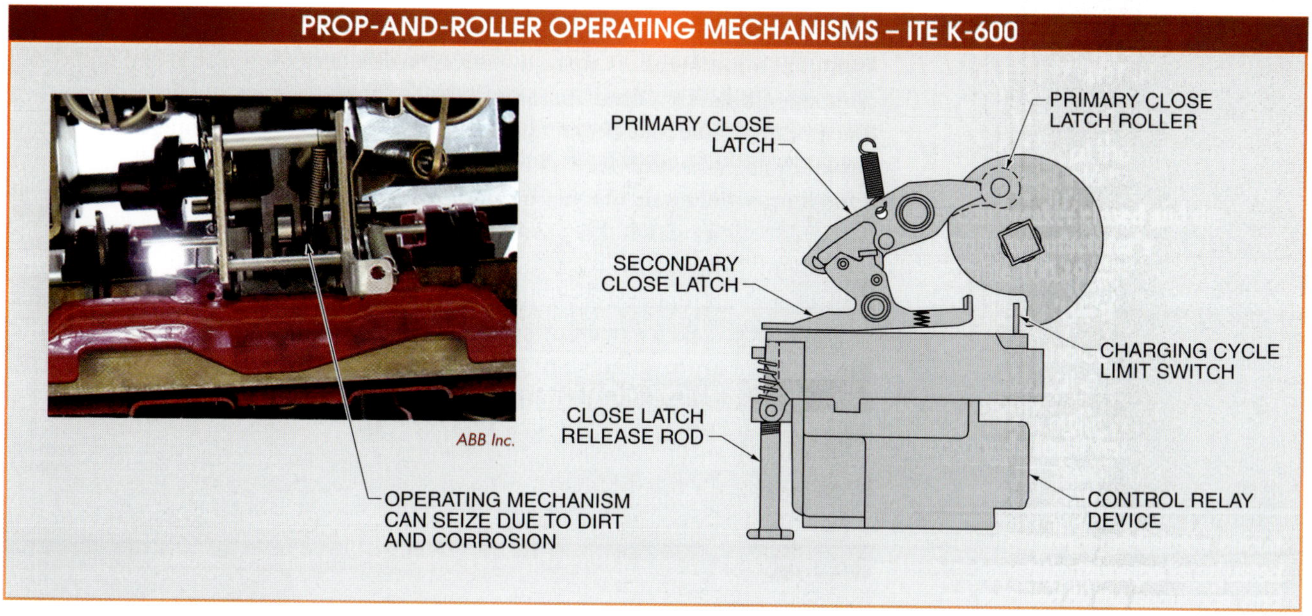

Figure 1-34. With an ABB/ITE K-600 circuit breaker, the small bearing in the main roller often seizes due to dirt or corrosion, which causes a trip-free condition or prevents the circuit breaker from tripping when the trip button is depressed.

Frames

The frame on a medium-voltage air or vacuum circuit breaker has the same function as the frame on an LVPCB. It supports the contact assemblies or vacuum bottles and the primary disconnects. The frame on medium-voltage air and vacuum circuit breakers is heavier due to the increased weight and spacing of the components. **See Figure 1-35.**

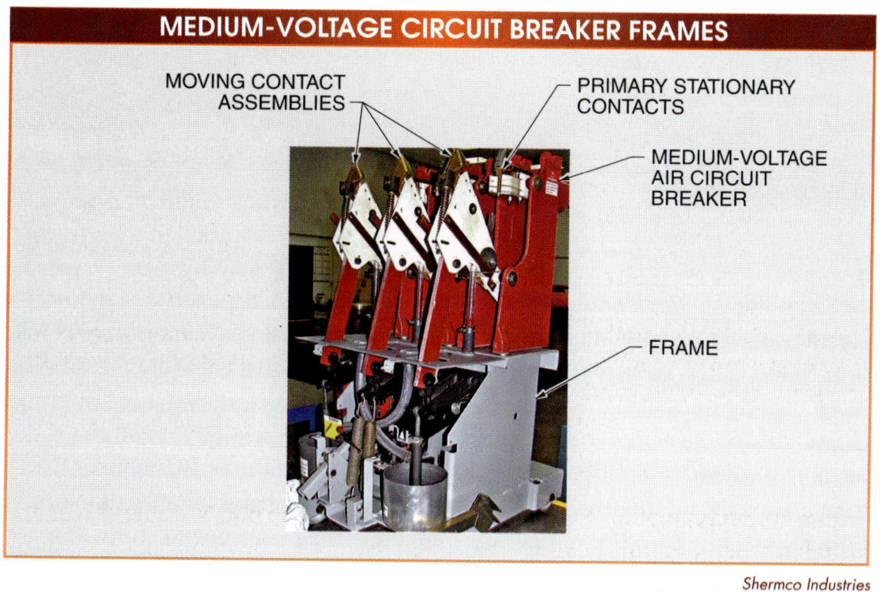

Figure 1-35. The frame on a medium-voltage circuit breaker supports the contact assemblies and primary disconnects.

Vacuum Circuit Breakers

Vacuum circuit breakers share the same basic operating mechanism and other components as air circuit breakers, except that vacuum bottles are used in place of arc chute and contact assemblies. The primary advantage of vacuum circuit breakers is their size. A vacuum circuit breaker is about half the size of a medium-voltage air magnetic circuit breaker. This means that there can be two per section of switchgear instead of just one.

Vacuum bottles also require little maintenance compared to air magnetic contact and arc chute assemblies. The contact moves only about ½″ in the vacuum bottle, and the opening springs are considerably lighter than the opening springs in an air-magnetic circuit breaker. This reduces wear on the assembly as well as their size and weight, since the use of heavy metal supports and a heavy frame can be reduced. **See Figure 1-36.**

VACUUM CIRCUIT BREAKERS

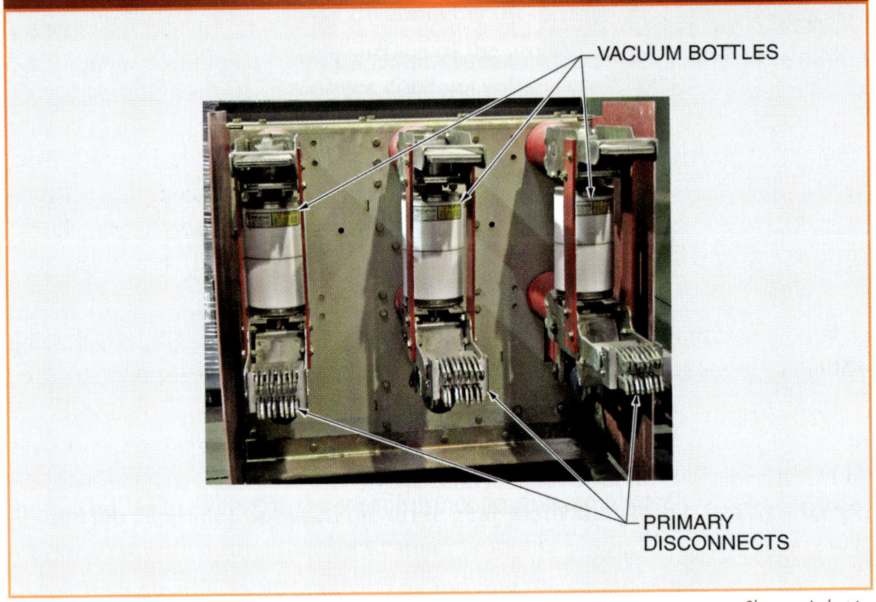

Shermco Industries

Figure 1-36. Vacuum circuit breakers have the same basic operating mechanisms as air circuit breakers.

Since the vacuum cannot be checked once the bottle is in service, it must be tested using an alternating current (AC) high-potential test set, or hipot tester, or a MAC tester. A *hipot tester* is a test instrument that measures insulation resistance by measuring leakage current. A magnetron atmospheric condition (MAC) tester measures the internal pressure of a vacuum bottle and helps determine the remaining life of a vacuum bottle. A vacuum bottle is most likely to fail when the circuit breaker is closed after being opened. The vacuum bottle does not usually fail during opening, since the load is removed from all three phases at the same time. However, when it is closed, especially when there is a large inrush of current or a fault, the bottle could fail violently. Most often it will not cause much damage, but there is a chance of personal injury.

Shermco Industries

Vacuum bottles can fail when closed, especially if exposed to large inrushes of current or faults.

CIRCUIT BREAKER SECONDARY COMPONENTS

Secondary components for a circuit breaker include current-limiting fuses, blown fuse indicators, undervoltage OCPDs, bell alarms, shutters, and inter-phase barriers. These devices are used to provide the circuit breaker with extra protection for short circuits, overloads, blown fuses, and undervoltage.

Current-Limiting Fuses

When a power system has higher available short-circuit current than the interrupting rating of the circuit breaker, current-limiting fuses are added to the line side of the circuit breaker. Current-limiting fuses vary somewhat in construction, but they typically have links or spiraled wires surrounded by silica. They also may have insulated end-caps to prevent accidental contact with energized components. **See Figure 1-37.**

CURRENT-LIMITING FUSES

Shermco Industries

SIEMENS LA

Shermco Industries

GE AKR

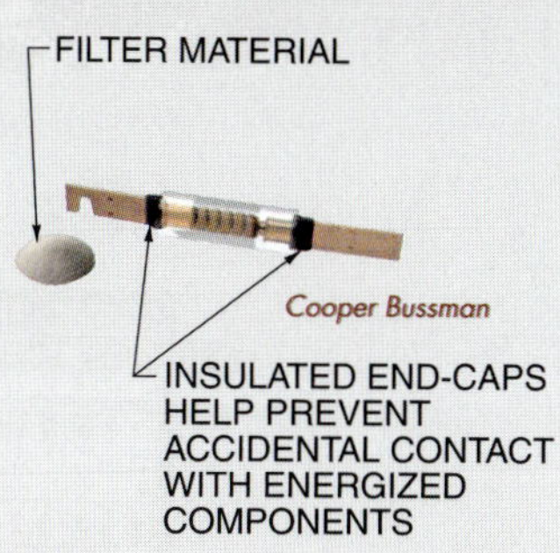

Cooper Bussman

INTERIOR COMPONENTS

Figure 1-37. Current-limiting fuses are used to provide extra protection to circuit breakers for short circuits, overloads, blown fuses, and undervoltage.

During normal loading and with overloads, these links or spiraled wires carry current without melting. During a short circuit, they fuse (melt) rapidly and cause the silica to fuse into glass. Glass, being an insulator, interrupts the current flow and opens the circuit. This process occurs so quickly that the interrupting rating of the circuit breaker is not exceeded.

Current-limiting fuses must be properly installed in the electrical power system to function correctly. The short-circuit current must exceed the threshold for the current-limiting characteristic of the fuse. If short-circuit current does not have a value high enough, a current-limiting fuse will act as a normal dual-element time delay fuse.

Blown Fuse Indicators and Lockouts

Blown fuse indicators are used on LVPCBs with current-limiting fuses. They indicate that a fuse has operated and the location of the fuse. They also ensure that the circuit breaker trips to prevent single phasing. Until a blown fuse is replaced, the circuit breaker cannot be reset if the device has a lockout feature. Each phase has a coil that is parallel with a fuse. **See Figure 1-38.**

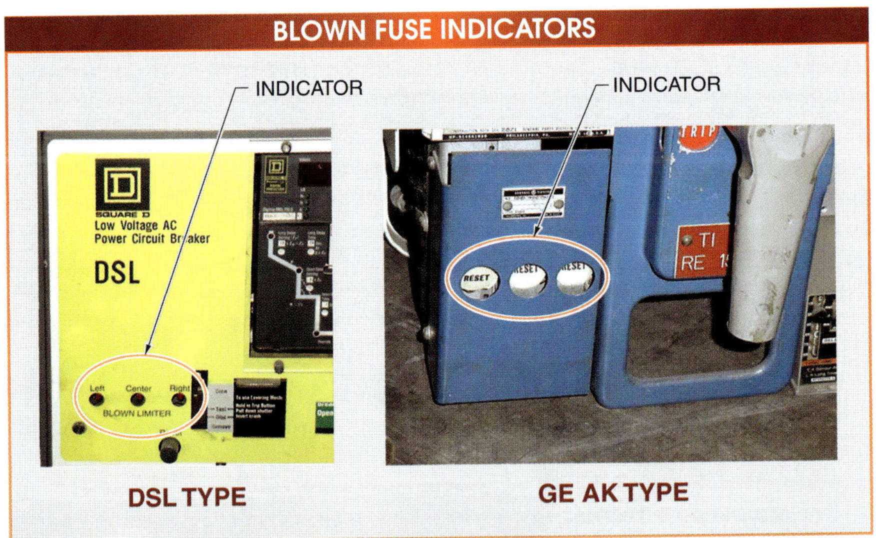

Shermco Industries

Figure 1-38. Blown fuse indicators indicate that a fuse has functioned and the fuse location. They also ensure that circuit breakers trip to prevent single phasing.

Undervoltage OCPDs

An *undervoltage OCPD* is a device that senses when the voltage of a circuit drops below a predetermined value. When the voltage drops below that value, the undervoltage OCPD actuates and causes the circuit breaker to open. Undervoltage OCPDs can be used on LVPCBs to protect motors, computers, and other voltage sensitive devices.

There are many applications for these devices. For example, without them, low voltage at motor starting increases the current required to a level that could damage the windings, while computers may fail their power supplies if the voltage drops too low. When testing an LVPCB that has an undervoltage OCPD, the undervoltage OCPD must be energized using the required control voltage. Medium-voltage air and vacuum circuit breakers use protective relays, rather than breaker-mounted undervoltage OCPDs.

Bell Alarms and Lockouts

A *bell alarm and lockout* is a device installed on LVPCBs to send an alarm and/or to lock out the circuit breaker when it trips. Some bell alarms are used to send an alarm when overcurrent reaches a certain level, which is often somewhat below the actual trip setting of the circuit breaker. This provides time to correct the overcurrent condition before it actually causes a trip condition. This option and the undervoltage OCPD are not used on many circuit breakers, since they are fairly specialized in their application. Some of the more common trade names for this device are the following:

- overcurrent trip switch (Siemens-Westinghouse)
- bell alarm and lockout device (General Electric)
- automatic trip indicator (ABB/ITE)

Shutters

Due to the high voltage found with medium-voltage metal-clad switchgear, shutters and barriers must be used to prevent accidental contact at the point where the circuit breaker connects to the main bus. Shutters are also found on some modern styles of low-voltage switchgear. In addition to the clamshell shutter, the most common type of device used on air-magnetic circuit breaker cubicles (and on some modern vacuum circuit breaker cubicles) is the shutter type. It consists of a flat fiberglass or aluminum sheet that drops over the energized bus connections when the circuit breaker is racked out. When the circuit breaker is racked in, the shutter is lifted by the racking mechanism so that the primary disconnects on the circuit breaker can contact the bus connections. **See Figure 1-39.**

Interphase Barriers

Since the arc chute assembly on a medium-voltage circuit breaker is energized, an interphase barrier is installed over the arc chute/contact assemblies. An *interphase barrier* is an electrical safety device that prevents contact with energized electrical components and protects against spark scatter. *Spark scatter* is the molten metal and debris that may be ejected from an arc chute when a circuit breaker interrupts a short circuit at or very near its rating.

SAFETY TIP

To avoid electrical shock, never attempt to repair a defective shutter mechanism while a bus is energized. If the shutter mechanism does not fully cover the energized bus when the circuit breaker is racked out of the cubicle, the circuit breaker should be reinserted and work cannot be performed until the shutter is repaired. This would require de-energizing the entire switchgear and bus so that it could be worked safely.

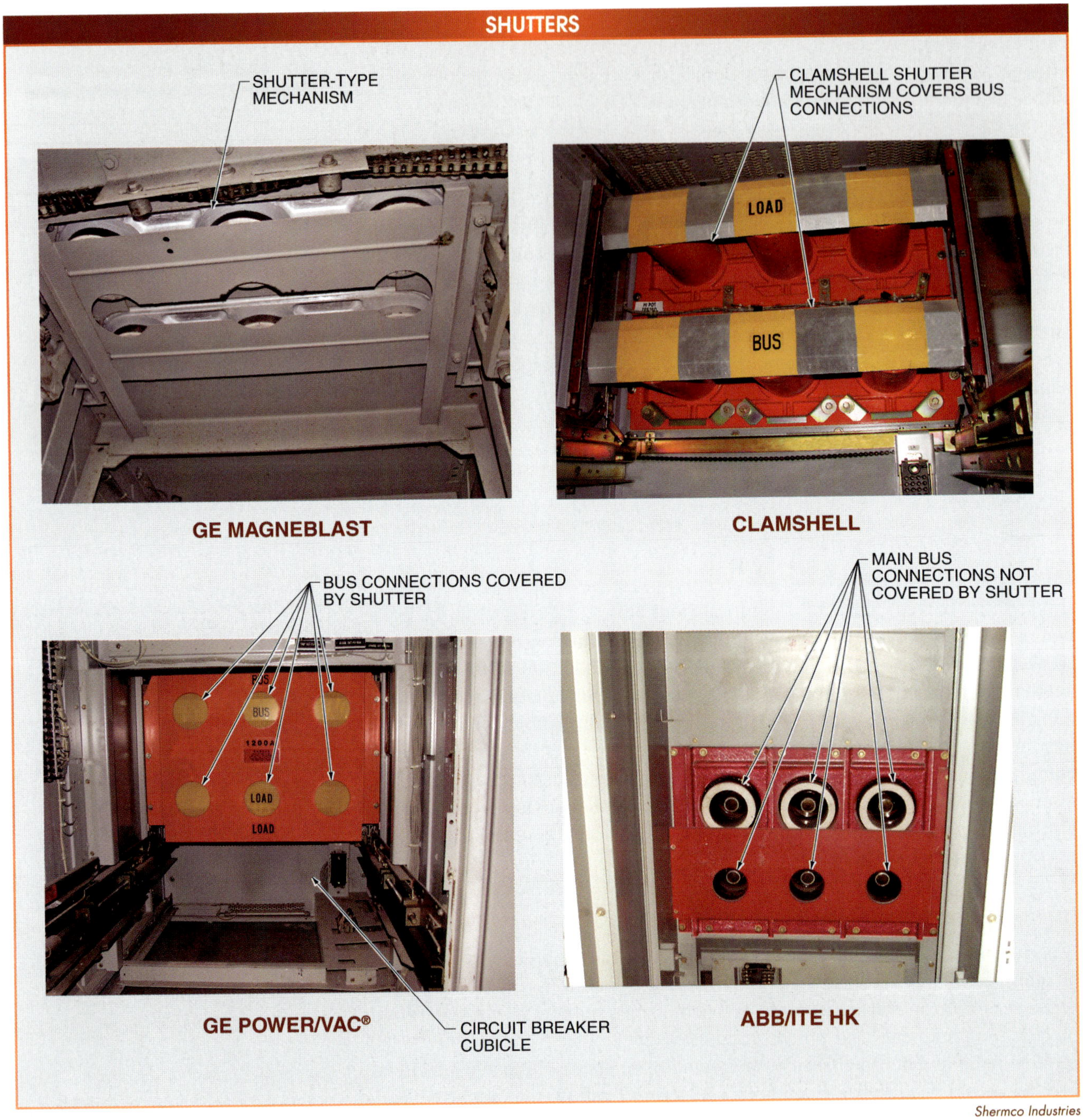

Shermco Industries

Figure 1-39. Due to the high voltage found on medium-voltage metal-clad switchgear, shutters (barriers) must be used to prevent accidental contact at the point where the circuit breaker connects to the main bus.

A handle located at the bottom front of the interphase barrier is used to pull the circuit breaker. LVPCBs also have interphase barriers, but they are smaller and only fit between the phase units. **See Figure 1-40.** *Note:* Never pull on the top or sides of an interphase barrier when moving a circuit breaker. It is held in place by nylon hardware and could break from its mounting.

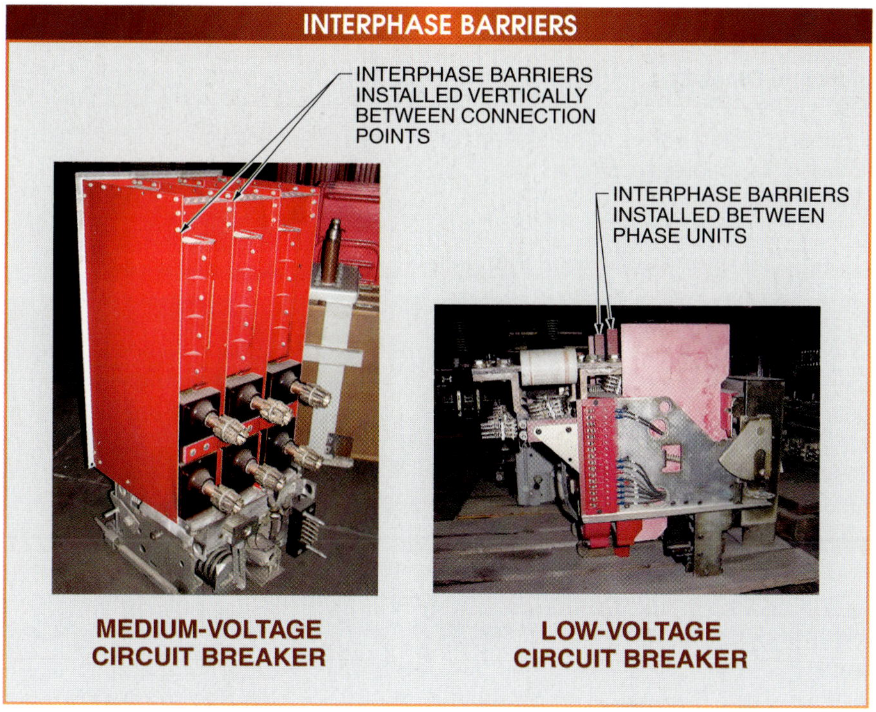

Shermco Industries

Figure 1-40. An interphase barrier is an electrical safety device used to prevent contact with energized electrical components and protect against short circuits and spark scatter.

INTERNAL SCHEMATIC DIAGRAMS

Many LVPCBs and virtually all medium-voltage metal-clad circuit breakers use electrically operated mechanisms. Electrically operated circuit breakers have trip and close circuits, a spring-charging motor, and indicator lights. Understanding how these control circuits function is important when troubleshooting circuit breakers. It is standard protocol to show the control circuits for a circuit breaker in a schematic diagram in a de-energized condition just as it would be sitting in a cubicle with no control power applied.

A *schematic diagram* is a drawing that shows electrical system circuitry with symbols that depict electrical devices and lines representing conductors. Only qualified persons should review schematic diagrams and perform work on circuit breakers. A *qualified person* is a person who has special knowledge, training, and experience in the installation, programming, maintenance, and troubleshooting of electrical circuit breaker equipment and also has the safety skills and knowledge that allow him/her to perform the task safely. Schematic diagrams are required for the setup and operation of common circuit breakers from OEMs such as ABB/ITE, General Electric, and Westinghouse/Cutler-Hammer.

INTERPRETING COMMON CIRCUIT BREAKER OEM SCHEMATIC DIAGRAMS

ABB/ITE K-Line Circuit Breaker Schematic Diagrams

With an ABB/ITE K-Line circuit breaker, the contact marked "CPS" (control power switch) on the schematic is used to disable the motor charging circuit and would normally be closed. When the circuit breaker is racked into a cubicle and is connected to control power, the following occurs:

- The spring charging motor immediately runs.
- The closing springs charge.
- Contact limit switches 1 and 3 (LS1 and LS3) open and shut off the motor.
- Contact LS2 closes and enables 52X (closing control coil).
- Contact 52b is closed because the circuit breaker is open.

The circuit breaker is now enabled to close and will remain in this condition until the CLOSE button is pressed. It is typical for the remote operating switch, identified as "Remote Close" on this drawing, to have a contact that stays closed when operated, as opposed to a momentary close contact. This means that control power is available to the circuit at all times that the control switch is in the closed position. When the remote close switch is operated, the following occurs:

- The circuit breaker main contacts close and discharge the closing springs.
- LS1 and LS3 close, but the motor circuit is disabled by 52b, which is now open.
- LS2 opens and disables the close circuit.
- LS3 closing energizes 52Y (anti-pump coil), which in turn closes 52Y/1 and opens 52Y/2.
- Contact 52Y/1 forms a parallel path, known as a seal-in circuit, around LS3. This keeps 52Y coil energized until control power is removed.
- Contact 52Y/2 remains open as long as the antipump (lockout) coil is energized.

The circuit breaker remains in this condition until it trips. When it does trip, the following occurs:

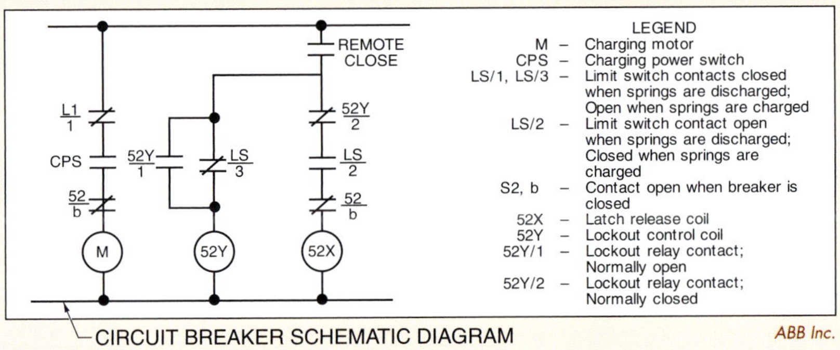

ABB Inc.

SCHEMATIC DIAGRAMS—ABB

- Contact 52b recloses.
- LS1 and LS3 close and LS2 opens.
- LS1 allows the spring-charging motor to charge the springs, causing LS1 and LS3 to open and LS2 to close.

During this sequence, 52Y remains energized due to the seal-in contact (52Y/1). Ordinarily, the circuit breaker would now be enabled to close, but 52Y prevents this by keeping the 52Y/2 contact open. To close the circuit breaker, control power would have to be removed by resetting the remote close switch to the open position. As a result, 52Y would be de-energized, which would close 52Y/2 and open 52Y/1.

General Electric

In a General Electric Power/Vac medium-voltage circuit breaker schematic diagram, the circuit breaker is shown de-energized with the circuit breaker open and no control power applied. When control power is applied, the following occurs:

- The spring charging motor (M) energizes.
- Coil 52Y energizes (prevents a closing operation until the springs are charged).
- When the springs are charged, $\frac{52}{SM/LS}$ locations open.
- Since the control switch (CS) is open, 52Y de-energizes.

The circuit breaker is now in position to close. To close the circuit breaker, the remote CS is turned to the closed position. In the closed position, the following occurs:

- Closing coil 52X becomes energized and releases the closing springs.
- 52 SM/LS locations close because the springs are discharged.
- Auxiliary contact 52b in the closing control circuit changes position as well as the 52a contacts in the trip circuit (they are mechanically driven).

continued on next page...

INTERPRETING COMMON CIRCUIT BREAKER OEM SCHEMATIC DIAGRAMS
(...continued from previous page...)

- Coil 52Y is energized. Three contacts change position, which open the circuit to 52X and close the 52Y seal-in contact.

When the breaker is fully set into the closed position, $\frac{52}{CL/MS}$ closes, allowing the spring-charging motor to operate. When the springs are charged, $\frac{52}{SM/LS}$ opens, stopping the spring-charging motor. As long as the remote control switch contacts remain closed, 52Y is energized, preventing repeated automatic reclosings (antipump). To trip or open the circuit breaker once the circuit breaker is closed, the following must occur:

- Both 52a contacts are closed, because the circuit breaker is closed.
- Terminals 9 (+) and 10 (–) are connected to the DC voltage source.
- Closing $\frac{CS}{T}$ energizes $\frac{52}{TC}$ (trip coil), opening the circuit breaker.

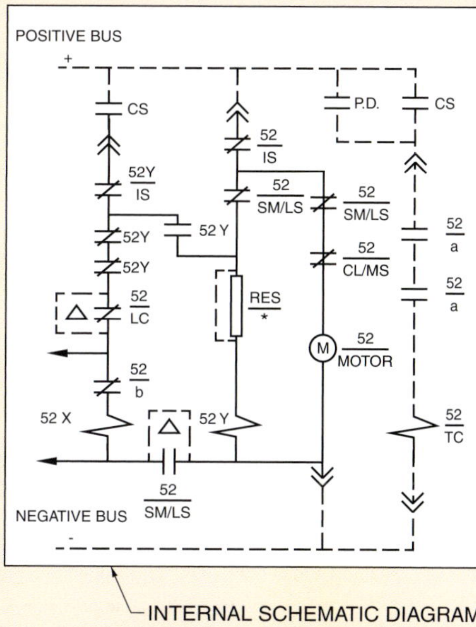

	CONTACT TYPE		DESCRIPTION
	N/O	N/C	
52/AUX SW	2	3	AUXILIARY SWITCH
52 X	6	15	CLOSING COIL (SPRG, REL. SOL'D)
52 Y	6	17	CONTROL RELAY
52/TC	5	3	TRIP COIL
52/IS	2	2	INTERLOCKING SWITCHES
52/CL/MS	6	13	CLOSING LATCH MONITORING SWITCH
52/SM/LS	6	7	POWER SWITCHES
52/LC	6	1	LATCH CHECKING SWITCH
52/MOTOR	2	11	SPRING CHARGING MOTOR

General Electric Company

SCHEMATIC DIAGRAMS—GENERAL ELECTRIC

Westinghouse/Cutler-Hammer

In a Westinghouse/Cutler-Hammer schematic diagram, the circuit breaker is shown open, and the closing springs are shown discharged. As soon as power is applied to the control circuit (terminals 7 and 4), the spring charging motor is energized, and the closing springs charge. Once the springs are charged, the limit switch (LS) changes from the closed position to the open position on wire 67, and the LS contact in wire 65/66 closes. This disables the spring charging motor and enables the closing circuit.

At this point, the antipump coil is short-circuited because the normally open LS contact closes. The "b" contacts are closed when the circuit breaker main contacts are open. When the CS is closed, the following occurs:

- The spring release (SR) coil is energized and allows the circuit breaker main contacts to close.
- Both limit switches change position. The spring charging circuit is enabled, and the spring release coil is disabled.
- Both "b" contacts open. One contact de-energizes the green light, while the other contact disables the spring charging motor, antipump coil, and spring release coil.
- The close contact in the CS remains closed, even when the handle is released, as it is a maintain contact.
- The "a" contact on wire 68 closes, which enables the shunt trip (ST) coil.
- The red indicator light illuminates.

The circuit breaker remains in this state until the trip coil is energized by a relay operation. When the trip coil is energized, the circuit breaker main contacts open, which closes both b contacts. The green light illuminates, and the spring charging motor is energized, which charges the closing springs.

When the springs are charged, both LSs change position. One LS de-energizes the motor, and the other energizes the antipump coil (Y). The spring release is not energized at this time because the Y contact opens and disables the spring release coil. The Y coil is energized because the CS closed contact is a maintain contact. A maintain contact stays closed even though the CS handle is released. If the CS is turned to the trip position, it will open the close/maintain contact in the CS, and the circuit breaker cannot close automatically.

The CS must be reset to the trip position at this time to reclose the circuit breaker. To trip the circuit breaker open, the trip coil (ST) is connected directly between the positive and negative power terminals. Once 52a contact closes (when the circuit breaker closes), the CS can be operated to the trip position. This will trip the circuit breaker and open the close/maintain contact on the CS.

continued on next page...

INTERPRETING COMMON CIRCUIT BREAKER OEM SCHEMATIC DIAGRAMS

(...continued from previous page)

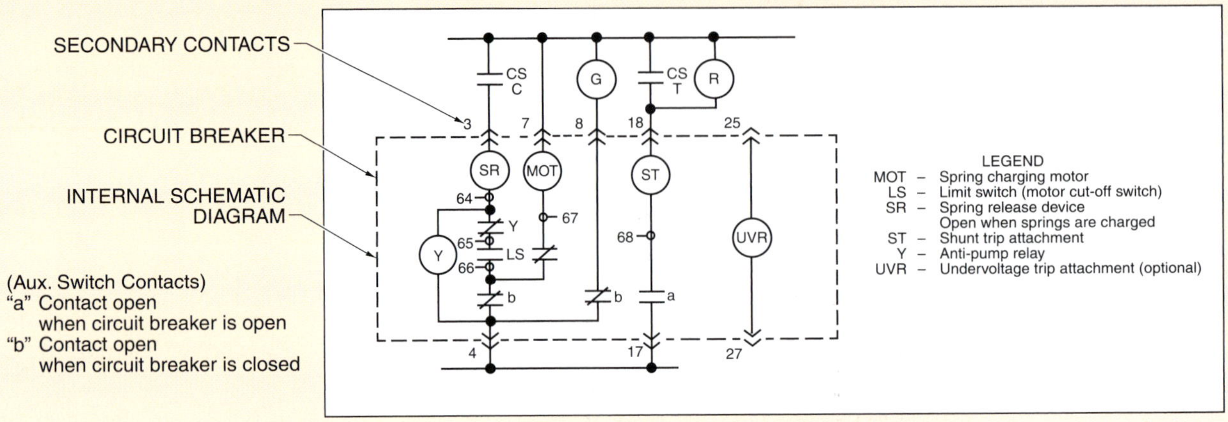

Cutler-Hammer

SCHEMATIC DIAGRAMS—WESTINGHOUSE/CUTLER-HAMMER

SUMMARY

The proper application of a circuit breaker, in accordance with its ratings, is required to ensure that it is not damaged in service. Exceeding any circuit breaker rating can cause failure. Knowledge and understanding of how a circuit breaker is constructed and operates is also required. Due to the size of circuit breakers and their complexity, technicians especially must be aware of safety hazards created by circuit breakers and how to avoid them.

The operating mechanism is one area that many technicians have difficulty understanding. Only qualified persons should perform work on circuit breakers. However, most OEMs provide drawings with four detailed views of the main components of a circuit breaker when in the four primary positions:

- open, springs discharged
- open, springs charged
- closed, springs discharged
- closed, springs charged

These drawings aid in the troubleshooting of circuit breakers. The electrical schematic is also important for troubleshooting circuit breakers. Knowing that an interlock could prevent the circuit breaker from closing or that a limit switch must be in a certain position will allow a logical troubleshooting path to be developed. A few minutes of working out the proper sequence will allow quick troubleshooting during unplanned outages and get the circuit breaker back into service as quickly as possible.

DIGITAL RESOURCES
ATPeResources.com/QuickLinks
Access Code: 223436

chapter 1

CIRCUIT BREAKER CONSTRUCTION AND OPERATION

CHAPTER REVIEW

Name ______________________ Date ______________

True-False

T F **1.** The actual voltage rating of the circuit breaker is the lowest voltage listed on the name-plate.

T F **2.** Medium-voltage circuit breakers rely on protective relays rather than built-in OCPDs to provide a trip signal.

T F **3.** When the circuit breaker closes, the arcing contacts make first.

T F **4.** The NEC® definition states that a circuit breaker will open a circuit *"without damage to itself when properly applied within its rating."*

T F **5.** Closing springs accelerate the contacts to the closed position and hold the contacts closed.

T F **6.** By standard, current-limiting fuses must operate in two cycles or less.

T F **7.** The only time the trip latch changes position is when the circuit breaker is tripped.

T F **8.** It is important not to change the shape of the contact faces, as this could cause the circuit breaker to malfunction.

T F **9.** Any part of an air-frame circuit breaker that is damaged or compromised can be repaired or replaced.

T F **10.** Settings above 100% of the continuous current rating do not increase the amount of current the device can safely carry.

T F **11.** Air magnetic contact and arc chute assemblies require little maintenance compared to vacuum bottles.

T F **12.** A shutter consists of a flat fiberglass or aluminum sheet that drops over the energized bus connections when the circuit breaker is racked out.

T F **13.** The main contacts in a circuit breaker are constructed of an alloy harder than arcing contacts.

T F **14.** A vacuum bottle is most likely to fail when the circuit breaker is opening.

T F **15.** Insulated-case circuit breakers have lower short-circuit current ratings than molded-case circuit breakers.

T F **16.** When a conductor becomes annealed, it does not carry the proper amount of current due to increased impedance.

T F **17.** Molded- and insulated-case circuit breakers have small, light contacts that can interrupt an arc quickly.

T F **18.** A megavolt ampere (MVA) rating is the rated voltage times the rated interrupting current of a circuit breaker.

T F **19.** It is standard protocol to show the control circuits for a circuit breaker in a schematic diagram in an energized condition.

T F **20.** Modern circuit breaker operating mechanisms are quick make, quick break.

T F **21.** Until a blown fuse is replaced, the circuit breaker cannot be reset if the device has a lockout feature.

T F **22.** Since the vacuum in vacuum bottles is a perfect vacuum, there is no arcing.

T F **23.** A thermal-magnetic overcurrent protective device uses a bimetallic strip placed in the current path.

T F **24.** Current-limiting fuses do not have a rated short-time current.

T F **25.** New generation current-limiting circuit breakers differ from standard molded- and insulated-case circuit breakers primarily by their contact structures.

Multiple Choice

______________ **1.** The ___ open and close circuits.
- A. backboard assemblies
- B. terminal connections
- C. contact assemblies
- D. arc chutes

______________ **2.** The voltage for low-voltage circuit breakers can range up to ___ V.
- A. 635
- B. 725
- C. 780
- D. 910

______________ **3.** Arc chutes in low-voltage power circuit breakers cool down the electrical arc and extinguish it in ___ cycles.
- A. four
- B. six
- C. eight
- D. ten

______________ **4.** A(n) ___ is an electrical safety device that prevents contact with energized electrical components and protects against spark scatter.
- A. shutter
- B. barrier
- C. undervoltage OCPD
- D. interphase barrier

_______________ **5.** The actual continuous current rating of a low-voltage circuit breaker is always determined by the rating of its ___.

A. overcurrent protective device
B. trip latch
C. arc dividers
D. arc chutes

_______________ **6.** ___ disconnects connect a circuit breaker to the primary bus and allows load current to flow through the circuit breaker.

A. Main
B. Secondary
C. Tertiary
D. Ground

_______________ **7.** Circuit breaker ratings are determined by the National Electrical Manufacturer's Association (NEMA) and the ___.

A. Occupational Safety and Health Administration (OSHA)
B. Institute of Electrical and Electronics Engineers (IEEE)
C. National Electrical Contractors Association (NECA)
D. National Fire Protection Association (NFPA)

_______________ **8.** The arcing contacts on low-voltage circuit breakers are often made of a silver and ___ alloy.

A. zinc
B. tungsten
C. copper
D. aluminum

_______________ **9.** A circuit breaker that uses current-limiting fuses will not have a(n) ___ rating but will use that of the fuse.

A. frequency
B. voltage
C. interrupting
D. frame

_______________ **10.** Circuit breakers are typically rated at 50 Hz or ___ Hz.

A. 60
B. 75
C. 80
D. 95

_______________ **11.** With ___ circuit breakers, arc interruption takes place in air and is usually assisted by arc chutes, blowout coils, and puffer tubes.

A. vacuum
B. gas
C. air-blast
D. air-magnetic

_______________ **12.** Outdoor circuit breakers and indoor circuit breakers without enclosures shall have ratings based on ___°C ambient temperature.
A. 20
B. 30
C. 40
D. 50

_______________ **13.** Bushings connect the ___ to the stationary contact assembly.
A. primary disconnects
B. secondary disconnects
C. arc chutes
D. backboard

_______________ **14.** A(n) ___ circuit breaker cannot be opened for maintenance or renewal because its UL rating will be nullified.
A. molded-case
B. sealed-case
C. vacuum
D. air-frame

_______________ **15.** A molded-case circuit breaker can only have an overcurrent protective device with a continuous current rating of ___% of the frame rating.
A. 80
B. 85
C. 90
D. 95

_______________ **16.** Vacuum circuit breakers share the same basic operating mechanism and other components as air circuit breakers, except that vacuum bottles are used in place of ___.
A. primary disconnects
B. overcurrent protective devices
C. arc chute and backboard assemblies
D. arc chute and contact assemblies

_______________ **17.** ___ contacts act as a second set of arcing contacts.
A. Main
B. Intermediate
C. Auxiliary
D. Secondary

_______________ **18.** Instead of zinc being used to harden the arcing contacts, ___ is often used for medium-voltage air circuit breakers.
A. aluminum
B. steel
C. copper
D. tungsten

_______________ **19.** Circuit breaker ___ are current-carrying components that connect the circuit breaker to the bus or other connections.
A. disconnects
B. frames
C. fuses
D. shutters

______________ **20.** When the voltage drops below a predetermined value, the ___ actuates and causes the circuit breaker to open.
- A. shutter
- B. fuse
- C. undervoltage OCPD
- D. interphase barrier

Completion

______________ **1.** A(n) ___ is a circuit breaker, fuse, or switch with an element that disconnects or discontinues current flow when the amount of current exceeds the design load.

______________ **2.** A(n) ___ is a metal plate attached to a device, such as a circuit breaker, that lists its technical specifications.

______________ **3.** A(n) ___ is a sustained discharge of electricity across a gap in a circuit or between electrodes, usually accompanied by the electrodes (contacts) being vaporized and/or melted by the extreme heat.

______________ **4.** A(n) ___ assembly is the section of a circuit breaker that supports stationary contacts, the primary line, and load disconnects.

______________ **5.** A(n) ___ is an electrical device that opens and closes a circuit by nonautomatic means and automatically opens a circuit when a predetermined overcurrent is reached, without damage to itself.

______________ **6.** A(n) ___-voltage power circuit breaker (MVPCB) is a circuit breaker that operates with voltages ranging from 2300 V to 69,000 V.

______________ **7.** A(n) ___ is a drawing that shows electrical system circuitry with symbols that depict electrical devices and lines representing conductors.

______________ **8.** A(n) ___ contact is the contact of a switching device on which an arc is drawn after the main contacts have parted.

______________ **9.** The ___ rating is the maximum short-circuit current a circuit breaker can safely clear without external damage.

______________ **10.** As current flows through a conductor, it creates heat, which is often referred to as ___.

______________ **11.** A(n) ___ circuit breaker is a circuit breaker that is similar in construction to a molded-case circuit breaker but typically uses an electronic or digital OCPD and has much higher interrupting ratings.

______________ **12.** The ___ rating is the continuous current rating of all current-carrying parts of a low-voltage circuit breaker, excluding its OCPD.

______________ **13.** The operating mechanism of a molded-case circuit breaker opens and closes the contact assemblies and has three positions: open, closed, and ___.

______________ **14.** Both arcing and main contacts can be butt- or ___-style contacts.

______________ **15.** A(n) ___ is a device installed on low-voltage power circuit breakers to send an alarm and/or to lock out the circuit breaker when it trips.

______________ **16.** Gas circuit breakers use ___ gas to interrupt arcs.

______________ **17.** Since the vacuum cannot be checked once a bottle is in service, it must be tested using an alternating current (AC) high-potential test set, or hipot tester, or a(n) ___ tester.

______________________ **18.** The connection to the ___ disconnect is the first connection made when racking a circuit breaker into the cubicle and is the last connection broken when racking it out of the cubicle.

______________________ **19.** The ___ switch is used to disable the circuit breaker during racking operations.

______________________ **20.** The ___ is the temperature rise of an electrical device or component above ambient, or surrounding air, temperature.

______________________ **21.** ___ is the molten metal and debris that may be ejected from an arc chute when a circuit breaker interrupts a short circuit at or very near its rating.

______________________ **22.** The short-time current rating of a circuit breaker is the maximum current it can safely carry for two ___ sec periods, with a 15 sec rest period between them at zero current.

______________________ **23.** A(n) ___ is an electrical overcurrent protective device with a fusible portion that is heated and broken by the passage of excessive current.

______________________ **24.** A low-voltage power circuit breaker is also referred to as a(n) ___ because the frame of the circuit breaker is only used to support the components and is not part of the insulation system.

______________________ **25.** Control power is brought into the circuit breaker through the ___ disconnects.

Short Answer

1. Explain how the arc chute of an MCCB or ICCB extinguishes an arc.

2. How do low- and medium-voltage circuit breaker arc chutes differ?

3. Why do contacts have a contoured surface?

4. Explain the function of current-limiting fuses.

5. Why is improper torqueing an issue with terminal lugs?

Matching

Circuit Breaker Nameplates

_______________ 1. Frame rating

_______________ 2. Interrupting rating

_______________ 3. Model number

_______________ 4. OEM

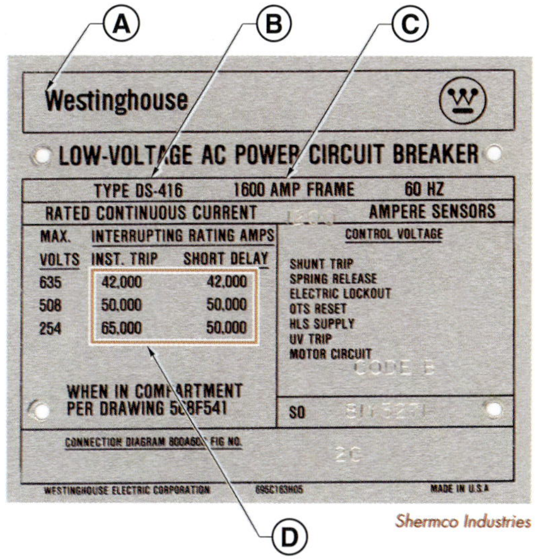

Shermco Industries

Medium-Voltage Power Circuit Breakers

_______________ 1. Air-blast

_______________ 2. Air-magnetic (ACB)

_______________ 3. Gas

_______________ 4. Oil (OCB)

_______________ 5. Vacuum (VCB)

A

B

C

D

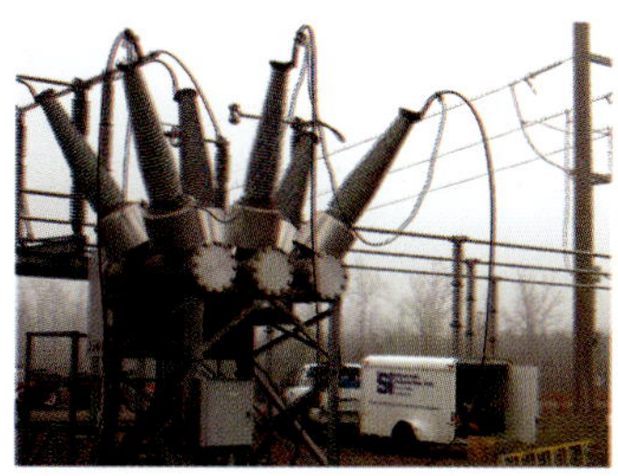

E

Shermco Industries

Cutler-Hammer DS Operating Mechanisms (Open, with Springs Charged)

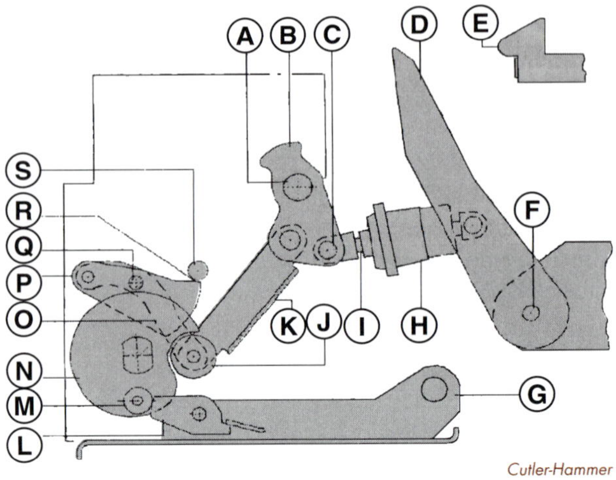

Cutler-Hammer

_______________ 1. Center pole lever

_______________ 2. Close cam

_______________ 3. Insulating link adjusting stud and locknut

_______________ 4. Insulating link

_______________ 5. Main drive link

_______________ 6. Main roller

_______________ 7. Mechanism side frame

_______________ 8. Moving contact arm

_______________ 9. Moving contact pivot pin

_______________ 10. Pivot pin

_______________ 11. Pole lever pin

_______________ 12. Pole shaft

_______________ 13. Roller constraining link

_______________ 14. Spring release latch

_______________ 15. Stationary arcing contact

_______________ 16. Stop roller

_______________ 17. Trip latch

_______________ 18. Trip shaft latching surface

_______________ 19. Trip shaft

chapter 2

LOW- AND MEDIUM-VOLTAGE ARC INTERRUPTION

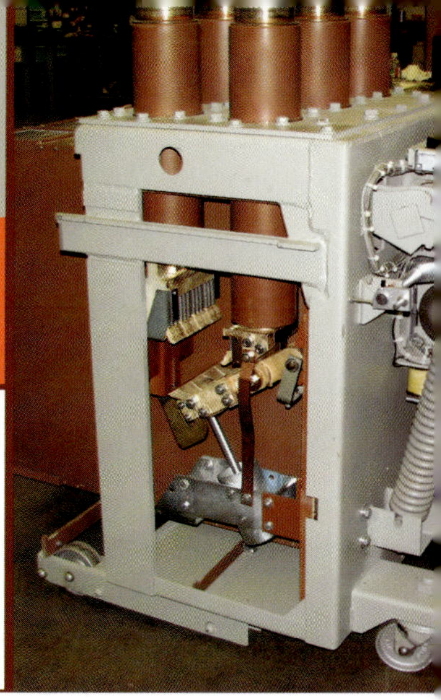

Installing and removing a circuit breaker from its bus creates an arc flash hazard. Technicians must follow safety procedures per original equipment manufacturer (OEM) instructions, understand the arc interruption process, and have knowledge of the equipment and components installed on circuit breakers to interrupt and extinguish electric arcs. The arc interruption process varies, depending on the voltage level and type of circuit breaker.

OBJECTIVES

- Define overcurrent and short circuit.
- Describe how air becomes ionized.
- List the four functions of an arc chute.
- Define restrike.
- State the effect improper maintenance can have on a circuit breaker's arc clearing time.
- State where in a sine wave an electric arc can be interrupted.
- List the components of low- and medium-voltage arc chutes.
- Describe how medium-voltage arc chutes differ from low-voltage arc chutes.
- List the components of a vacuum bottle.
- Describe how an arc is interrupted in a vacuum circuit breaker.

ARC CONDITIONS

Circuit breakers are switches designed to open and close circuits. They may be operated manually or automatically through the use of overcurrent protective devices (OCPDs). The automatic operation of a circuit breaker occurs when there is an abnormal condition such as an overcurrent or short circuit.

For example, an overcurrent condition can occur when a coupling between an electric motor and a pump is misaligned. The misalignment causes the motor to work harder to overcome the increased mechanical resistance. As a result, the motor draws more than the rated current. If allowed to continue, this overcurrent condition can cause the motor to overheat and the insulation to fail due to thermal degradation. These conditions create current that is several times the full-load rating of the circuit breaker.

Overcurrent is electrical current in excess of the equipment limit, total amperage load of a circuit, or conductor or equipment rating. An overcurrent condition is not a short circuit. A *short circuit* is an unintentional connection of two ungrounded conductors that have a potential difference between them. A short circuit occurs when there is an unintentional grounding of an energized conductor, usually resulting in an arc flash. Short circuits can be four or more times the full-load current of a circuit.

When a circuit breaker opens to clear a system short circuit, an arc is also created between the opening contacts, which are located inside the arc chute. The arc between the circuit breaker contacts occurs due to the ionization of air, just as the air is ionized during a system short circuit. In short-circuit conditions, the arc flows from an energized conductor/component to ground or possibly

KEY TERMS

- restrike
- maximum total clearing time
- zero crossover point
- persistence

phase-to-phase. It remains a conductor, and short-circuit current continues to flow through it as long as the air is ionized. This ionization of air occurs when electrons are knocked out of their orbits around the molecules of air by the extreme heat of an electric arc. Since the electrons are no longer attached to air molecules, the electricity uses the free electrons as a path across the arc chute. Circuit breakers use arc chutes to deionize the air and interrupt the flow of short-circuit current.

ARC INTERRUPTION

An arc chute (arc extinguisher) contains, cools, stretches, and deionizes and extinguishes the arcs between contacts. This occurs within 1⁄10 of a second (six cycles) or less and is critical to the safe operation of the circuit breaker and power system. A circuit breaker can fail to clear (extinguish) an arc due to the following reasons:

- arc chute component is defective
- arc chute is contaminated with moisture or carbon residue from previous operations
- mechanical failure of the operating mechanism in that it opens too slowly or even fails to open

When a circuit breaker does not clear the arc, the arc re-establishes itself and a restrike occurs. A *restrike* is a resumption of current between the circuit breaker's contacts if the circuit breaker fails to clear an arc. **See Figure 2-1.**

RESTRIKES

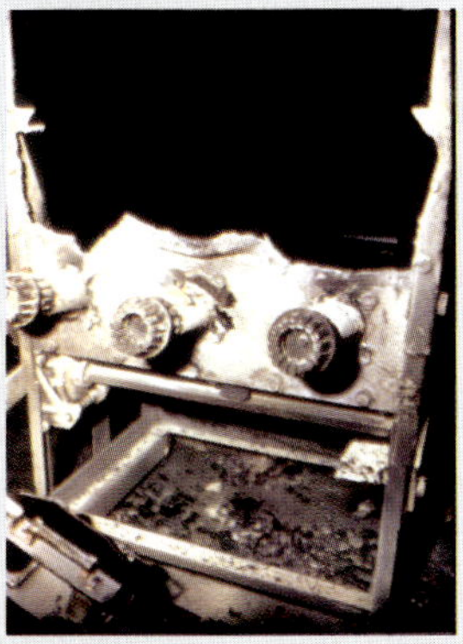

15 kV AIR CIRCUIT BREAKER WITH RESTRIKE DAMAGE

Figure 2-1. Circuit breakers can become permanently damaged by restrikes.

One common problem, especially with older arc chutes, is that arc chutes can absorb moisture. Older arc chutes are made of materials such as ceramic, asbestos, and celluloid or phenolic plastic resins that tend to absorb moisture.

When a circuit breaker attempts to interrupt an arc, the moisture absorbed by the arc chute is released into the arc chute as steam. Small amounts of moisture can cause catastrophic restrikes. To help prevent restrikes, arc chutes can be dried by placing heat lamps around them while they are covered with a tarp or by using an industrial oven. Both methods take several hours or days, depending on the saturation of the arc chute.

Maximum total clearing time is the time required to interrupt an arc and is considered to be from when the arc starts until it is completely extinguished. This characteristic is used to properly coordinate power systems so that they will trip in the right sequence (selective tripping). During selective tripping, the device closest to the load operates first. If that device does not operate, due to mechanical or electrical failure, then the next device upstream operates to clear the fault. If that device does not operate, then the next device upstream operates and so on. This process is sometimes referred to as "cascading."

Improper maintenance can also cause circuit breakers to malfunction, either by slowing their operating speed or by preventing them from operating. When this occurs, arc flash studies and their labels do not correctly reflect the true incident energy that would be received by a worker if exposed to an arc flash.

Typically, incident energy is proportional to time. If the duration of operation doubles from two cycles to four cycles, the incident energy received by the technician also doubles. The personal protective equipment (PPE) and arc-rated clothing will be inadequate. The situation may result in serious injuries or a fatality. This aspect of incident energy makes the maintenance of OCPDs a matter of safety not just maintenance.

The maximum total clearing time includes the amount of time it takes for the circuit breaker to clear the arc, the operating time of the OCPD and any other auxiliary relay that may be part of the circuit. The incident energy is directly affected by the speed of the circuit breaker's operation, the distance which its contacts must travel and the ability of the arc chutes to extinguish the arc according to the manufacturer specifications. Changing one of these factors from its design specifications can cause major problems when the circuit breaker must operate at or near its interrupting rating.

Another factor that must be considered for arc interruption is the zero crossover point of the alternating-current sine wave. The *zero crossover point* is the point at which current alternates and begins to change direction. **See Figure 2-2.**

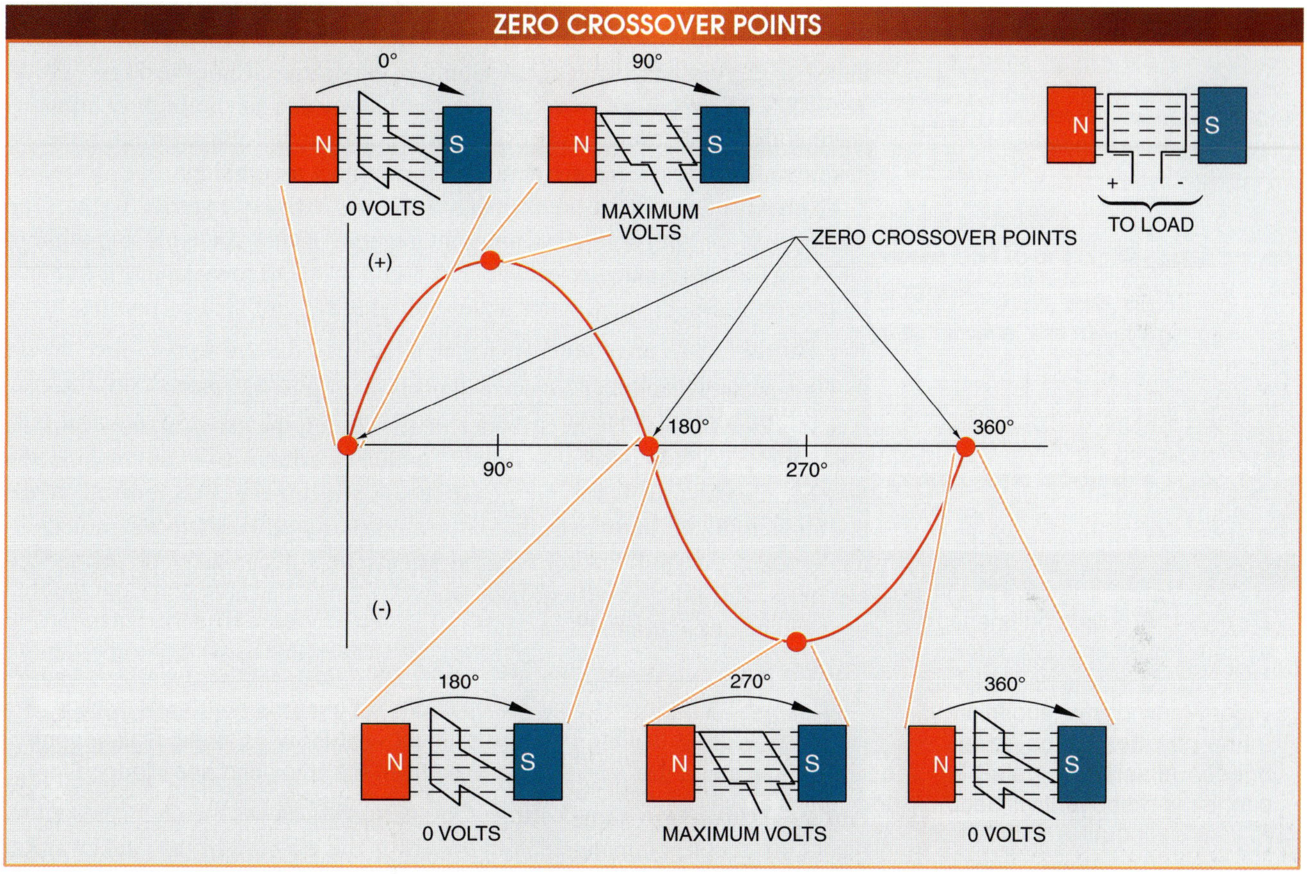

Figure 2-2. The zero crossover point is the point where current alternates and begins to change direction and the only point where an arc can be interrupted.

As the generator armature in Figure 2-2 begins to rotate, it is at a zero crossover point (0°). Rotating the armature to the 90° position causes the number of lines of flux cutting the armature to increase to its maximum amount, which produces the maximum voltage and current. As the armature continues to rotate, the number of lines of flux being cut decreases until it is at the 180° position. At this position, no lines of flux are being cut, and it is again at a zero crossover point.

It is only at a zero crossover point that the current can be interrupted, so the arc must be stretched, cooled, and then, at zero crossover, extinguished. The approximate maximum total clearing times for a circuit breaker may vary, depending on the circuit breaker type. **See Figure 2-3.**

ARC CHUTE MAXIMUM TOTAL CLEARING TIMES		
OCPD Type	**Cycles**	**Seconds**
Molded-case	1.5–2	0.025–0.033
Low-voltage power	3–5	0.05–0.08
Medium-voltage air	5–8	0.08–0.0133
Medium-voltage vacuum	2–3	0.033–0.05

Figure 2-3. The approximate maximum total clearing times for a circuit breaker may vary, depending on the circuit breaker type.

Low-Voltage Circuit Breaker Arc Chute Components

SAFETY FACT

With alternating current (AC) circuits, arcs are extinguished at a current of zero (0 V) when the AC voltage applied across the arcing contacts reverses polarity.

During arc interruption, as the contacts of a circuit breaker begin to open, the hot arc rises. Most air circuit breakers have arc chutes positioned above the contact assemblies since arcs have a tendency to rise. Most circuit breakers have one set of contacts per phase. Some larger-frame circuit breakers (3000 A and 4000 A) may have two contacts per phase, due to the high continuous current and interrupting ratings. Therefore, some low-voltage circuit breakers have two openings in the arc chute for the contact assemblies.

A low-voltage circuit breaker arc chute has a heavy exterior frame. The frame must contain, at its maximum interrupting rating, an explosion without allowing the arc products to escape. The components of low-voltage arc chutes include arc runners, arc divider plates, and muffler and vent assemblies.

Arc Runners. An arc runner provides a path for the arc to rise into the arc chute. It also helps reduce the amount of time required to stretch out the arc.

Arc Divider Plates. The arc divider plates stretch the arc as it rises into the arc chute. Stretching the arc cools it, which is critical to extinguishing the arc. The arc can travel through the air because the air becomes ionized. When the air is ionized, the electrons that are normally attached to air molecules are knocked out of their orbits by the intense heat of the arc. The unattached electrons facilitate the current flow from the stationary contact to the moving contact and act as conductors.

Muffler and Vent Assemblies. The muffler and vent assembly on the end of an arc chute serves multiple purposes. It reduces the sound of arc interruption and allows combustible gases that are created during the arc to safely escape. It also captures any molten metal that may try to escape from the arc chute. **See Figure 2-4.**

Arc Interruption in Low-Voltage Circuit Breakers. As the arc is created by the arcing contacts moving apart, it travels up the arc horn of the contact and is transferred to the arc runner. The arc runner facilitates the transfer since the arc has a magnetic field and is conductive. The heat of the arc causes the arc to move up into the arc chute. Up to this point, arc extinguishing on low- and medium-voltage air circuit breakers is the same. However, low-voltage circuit breakers rely totally on the arc runners and arc divider plates to stretch out the arc and cause it to cool.

As the arc is cooled, it becomes deionized. As the current tries to alternate through a zero crossover point, the arc is extinguished because there is no ionized path through the air. In low-voltage circuit breakers this all occurs in two cycles (molded- and insulated-case) to four cycles (low-voltage power). The larger the circuit breaker, the slower the arc is extinguished. Other factors include the system voltage and short-circuit current that is being interrupted, but larger circuit breakers are typically slower at extinguishing an arc than smaller ones. **See Figure 2-5.**

LOW-VOLTAGE ARC CHUTE COMPONENTS

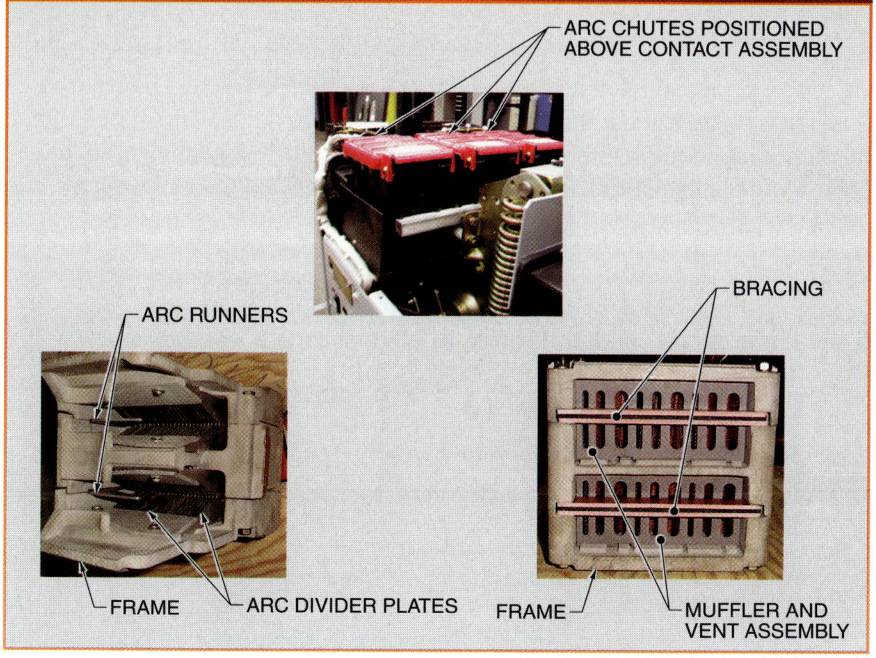

Figure 2-4. The components of low-voltage arc chutes include arc runners, arc divider plates, and muffler and vent assemblies.

ARC INTERRUPTION IN LOW-VOLTAGE CIRCUIT BREAKERS

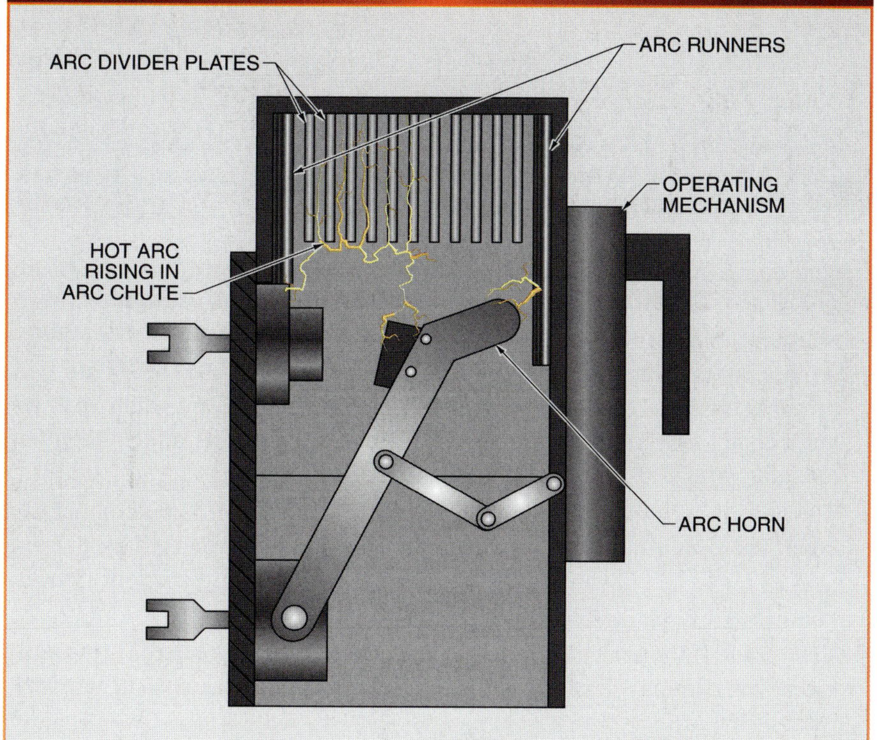

Figure 2-5. Low-voltage circuit breakers rely totally on arc runners and arc divider plates to stretch out an arc. This causes the arc to cool, become deionized, and then be extinguished.

Medium-Voltage Air-Magnetic Circuit Breaker Arc Chute Components

In medium-voltage air circuit breakers, the arc is moved through the extinguishing process by various additional components such as puffers and blowout coils. Some types of medium-voltage circuit breakers have arc chutes positioned at the rear of the breaker assembly. **See Figure 2-6.** The puffer assembly is not an arc chute component, but it works with the arc chute to quickly extinguish the arc.

MEDIUM-VOLTAGE CIRCUIT BREAKER ARC CHUTES

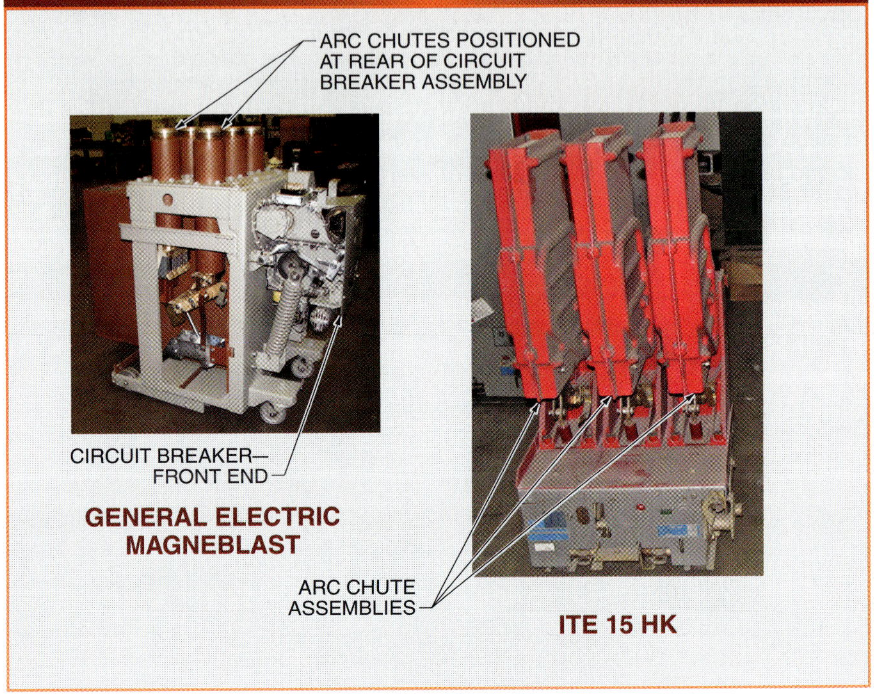

Figure 2-6. Some types of medium-voltage circuit breakers have arc chutes positioned at the rear of the breaker assembly.

PUFFER ASSEMBLIES

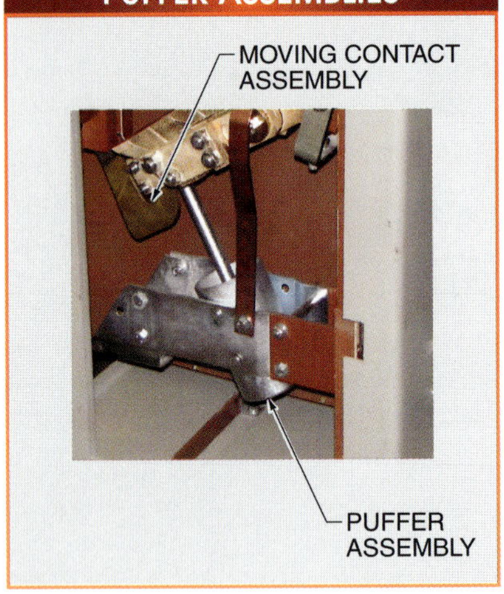

Figure 2-7. The puffer assembly is a bellows or piston/tube assembly that creates a strong puff of air aimed at the area where the contacts are parting.

The components involved in arc interruption on a medium-voltage air-magnetic circuit breaker include arcing contacts, puffer assembly, throat, arc runners, blowout coils, pole pieces, and arc dividers.

- Arcing contacts are designed to withstand the heat created by an arc with minimal erosion. The arcing contacts are made of an alloy, often zinc, silver, copper, and sometimes tungsten. The arcing contacts protect the main contacts, which are made of a softer alloy.
- The puffer assembly is a bellows or piston/tube assembly that is compressed when the circuit breaker contacts open. The compression creates a strong puff of air to be blown directly into the arc path, forcing it up into the arc chute. **See Figure 2-7.**
- The throat of the arc chute is where the arc is initially created. It is usually lined with ceramic or asbestos to allow it to withstand the extreme temperatures of the arc.
- Arc runners are conductive strips, usually copper, and run from the bottom of the arc chute to the top on each side.
- Blowout coils are magnetic coil assemblies that are energized when the arc

travels up the arc runner. The blowout coil assemblies create a magnetic field that attracts the arc, accelerating it up into the arc chute.

- Pole pieces are mounted on the outside surface of the arc chute and are made of layers of laminated sheet silicone steel, similar to a transformer core. As the arc rises in the arc chute, the pole pieces are energized and create a strong counter electromagnetic force that helps contain and force the arc up into the arc chute.
- Arc dividers have the same purpose in medium-voltage air magnetic circuit breakers as they do in low-voltage circuit breakers. They stretch, cool, and deionize the arc so it can be extinguished at the zero crossover point.

All these components must function properly in order to extinguish the arc. **See Figure 2-8.**

MEDIUM-VOLTAGE ARC CHUTE COMPONENTS

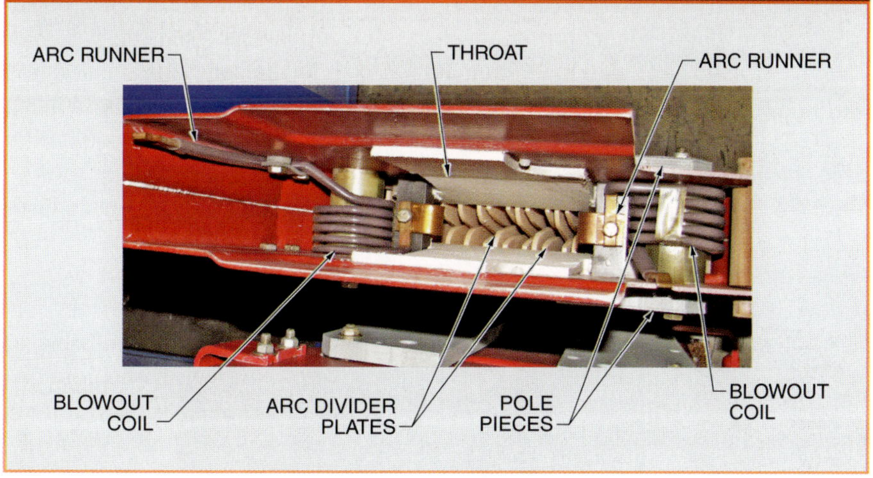

Figure 2-8. Medium-voltage arc chute components include arc runners, throats, blowout coils, pole pieces, and arc divider plates.

Arc Interruption in Medium-Voltage Air-Magnetic Circuit Breakers. With a medium-voltage air circuit breaker, current flows into the line-side primary disconnects, through the contact structure, and out the load-side primary disconnects. Circuit breaker contacts are designed so that the circuit resistance through the contacts is low and no current flows through the blowout coils under normal operation. When the breaker is opened, the extremely strong opening springs immediately begin to open the contacts. The circuit resistance increases as the contacts open. **See Figure 2-9.**

The increased resistance forces some of the current to travel up the arc horn and runners, where the blowout coils are energized. The arc horn (on the backside of the arcing contact) helps transfer the arc to the arc runners. At the same time that the contacts begin to open, the puffer assembly is depressed and causes a puff of air to be forced through the contacts.

As the puffer piston descends into the puffer tube, the arc continues to be forced through the parting contacts. The contacts are further parted and resistance increases to the point where the arc now fully follows the arc runners up into the arc chute. More blowout coils are energized and create a stronger magnetic pull on the arc. The magnetic pole pieces are also energized at this time and work to both contain the force of the arc interruption and assist in driving the arc into the arc chute. The arc is stretched by the arc divider plates and is cooled and deionized.

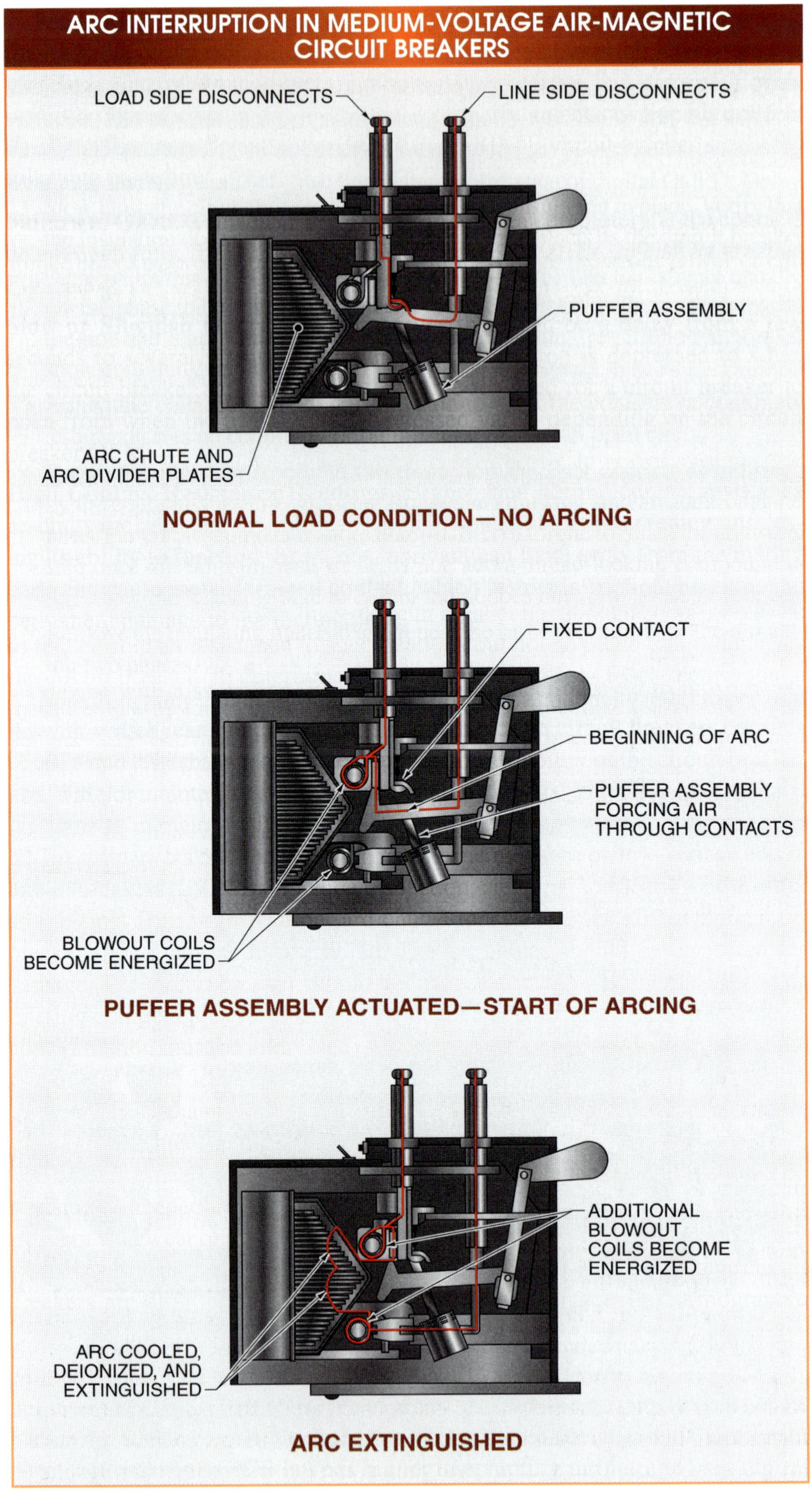

Figure 2-9. With medium-voltage air-magnetic circuit breakers, the combination of the arc being stretched, contact gap increasing, and arc being cooled and deionized allows the arc to be interrupted and extinguished.

As the contact assembly continues to part, the arc is forced to stretch longer. The combination of the arc being stretched, the contact gap increasing, and the arc being cooled and deionized allows the arc to be interrupted and extinguished.

Medium-Voltage Circuit Breaker Vacuum Bottle Components

Vacuum bottles serve the same purpose in a circuit breaker as the contact assemblies and arc chutes. Vacuum bottles extinguish the arc through the absence of air. Vacuum bottles require little maintenance compared to air magnetic contact and arc chute assemblies. **See Figure 2-10.** The primary components of a medium-voltage circuit breaker vacuum bottle include a bottle, flexible metal bellows, bellows shield, contacts, and metal vapor condensing shield.

VACUUM BOTTLE COMPONENTS

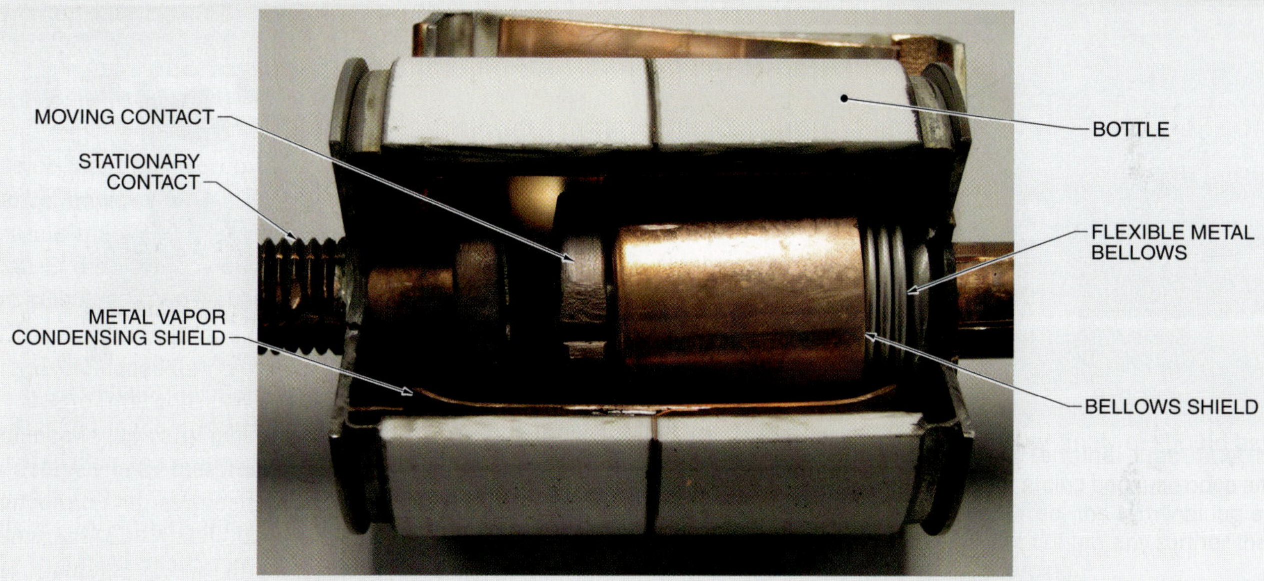

Figure 2-10. The primary components of a medium-voltage circuit breaker vacuum bottle include a bottle, flexible metal bellows, bellows shield, contacts, and metal vapor condensing shield.

Bottles. The bottle is composed of extremely hardened ceramic or glass. It must contain the explosive force of an arc.

Flexible Metal Bellows. The flexible metal bellows is soldered or welded to the moving contact stem on one end and the bottle end plate on the other. It maintains the seal between the moving contact and the bottle. There are no other seals or gaskets. The bushing around the moving contact can be mistaken for the seal, but it is the bellows that acts as the seal. The typical failure mode for vacuum bottles is cracking of the flexible metal bellows, which allows air into the bottle as a result. The bellows can crack due to wear. This results in hardening of the part that flexes. Since it is metal, it can only flex a certain number of times before it fails. Most bellows manufacturers specify the bellows to operate properly about 10,000 to 20,000 times before it fails.

Bellows Shields. A bellows shield protects the metal bellows from the intense heat of an arc. Because it does not have a perfect vacuum, there will be some arcing inside the bottle.

Contacts. Vacuum circuit breakers have only one set of contacts. When an arc is interrupted, some of the metal is vaporized. Most metal recollects onto the contact surface, while some drifts toward the inside of the bottle. The contact face design of medium-voltage vacuum circuit breakers is based on the application. **See Figure 2-11.**

CONTACT FACES

Contact Structure	Flat Contact	Spiral Contact	Axial Magnetic Field Contact
Appearance		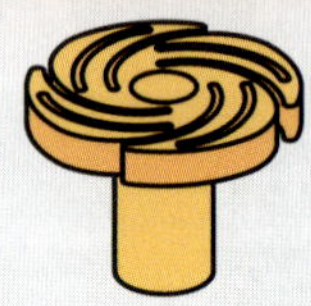	
Application	VCB (General) Contactor	VCB (Large capacity)	VCB (More interruption) (Low surge)
Materials	Chromium-Copper alloy Silver Tungsten Carbide alloy Copper-Tungsten alloy	Chromium-Copper alloy	Chromium-Copper alloy Silver Tungsten Carbide alloy

Figure 2-11. Contact face designs of medium-voltage vacuum circuit breakers are based on the application.

Metal Vapor Condensing Shields. Since a small part of the contact face does not recombine onto the contact face, it moves toward the bottle wall. In earlier design-phase vacuum bottles, repeated arcing caused a conductive path to form due to the debris from arc interruption. The metal vapor condensing (arc) shield is designed so that both ends are open and do not contact the bottle. Any metal vapor that drifts to it cannot make a short circuit between the contacts.

Arc Interruption in Medium-Voltage Vacuum Circuit Breakers. An alternative to medium-voltage air circuit breakers and their large mechanisms and arc extinguishers is medium-voltage vacuum circuit breakers. Vacuum circuit breakers interrupt the arc by denying it air. In a pure vacuum, there can be no arc. Even though the vacuum is in a good vacuum bottle, it is not perfect. Therefore, some small amount of arcing takes place. Because the arcing is small, it is interrupted quickly, usually in two to three cycles, depending on the application.

Unlike air and oil circuit breakers, which can only interrupt the current at zero crossover point, vacuum circuit breakers can interrupt the arc at any point in the sine wave. This characteristic caused issues for electric motors when the technology was first implemented in the field.

Vacuum circuit breakers used as across-the-line motor starters cut the current waveform so quickly that it caused an inductive voltage spike, which damaged the motor windings. Medium-voltage vacuum motor starters have contacts that use a softer alloy and will not cut off the arc as quickly as a vacuum circuit breaker. This increases a characteristic known as persistence.

Persistence is the amount of time vaporized metal remains a vapor between the contact faces in a medium-voltage circuit breaker. By increasing the persistence of a contact, the voltage is allowed to bleed off and damage is eliminated. **CAUTION:** Never interchange vacuum bottles between circuit breakers and vacuum starters. Personal injury or equipment damage could result.

Large circuit breakers are often installed at electrical utility substations.

CASE STUDY—VACUUM BOTTLE REMOVAL AND INSTALLATION SAFETY

Due to the dielectric stresses and incident energy that can be higher in medium-voltage equipment than low-voltage equipment, medium-voltage circuit breakers present somewhat higher risks when being operated. Arc-rated PPE and clothing are required when racking or operating medium-voltage circuit breakers. Vacuum circuit breakers typically do not fail when they are opened because all three phases open the circuit at once. Vacuum circuit breakers can fail, however, when they are reclosed into circuits with high starting loads, or if there is a fault, as was the case in the example shown. Although the damage to the circuit breaker and cubicle was severe, the worker was wearing the proper PPE and suffered no injuries. NFPA 70E, *Standard for Electrical Safety in the Workplace,* must be consulted for guidelines on choosing arc-rated clothing and PPE.

FAILED VACUUM BOTTLE

DESTROYED CIRCUIT BREAKER

DESTROYED CUBICLE

SUMMARY

Arc interruption must take place quickly and without failure. If components within a circuit breaker do not function as designed, there is the possibility of a restrike. Restrikes typically have a catastrophic effect on circuit breakers and surrounding enclosures. They can destroy equipment and cause severe injuries or fatalities.

Low- and medium-voltage air circuit breakers interrupt arcs in a similar manner. The hot arc rises in the arc chute, which stretches the arc, cools it, and, at a zero crossover point, extinguishes it. This can take from ½ to 8 cycles, depending on the type of circuit breaker, the voltage, and the short-circuit current.

Arcs from medium-voltage air circuit breakers differ from low-voltage circuit breakers in that the arc is more persistent and harder to extinguish. Puffer assemblies, blowout coils, and pole pieces are used to increase the speed of arc extinguishing at higher voltages. The practical limit for air circuit breaker technology is 15,000 V since the size of the circuit breaker makes working with higher voltages infeasible. Air circuit breakers and their contact/arc chute assemblies require a high level of maintenance and large circuit breakers. For these reasons, manufacturers developed medium-voltage vacuum circuit breakers.

Vacuum circuit breakers are much smaller than air circuit breakers of the same voltage and current rating. They also require less maintenance. Typically, a vacuum bottle can operate properly about 10,000 to 20,000 times. Once the bottle fails or reaches the wear limit, it must be replaced with a new vacuum bottle. Vacuum circuit breakers are also more compact than air circuit breakers. Two vacuum circuit breakers can fit in the same space as one air circuit breaker. Therefore, a smaller amount of space is required, which greatly reduces installation costs.

DIGITAL RESOURCES
ATPeResources.com/QuickLinks
Access Code: 223436

chapter 2

LOW- AND MEDIUM-VOLTAGE ARC INTERRUPTION

CHAPTER REVIEW

Name ______________________________ Date ______________

True-False

T F **1.** Most circuit breakers have three sets of contacts per phase.

T F **2.** The larger the circuit breaker, the quicker the arc is extinguished.

T F **3.** The incident energy is directly affected by the speed of the circuit breaker's operation.

T F **4.** The typical failure mode for vacuum bottles is cracking of the flexible metal bellows.

T F **5.** Improper maintenance can cause circuit breakers to malfunction by speeding up their operating speed.

T F **6.** Circuit breaker contacts are designed so that the circuit resistance through the contacts is low.

T F **7.** An overcurrent condition is not a short circuit.

T F **8.** The arc between the circuit breaker contacts occurs due to the ionization of air.

T F **9.** Vacuum circuit breakers can interrupt the arc at any point in the sine wave.

T F **10.** In low-voltage air circuit breakers, the arc is moved through the extinguishing process by various additional components such as puffers and blowout coils.

T F **11.** Muffler and vent assemblies can capture any molten metal that may try to escape from an arc chute.

T F **12.** The automatic operation of a circuit breaker occurs when there is an abnormal condition such as an overcurrent or short circuit.

T F **13.** Small amounts of moisture in the arc chute cannot cause restrikes.

T F **14.** A metal vapor condensing (arc) shield is designed so that both ends are closed and seal the vacuum bottle.

T F **15.** During selective tripping, the device closest to the load operates first.

T F **16.** A vacuum bottle is composed of extremely hard metal alloy.

T F **17.** Circuit breakers are switches designed to open and close circuits.

T F **18.** Vacuum bottles require little maintenance compared to air magnetic contact and arc chute assemblies.

T F **19.** The approximate maximum total clearing times for a circuit breaker does not vary, even with different circuit breaker types.

T F **20.** A low-voltage circuit breaker arc chute has a light exterior frame.

T F **21.** In a pure vacuum, there can be no arc.

T F **22.** When a circuit breaker opens to clear a system short circuit, an arc is also created between the opening contacts.

T F **23.** A puffer assembly is an arc chute component.

T F **24.** With medium-voltage circuit breaker vacuum bottle components, the flexible metal bellows acts as a seal on the vacuum bottle.

T F **25.** Older arc chutes are made of materials such as ceramic, asbestos, and celluloid or phenolic plastic resins that tend to repel moisture.

Multiple Choice

______ **1.** A(n) ___ contains, cools, stretches, deionizes, and extinguishes the arcs between contacts.
- A. arc runner
- B. arc chute
- C. puffer assembly
- D. blowout coil

______ **2.** Vacuum circuit breakers interrupt the arc by ___.
- A. denying it air
- B. stretching it between contacts
- C. applying a counter electromagnetic force
- D. deionizing the air

______ **3.** ___ is electrical current in excess of the equipment limit, total amperage load of a circuit, or conductor or equipment rating.
- A. A restrike
- B. A short circuit
- C. Ionization
- D. Overcurrent

______ **4.** The zero crossover point is the point at which current ___.
- A. cannot be extinguished
- B. begins to decrease
- C. reaches its peak voltage
- D. alternates and begins to change direction

______ **5.** A(n) ___ provides a path for the arc to rise into the arc chute.
- A. arc runner
- B. arc divider plate
- C. pole piece
- D. vacuum bottle

_______________ **6.** Most flexible metal bellows manufacturers specify that the bellows operates properly at least ___ times before it fails.
A. 5000
B. 10,000
C. 25,000
D. 50,000

_______________ **7.** An arc chute is also referred to as an arc ___.
A. containment device
B. extinguisher
C. pipe
D. tube

_______________ **8.** Arc runners are conductive strips and usually composed of ___.
A. copper
B. sheet silicone steel
C. tungsten
D. zinc

_______________ **9.** Arc chute maximum clearing times for a medium-voltage air circuit breaker typically ranges from ___ sec to ___ sec.
A. 0.08; 0.0133
B. 0.025; 0.033
C. 0.033; 0.05
D. 0.05; 0.08

_______________ **10.** Short circuits can be ___ or more times the full-load current of a circuit.
A. four
B. five
C. ten
D. forty

Completion

_______________ **1.** ___ is the time required to interrupt an arc and is considered to be from when the arc starts until it is completely extinguished.

_______________ **2.** A(n) ___ is a resumption of current between the circuit breaker's contacts if the circuit breaker fails to clear an arc.

_______________ **3.** Ionization of air occurs when ___ are knocked out of their orbits around the molecules of air by the extreme heat of an electric arc.

_______________ **4.** The ___ stretch the arc as it rises into the arc chute.

_______________ **5.** A(n) ___ is an unintentional connection of two ungrounded conductors that have a potential difference between them.

_______________ **6.** ___ is the amount of time vaporized metal remains a vapor between the contact faces in a medium-voltage circuit breaker.

_______________ **7.** The ___ of the arc chute is where the arc is initially created.

______________________ **8.** The ___ assembly is a bellows or piston/tube assembly that is compressed when the circuit breaker contacts open.

______________________ **9.** The arc ___ (on the backside of the arcing contact) helps transfer the arc to the arc runners.

______________________ **10.** ___ are magnetic coil assemblies that are energized when the arc travels up the arc runner.

______________________ **11.** The muffler and vent assembly on the end of an arc chute reduces the ___ and allows combustible gases that are created during the arc to safely escape.

______________________ **12.** A bellows ___ protects the metal bellows from the intense heat of an arc.

______________________ **13.** ___ serve the same purpose in a circuit breaker as the contact assemblies and arc chutes.

______________________ **14.** Circuit breakers use arc ___ to deionize the air and interrupt the flow of short-circuit current.

______________________ **15.** An arc chute (arc extinguisher) contains, cools, stretches, deionizes, and extinguishes the arcs between contacts within ___ of a second (six cycles) or less.

______________________ **16.** Pole pieces are mounted on the outside surface of the arc chute and are made of layers of laminated ___, similar to a transformer core.

______________________ **17.** When a circuit breaker does not clear the arc, the arc re-establishes itself and a ___ occurs.

______________________ **18.** A ___ circuit occurs when there is an unintentional grounding of an energized conductor, usually resulting in an arc flash.

______________________ **19.** When a circuit breaker attempts to interrupt an arc, the moisture is absorbed by the arc chute and is released into the arc chute as ___.

______________________ **20.** A(n) ___ maintains the seal between the moving contact and a vacuum bottle.

Short Answer

1. Explain why, even with good vacuum in a vacuum bottle, that some small amounts of arcing takes place inside the bottle.

2. Briefly explain the process referred to as cascading.

3. Explain why vacuum bottles must never be exchanged between circuit breakers and vacuum starters.

4. List five components involved in arc interruption on a medium-voltage air-magnetic circuit breaker.

5. How can incident energy have an effect on the type of PPE required for a particular job?

Matching

Low-Voltage Arc Chute Components

_______________ 1. Arc chutes

_______________ 2. Arc divider plates

_______________ 3. Arc runners

_______________ 4. Bracing

_______________ 5. Frame

_______________ 6. Muffler and vent assembly

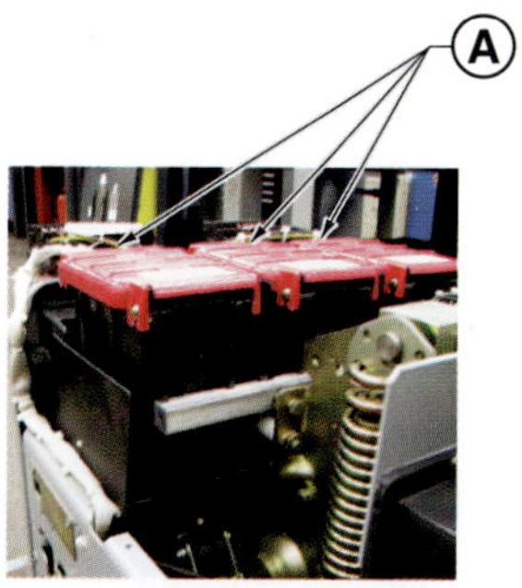

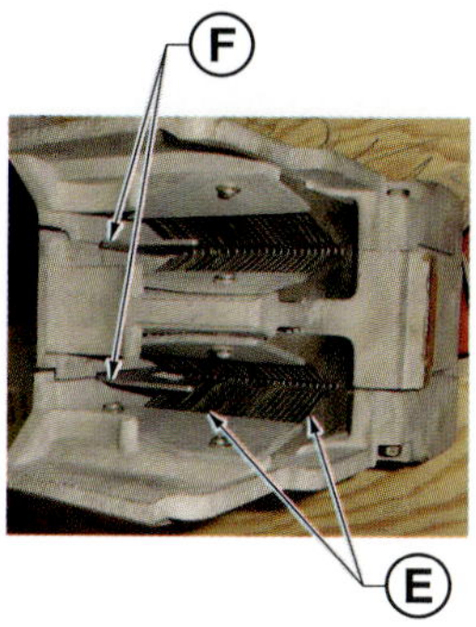

Medium-Voltage Arc Chute Components

_______________ **1.** Arc divider plates

_______________ **2.** Arc runner

_______________ **3.** Blowout coil

_______________ **4.** Pole pieces

_______________ **5.** Throat

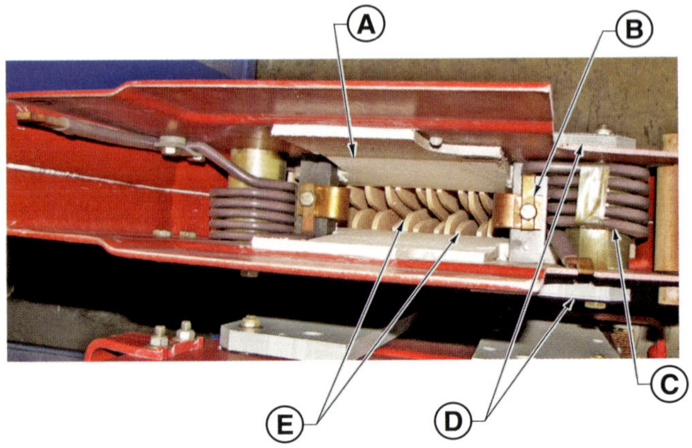

Vacuum Bottle Components

_______________ **1.** Bellows shield

_______________ **2.** Flexible metal bellows

_______________ **3.** Bottle

_______________ **4.** Moving contact

_______________ **5.** Stationary contact

_______________ **6.** Metal vapor condensing shield

chapter 3

CIRCUIT BREAKER RACKING

One of the more hazardous tasks involved in maintaining low- and medium-voltage circuit breakers is the process of racking (inserting or removing) them in and out of their cubicles. An arc flash can occur if proper racking and safety procedures are not adhered to. In addition to following original equipment manufacturer (OEM) racking procedures, specific Occupational Safety and Health Administration (OSHA) and National Fire Protection Agency (NFPA) 70E regulations must be followed. The safety precautions and racking procedures for low- and medium-voltage circuit breakers are similar with minor differences in the requirements for each type of circuit breaker.

OBJECTIVES

- State the primary hazard involved in racking a draw-out circuit breaker.
- List the shock approach boundaries for 480 V and 15 kV class circuits.
- Identify one method of calculating an arc flash boundary.
- Identify the PPE and arc-rated clothing required to perform the racking operation on a power circuit breaker.
- List the four racking positions of a draw-out circuit breaker.
- Understand safe racking procedures for low- and medium-voltage circuit breakers.
- Identify safety concerns of the racking procedure.
- State how interlocks can malfunction and the possible effects.
- Identify the differences in racking low-voltage and medium-voltage circuit breakers.

SAFETY PRECAUTIONS

Low- and medium-voltage draw-out circuit breakers are racked into and out of their cubicles in the same manner. There are differences from manufacturer to manufacturer, but the main process is usually the same.

The task of racking a circuit breaker is inherently hazardous because the circuit breaker's primary disconnects are connected and disconnected to an energized bus and there are no arc extinguishers (arc chutes). If an arc were to start, it would be at bus-to-breaker connections and would have to be extinguished by the breaker (upstream) that is feeding the affected bus. The upstream circuit breaker may not read the short circuit as a short circuit due to its overcurrent device settings. It may not process it as a large overcurrent instead.

The upstream circuit breaker may not process to the short circuit as a short circuit due to its overcurrent device settings and may see it as a large overcurrent instead. It will trip quickly, probably in 1 sec or 2 sec, but it may not use its instantaneous function to trip the circuit breaker open, which would take no more than 6 cycles (0.10 second). One or two seconds is considered a long time by arc flash protection standards. Since incident energy is proportional to time, the incident energy from such a situation can be much greater than expected. **See Figure 3-1.**

Due to the hazardous conditions involved with racking circuit breakers, shock and arc flash approach boundaries must be observed. An arc flash boundary is required, even if the door is closed, because the racking task is considered interacting with the equipment in a manner that could cause failure (refer to the definition of an Arc Flash Hazard in Article 100, NFPA 70E Informational Note No. 1).

KEY TERMS

- energized electrical work permit (EEWP)
- personal protective equipment (PPE)
- willful violation
- serious violation

ARC INCIDENT ENERGY

Figure 3-1. The task of racking a circuit breaker is inherently hazardous because there is making and breaking of a load-carrying electrical connection and there are no arc chutes.

The shock approach boundaries for 480 V low-voltage power circuit breakers (LVPCBs) are the following:

- limited—42″
- restricted—12″
- prohibited—1″

The shock approach boundaries for 208 V LVPCBs are the following:

- limited—42″
- restricted—avoid contact
- prohibited—avoid contact

The arc flash boundary can range from 3′ to 7′ or be greater than 7′, depending on the available short-circuit current and the trip function being used for short-circuit protection (short-time delay or instantaneous). Since the circuit breaker is installed in a metal box (cubicle), the arc flash is reflected off the back and sides of the cubicle and is focused toward the front opening, which is where a worker would typically be positioned. **See Figure 3-2.**

Operating time affects the arc flash boundary for LVPCBs. For example, assume that a low-voltage power (draw-out) circuit breaker has 42,000 A short-circuit current available with an instantaneous trip (4 cycles or 0.07 seconds). The arc flash boundary for this application would be 30.5″. However, if that same circuit breaker is using a short-time delay function of 0.33 seconds (nominal),

TECH TIP

There are a variety of tools available to calculate arc flash boundaries, such as the Cooper Bussmann Arc-Flash Calculator, which is available at the Cooper Bussmann website, www.cooperbussmann.com.

the arc flash boundary increases to 66.3″. The short-time delay function can have a nominal delay of up to 0.5 seconds, which causes the arc flash boundary to increase further to 81.6″.

ARC FLASH AND SHOCK APPROACH BOUNDARIES

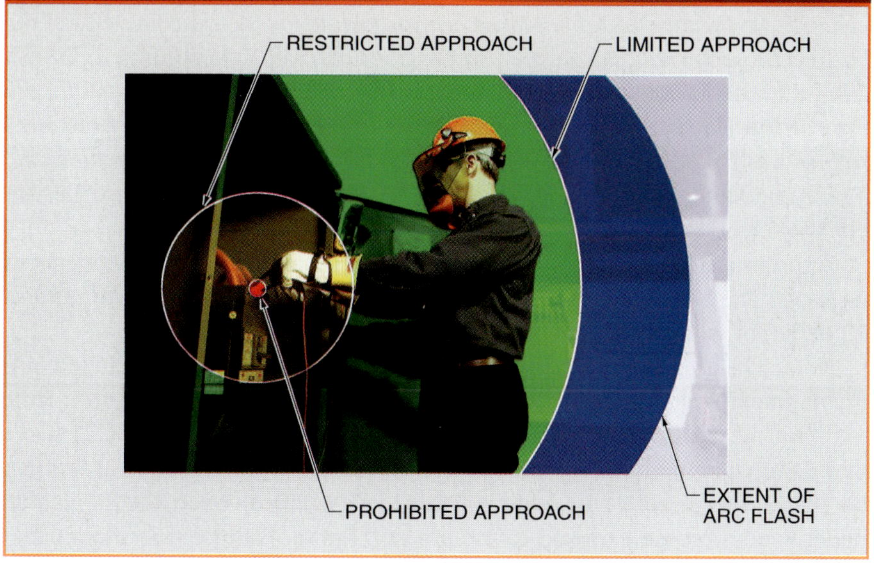

Figure 3-2. Due to the hazardous conditions involved with racking circuit breakers, arc flash and shock approach boundaries must be observed.

Since racking circuit breakers can be extremely hazardous, it is recommended that an energized electrical work permit be completed for this task. An *energized electrical work permit (EEWP)* is a document issued by an employer to allow specific work to be performed on energized electrical equipment with various limitations. Since the risks and hazards of racking a specific circuit breaker (or circuit breakers in a lineup of switchgear) are known, a standing EEWP can be written for a specific length of time but usually no more than 90 days. At the end of the 90-day period, the EEWP should be reviewed and any near-misses or problems must be addressed. At that time, it can be decided whether a new 90-day EEWP can be issued.

Personal protective equipment (PPE) is clothing, glasses, gloves, hard hats, respirators, or other safety devices designed to protect workers against safety hazards in the work area and from injury. The recommended PPE for racking circuit breakers includes the following:

- total arc-rated clothing system with a minimum rating of 40 cal/cm^2, including arc-rated flash suit, pants, and hood worn over arc-rated daily work wear
- class E hard hat (high-voltage protection) with arc-rated liner
- safety glasses or goggles (UV-rated)
- hearing protection (foam ear inserts)
- arc-rated gloves or rubber insulating gloves with leather protectors **See Appendix.**
- insulating sleeves and rubber blankets, as required
- heavy-duty leather footwear

Eye protection must be UV-rated to comply with OSHA 29CFR1910.133(a)(1), which states, *"The employer shall ensure that each affected employee uses appropriate eye or face protection when exposed to eye or face hazards from flying particles, molten metal, liquid chemicals, acids or caustic liquids, chemical gases or vapors, or potentially injurious light radiation."* According to OSHA 29CFR1910.335(a)(1)(i), *"Employees working in areas where there are potential electrical hazards shall be provided with, and shall use, electrical protective equipment that is appropriate for the specific parts of the body to be protected and for the work to be performed."* According to OSHA 29CFR1910.335(a)(1)(v), *"Employees shall wear protective equipment for the eyes or face wherever there is danger of injury to the eyes or face from electric arcs or flashes or from flying objects resulting from electrical explosion."* These regulations indicate the need for UV-rated safety glasses or goggles for electrical workers.

Appropriate footwear includes standard, industrial-grade leather work shoes or boots. The NFPA 70E requirement is leather footwear with a minimum of 12 oz/yd^2 of weight.

Note: The effective arc rating of the clothing system does not include flammable underlayers of any type, whether they are cotton, wool, or silk. Rule 410A of the 2012 NESC (National Electrical Safety Code/ANSI C2) states that flammable, nonmelting underlayers can increase the effective arc rating, but that is not correct. If the outer arc-rated layer(s) break open, the flammable underlayers become exposed to the arc and ignite. Natural-fiber flammable underlayers reduce the heat to the body and help reduce the risk of a burn but cannot increase the effective arc rating of the outer layers. Because of these clothing concerns, NFPA 70E 2012 edition added Section 130.7(C)(9), which states, *"Garments that are not arc rated shall not be permitted to be used to increase the arc rating of a garment or of a clothing system."*

CAUTION: Always apply specific safety guidelines to racking operations. The risk of serious injury is much higher with this task than many other tasks.

RACKING LVPCBs

The four positions LVPCBs must go through as they are being inserted or removed from their cubicles include connected, test, disconnected, and removed (withdrawn). **See Figure 3-3.**

- connected—The breaker is connected to the bus, carries load current, and operates electrically (on electrically operated circuit breakers).
- test—The primary stabs are disconnected. The secondary stabs, which carry control voltage, remain connected. The breaker does not carry load current, but operates electrically, if so equipped. On medium-voltage circuit breakers, it is common to be required to connect a separate secondary contact block in order to put the breaker in the test position. **See Figure 3-4.** *Note*: To engage a medium-voltage circuit breaker into the test position, grasp the T-handle, straighten, and push in firmly while the circuit breaker is in the disconnected position.
- disconnected—Both primary and secondary stabs are disconnected. The circuit breaker does not carry load current or operate electrically. The circuit breaker remains racked within the cubicle and the door can be closed.
- removed (withdrawn)—The circuit breaker is removed from its cubicle and placed either on extension arms or on the floor.

LOW-VOLTAGE POWER CIRCUIT BREAKER (LVPCB) RACKING POSITIONS

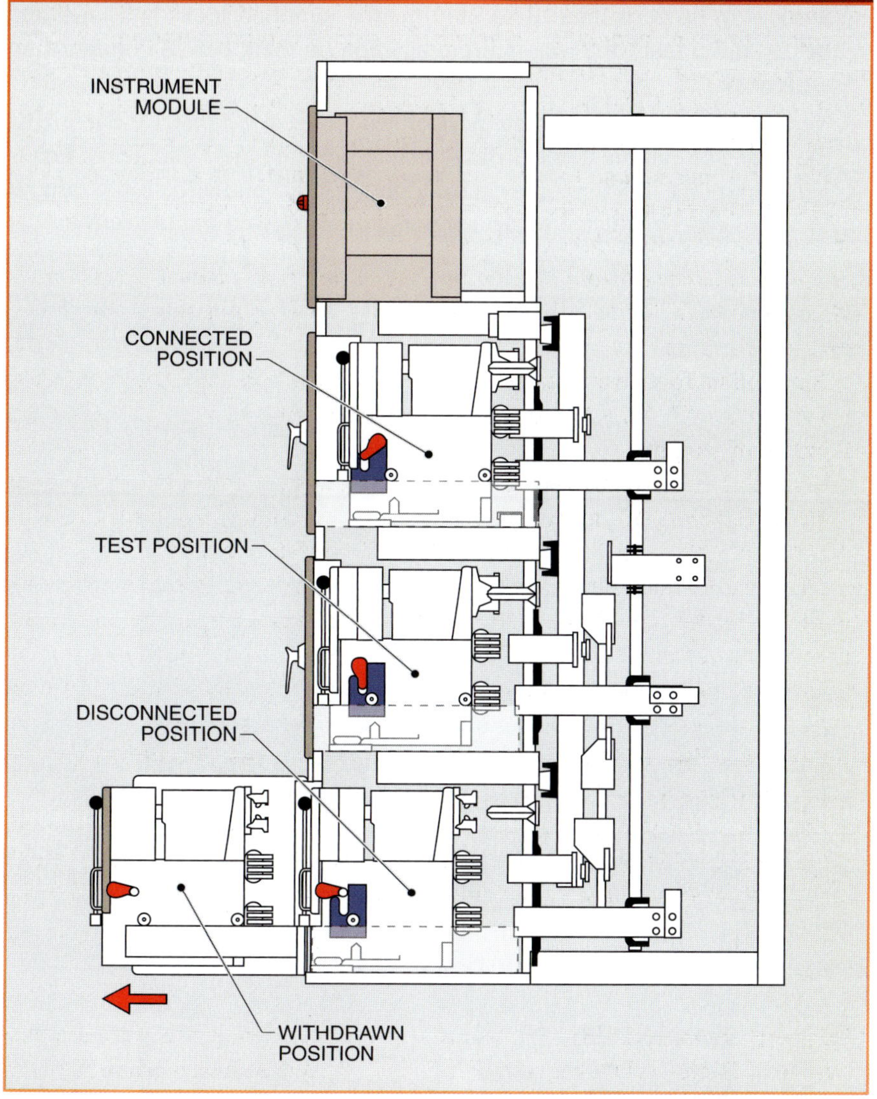

Figure 3-3. The four positions of LVPCBs as they are being inserted or removed from their cubicles include connected, test, disconnected, and removed (withdrawn).

The following procedure is a general guideline for racking most low-voltage draw-out circuit breakers. *Note:* This procedure may require adjustment, depending on the circuit breaker OEM, circuit breaker design, and enclosure. OEM instructions must always be followed. HRC level 4 PPE is required for protection from arc flash hazards when racking 480 V circuit breakers. Refer to NFPA 70E Tables 130.7(C)(15)(a) and 130.7(C)(16) for proper PPE requirements. **See Figure 3-5.**

To rack a low-voltage draw-out circuit breaker, employ the following procedure:

1. Obtain the proper clearances before attempting to take a power circuit breaker out of service. Clearances are notification to facility departments that certain circuits will be de-energized and that plans must be made to accommodate the outage.

2. Follow proper lockout/tagout procedures as specified by the facility where work is to be performed. Use only facility-supplied locks and tags, and be certain to mark the required information on them before commencing with work.
3. Verify that the circuit breaker to be operated is the correct one. De-energizing the wrong circuit breaker can cause major, possibly fatal, accidents. Emergency systems, life-support systems, lighting, and ventilation for hazardous locations can be affected by deenergizing the wrong circuit breaker.

ENGAGING MEDIUM-VOLTAGE CIRCUIT BREAKERS INTO TEST POSITION

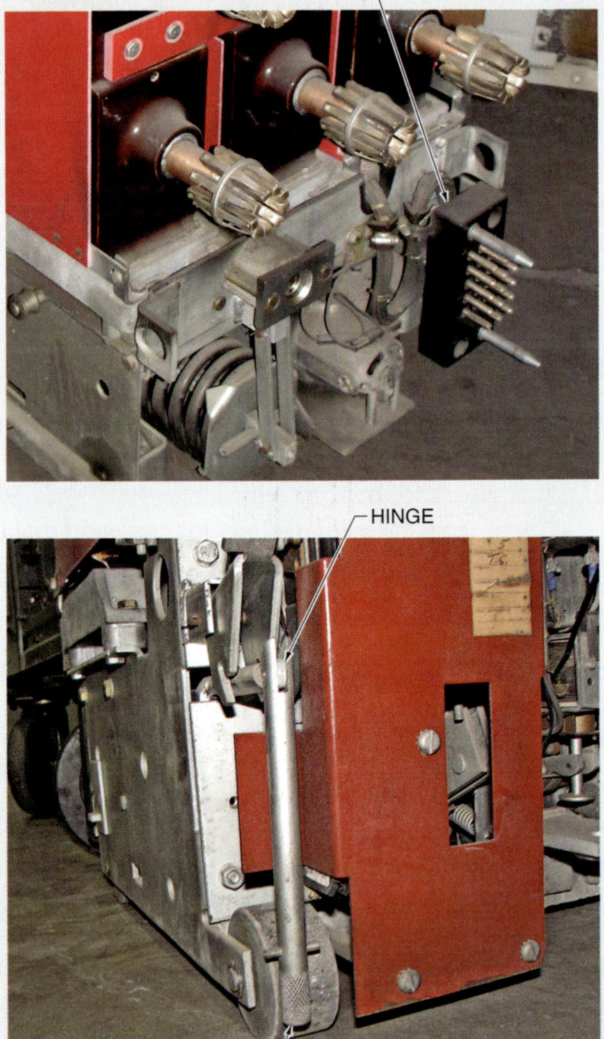

Figure 3-4. On medium-voltage circuit breakers, it is common to connect a separate secondary contact block in order to put the breaker in the test position, which is often accomplished with a T-handle located on the circuit breaker.

4. Disconnect load fed by the circuit breaker before operating. This reduces the chance of failure. Circuit breakers are switches and are designed to interrupt load current. However, circuit breakers are primarily a mechanical device, and mechanical devices fail. The best safe work practice is to de-energize the load before opening the circuit breaker.
5. Note the position of the open/close indicator (it should indicate "OPEN"). *Note:* Even if the indicator shows that the circuit breaker is open, it does not mean that it is. Many less modern models of circuit breakers have indicators that are spring-loaded to the open position. If the linkage comes apart or there is some other mechanical failure, it could indicate the circuit breaker was open when it is actually closed. **See Figure 3-6.**

Figure 3-5. The PPE for performing work on circuit breakers must be resistant to arc flashes and can be determined by referencing NFPA 70E Tables 130.7(C)(15)(a) and 130.7(C)(16).

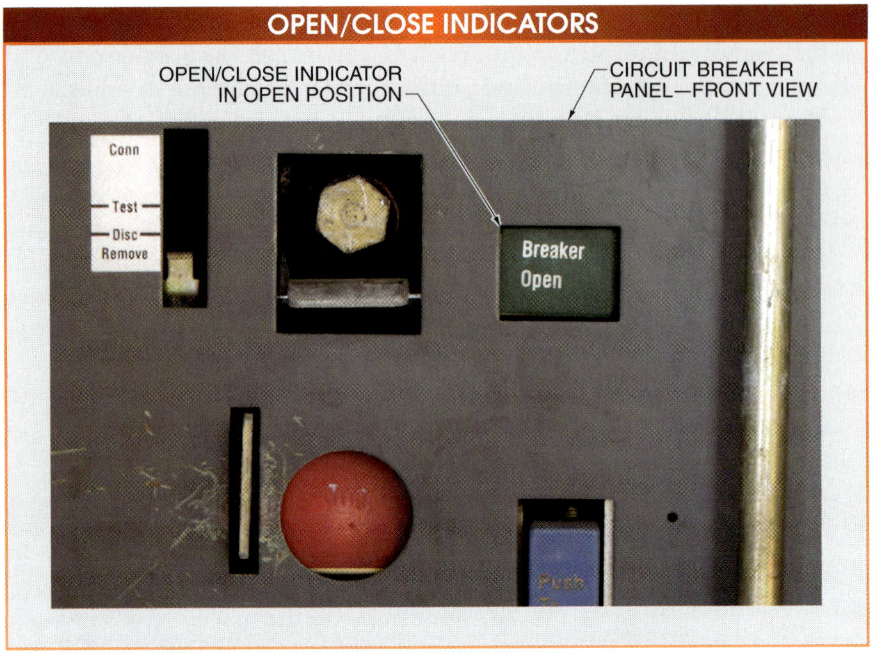

Figure 3-6. Circuit breakers typically have an open/close indicator located on the front of the circuit breaker panel.

6. Stand to the hinge-side of cubicle, extend arm and manually press trip button (even if indicator shows "OPEN"). If there is a failure, the handle side of cubicle should open, while the hinge side remains connected to the cubicle. The door can provide some amount of shielding from an arc flash, although not much if the door has louvers or vents. It should still divert some heat of the arc flash away from operator.
7. Rack circuit breaker with the door closed. *Note:* On some less modern circuit breakers, the cubicle door can be modified by using a metal saw and circular or rectangular metal template to cut a hole in the door to allow access to the racking screw.
8. Insert the racking handle while observing that the safety interlock device depresses the trip button. This is a common feature of all but the oldest circuit breakers. A *safety interlock device* is a device used to ensure that a circuit breaker does not get racked while the circuit breaker is in the closed position. **See Figure 3-7.**
9. Slowly rotate the racking handle counterclockwise to remove the circuit breaker. Perform this task slowly. It is best practice to listen for any arcing that may occur when the primary disconnects on circuit breaker just begin to part from the bus. **WARNING:** If the circuit breaker does not trip open and an arcing sound is heard, discontinue racking out immediately and rack circuit breaker back in.
10. Once the circuit breaker stabs are clear of the bus, the arc flash hazard is almost nonexistent. Rack the circuit breaker through the test position, unless there is an operational problem with the circuit breaker. The test position is used to check for proper functioning of the circuit breaker. Continue racking until the circuit breaker is in the disconnected position. The door may be opened at this point and extension arms (LVPCBs and some medium-voltage vacuum circuit breakers) installed in or extended from the cubicle.

Figure 3-7. A safety interlock device is used to ensure that a circuit breaker is not racked while the breaker is in the closed position.

11. Use a flashlight or similar lighting device to effectively view the circuit breaker rollers to verify that they are resting squarely on the extension arms with all the components properly in place.
12. Slowly pull the circuit breaker out of cubicle and onto extension arms. Pull until the circuit breaker rear rollers can be seen by using a lighting device. Verify that the rollers are resting squarely on the extension arms.
13. Slowly pull the circuit breaker fully out of the cubicle to the stops on the extension arms.
14. A *spreader bar* is a rigging device that is suspended from a crane hook to support lifting lines at various locations. Spreader bars are used to evenly lift long or wide loads. When removing a circuit breaker from extension arms, use the OEM's spreader bar and verify that the spreader bar is properly attached to the circuit breaker frame. **See Figure 3-8.** Often, one side will come loose when cable is tightened, which can cause the circuit breaker to fall to the floor.

    WARNING: Do not use rope or wire slings strung through the circuit breaker frame. They will push in the sides of the frame and possibly damage the arc chutes and/or operating mechanism. If using a cable and pulley mechanism, be certain to check the cable reel before lifting the circuit breaker for any crossed turns. Crossed cables on the reel can cause a circuit breaker to "bounce" when lowering and cause the cable to snap under the instantaneous weight/force increase. **See Figure 3-9.**
15. If using a lift platform to remove a circuit breaker, verify that it is completely positioned under the frame before lifting the circuit breaker.
16. Slowly lower the circuit breaker onto the floor and remove the spreader bar.

SPREADER BARS

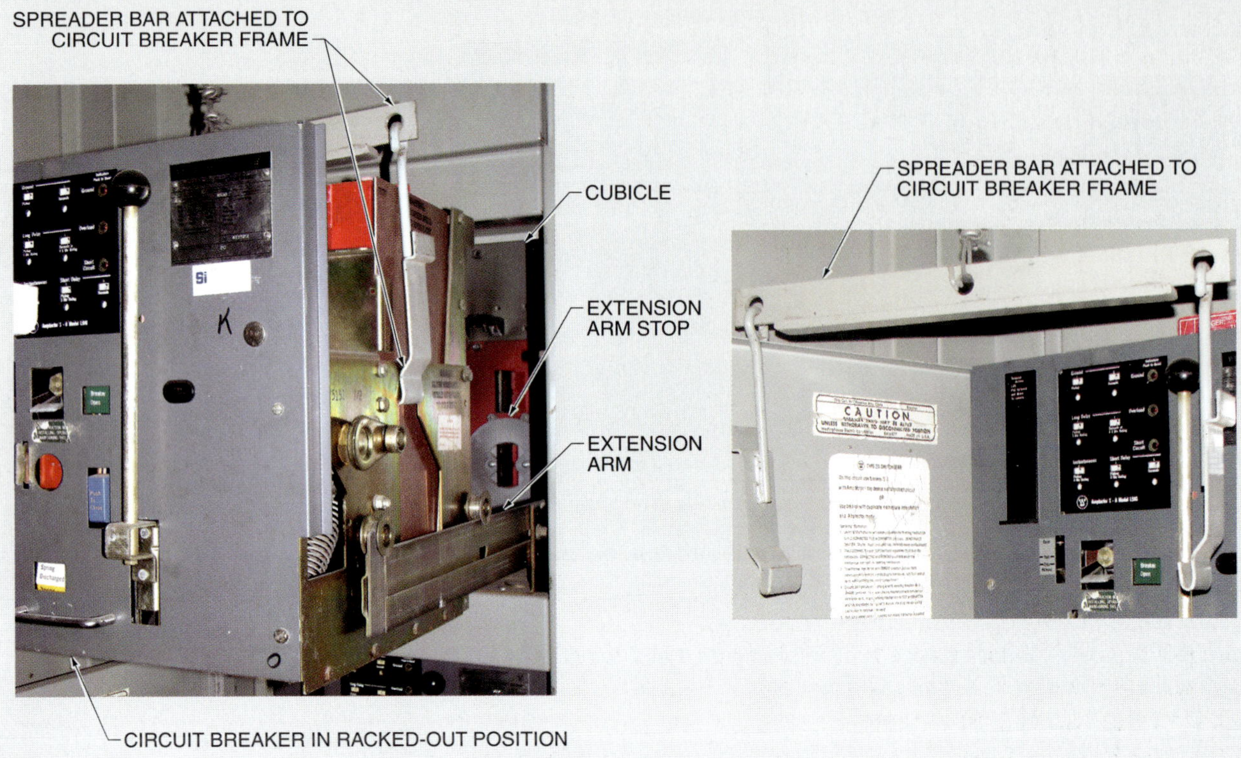

Figure 3-8. When removing a circuit breaker from extension arms, a spreader bar must be properly attached to the circuit breaker frame.

EFFECTS OF IMPROPER LIFTING AND RIGGING

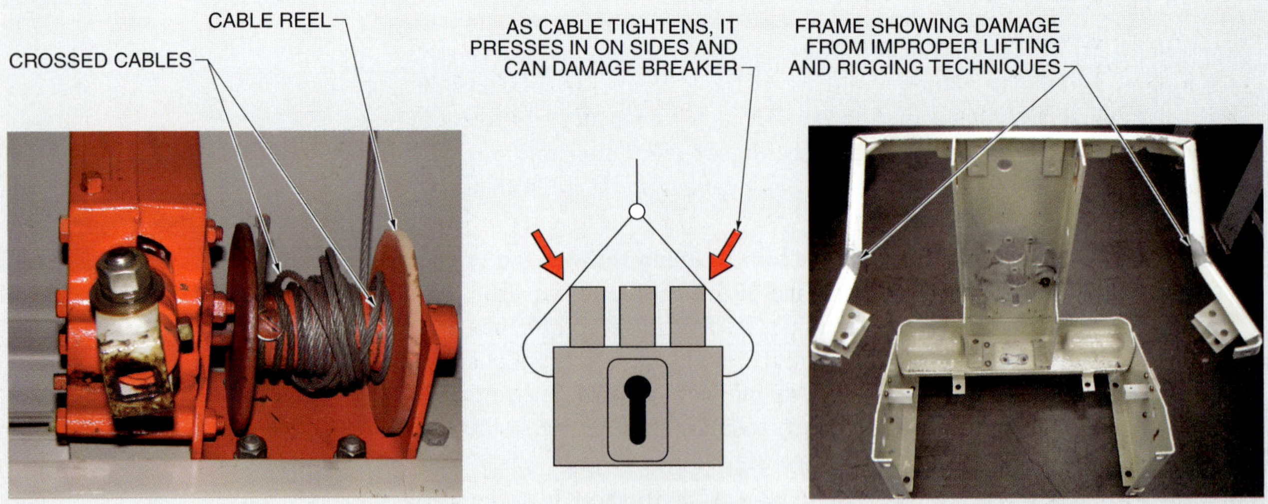

Figure 3-9. Improper lifting and rigging techniques can cause extensive damage to circuit breakers and related components.

INTERLOCKING DEVICES

On medium-voltage circuit breakers, it is not uncommon for a slight buzzing sound to be heard when the primary stabs part. This is much different from an arcing sound, which indicates the breaker is still closed. If in doubt, stop racking and reinsert the breaker to the bus. If the circuit breaker cannot be removed from its cubicle, the entire section of switchgear must be de-energized before the problem can be corrected. Never try to remove a circuit breaker if there is any doubt that it has actually opened.

An interlock device is commonly installed on most circuit breakers to prevent them from being inserted or removed from a cubicle while they are closed. As the circuit breaker is being racked in, the interlock device is designed to open the circuit breaker before the bus connections (stabs or primary disconnects) make contact. When a circuit breaker is racked out, an interlock device causes the circuit breaker to trip before the bus connections separate. Unfortunately, it is common for circuit breakers to have interlock devices that do not function properly. Always verify that the interlock devices function properly prior to performing any work.

To re-insert a circuit breaker into its cubicle, apply the following procedure:

1. When replacing a circuit breaker onto extension arms, verify that the spreader bar is secure prior to attempting a lift.
2. Raise the circuit breaker in a slow, smooth manner and position it so that rollers contact extension arms squarely. Often, the circuit breaker can rotate due to cable tension and must be continuously guided until it is completely resting on the extension arms.
3. Once the circuit breaker is firmly on the extension arms, remove the lifting device and spreader bar.
4. Push the circuit breaker into the cell until it is in the disconnected position. Manually trip the circuit breaker using the trip button, even if circuit breaker indicator shows that it is open.
5. Close and secure the cubicle door.
6. Rack the circuit breaker through the test position (unless it is necessary to test operate the circuit breaker). Be certain to depress the trip button before continuing to rack the circuit breaker in from this point.
7. Rack the circuit breaker into the connected position. Do not overtorque the racking mechanism. Once firmly seated onto the bus connections, discontinue racking.

   CAUTION: An overtorqued racking mechanism can fail mechanically. Overtorquing the racking mechanism can bend and eventually shear the stop pin. If the stop pin is sheared off, the circuit breaker cannot position itself correctly into the connected position and will rack through the connected position and begin to rack out again. This condition can create an extremely hazardous condition.
8. Firmly secure all fasteners on the cubicle door when finished. With the door closed and properly secured, there are no perceived hazards. This does not mean that the equipment will not fail, but failures in normally operating equipment are rare.

RACKING MEDIUM-VOLTAGE AIR AND VACUUM CIRCUIT BREAKERS

Medium-voltage draw-out circuit breakers have a similar racking operation to low-voltage circuit breakers. However, there are some equipment differences between the two types, including devices such as floor trippers, racking screws, and shutter mechanisms installed on medium-voltage circuit breakers. **See Figure 3-10.**

- Medium-voltage draw-out circuit breakers have floor trippers installed in the cubicle, which open the breaker and discharge the closing springs when racking the circuit breaker in or out of its cubicle.
- The floor tripper must trip the circuit breaker before the bus connection (primary disconnects) touch while racking in and also trip it before the bus connection (primary disconnects) part when racking out.
- On some medium-voltage vacuum circuit breakers, such as the General Electric (GE) Power/Vac®, there are two racking screws that cause the circuit breaker to rack in or out. If these racking screws are not cleaned and lubricated as part of a regular maintenance program, the circuit breaker can bind when racking. This binding action is caused by the circuit breaker not racking into the cubicle squarely because the racking screws cannot push the circuit breaker forward evenly but allow it to tilt in the cubicle. When racking screws are properly maintained, there are fewer problems. This issue is common to all circuit breakers that use two racking screws.
- Medium-voltage circuit breakers use a shutter, or barrier, to cover the energized bus connections when the breaker is racked out. Most facilities do not allow work to be performed within the cubicle even if the shutter is covering energized bus connections, due to the extremely high incident energies that may be generated if there is an arc. If the shutter does not completely cover the bus connections when the circuit breaker is racked-out, any attempt to remove the circuit breaker from its cubicle or perform any other task should not be made. The circuit breaker must be racked back into place and left undisturbed. Under no circumstances should any attempt be made to correct the problem until the entire bus can be de-energized and placed in an electrically safe work condition.

Some modern low-voltage draw-out circuit breakers are also equipped with shutter mechanisms. The same precautions must be taken when racking them out. Operating these circuit breakers when the shutter mechanism is defective or inoperable violates the OEM instructions and would constitute an unsafe act. The shock approach boundaries for medium-voltage draw-out power circuit breakers (2.3 kV to 15 kV) are the following:

- limited—60″
- restricted—26″
- prohibited—7″

An arc flash boundary can be up to several feet and most medium-voltage circuit breakers in this voltage range tend to be about 30 cal/cm^2 to 40 cal/cm^2, but these are often feeder circuit breakers. There are medium-voltage circuit breakers that have extremely high incident energies, especially circuit breakers feeding loads such as arc-furnace transformers or other high-demand loads.

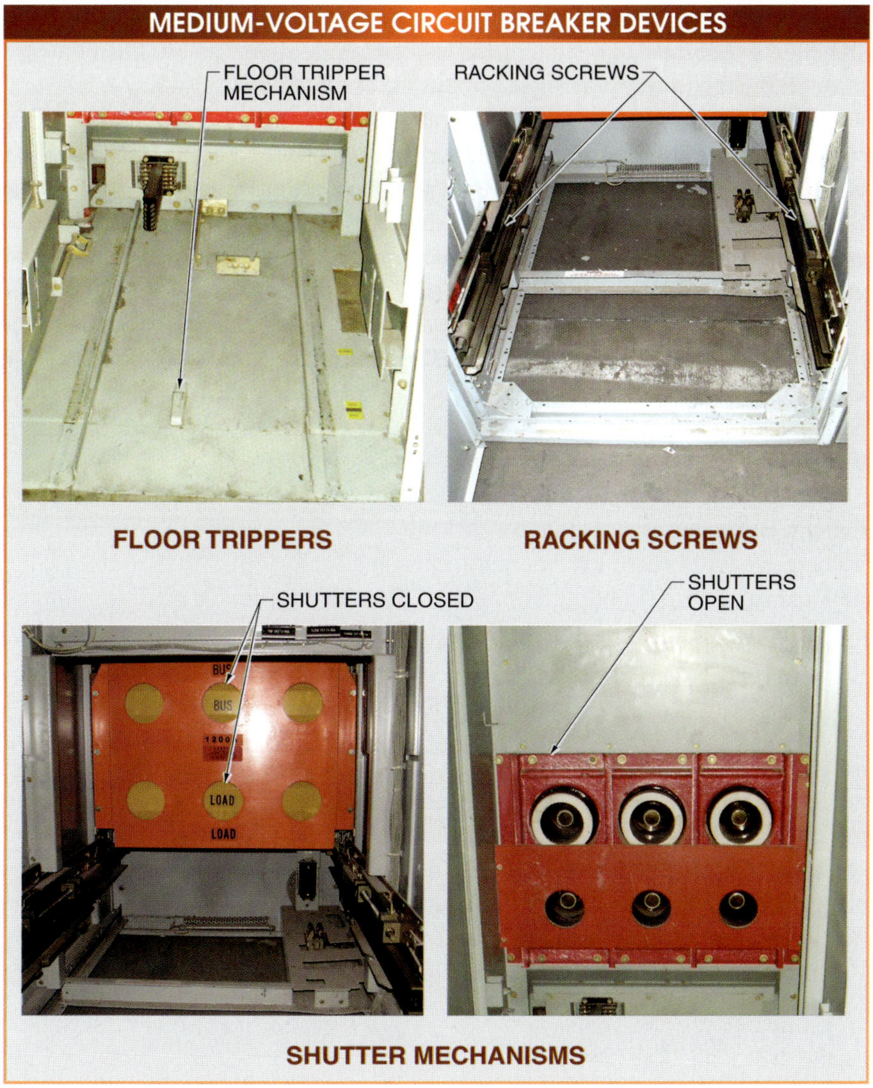

Figure 3-10. Devices such as floor trippers, racking screws, and shutter mechanisms are typically installed on medium-voltage circuit breakers.

Medium-voltage circuit breakers are often used in wind farm substations.

Also, arc-resistant switchgear is an alternative to regular NEMA-rated switchgear. Arc-resistant switchgear is designed to divert an arc flash upward through the top of the switchgear where the arc flash can be safely exhausted.

When the doors are properly closed and secured, there is no arc flash PPE or clothing required to rack circuit breakers in or out of their cubicles. According to NFPA 70E Table 130.7(C)(15)(a), if the doors are opened, HRC-4 arc-flash PPE and clothing are required. Arc-resistant switchgear can cost 15% to 20% more than standard switchgear, but it eliminates the PPE requirement and greatly reduces risks for hazardous equipment and tasks.

Note: NFPA 70E Table 130.7(C)(15)(a) specifies that arc-resistant switchgear in compliance with IEEE C37.20.7, *IEEE Guide for Testing Medium-Voltage Metal Enclosed Switchgear for Internal Arcing Faults,* must be Type 1 or Type 2 and have a clearing time of 0.5 seconds or less. Fault current cannot exceed the rating of the switchgear.

The PPE required for racking medium-voltage power circuit breakers (MVPCBs) is the same as that for racking LVPCBs. However, some medium-voltage systems can have incident energies well above that found at the 480 V level. The 2012 edition of NFPA 70E added limits for medium-voltage equipment. These limits are maximum 35 kA short-circuit current available, up to 0.2 second fault clearing time, and a 36″ working distance.

The total arc-rated clothing system must have a minimum rating of 40 cal/cm^2. Typically, when arc-rated daily work wear is worn, a minimum 40 cal/cm^2 arc-rated flash suit, pants, and hood are worn over it. The total arc-rating of the PPE system can be increased by layering arc-rated clothing and PPE, such as arc-rated daily work wear coveralls or shirt and pants, and wearing arc-rated underclothing. Even though it would not increase the arc-rating of the PPE system, wearing flammable, nonmelting underlayers (t-shirt and shorts) can decrease the heat that may reach the body and decrease the likelihood of a burn. Additional PPE requirements for racking medium-voltage circuit breakers include the following:

- Class E hard hat (high-voltage protection) with arc-rated liner.
- safety glasses or goggles (These must be UV-rated to comply with OSHA 29CFR1910.133(a)(1), which states, *"The employer shall ensure that each affected employee uses appropriate eye or face protection when exposed to eye or face hazards from flying particles, molten metal, liquid chemicals, acids or caustic liquids, chemical gases or vapors, or potentially injurious light radiation."*)
- hearing protection (foam ear inserts)
- arc-rated gloves or Class 2 rubber insulating gloves and leather protectors
- heavy-duty leather work shoes that meet the NFPA 70E minimum requirement of 12 oz/yd^2 weight

One additional safety device that is used when racking in medium-voltage circuit breakers is the stab-finger cluster (connector) assemblies, sometimes referred to as tulips, which connect the circuit breaker to the main bus. **See Figure 3-11.** Stab-finger clusters are initially lubricated by the OEM and require periodic lubrication as part of routine maintenance. If installed without lubricant, they can tilt and bind when sliding onto the bus. If they are not lubricated, they cannot slide on to the bus as designed and can break apart, which can cause an arc flash at the bus-to-breaker connection.

STAB-FINGER CLUSTER ASSEMBLIES

Figure 3-11. A stab-finger cluster (connector) assembly is a safety device that connects the circuit breaker to the main bus and is sometimes referred to as a tulip.

CASE STUDY—THE IMPORTANCE OF PROPER LUBRICATION

A 15 kV-class circuit breaker failed during insertion (racking). The fingers on the primary disconnects (stabs) were not lubricated prior to the racking operation. When the primary disconnects on the circuit breaker hit the bus connection, the fingers were cocked, causing them to break apart. A short circuit was created that damaged four other circuit breakers. The worker was wearing properly-rated arc-flash PPE and clothing and suffered no injuries.

This is an example of where a few minutes spent applying a lubricant would have saved this company tens of thousands of dollars in repair costs and lost production. Not only were the circuit breakers damaged but also the switchgear.

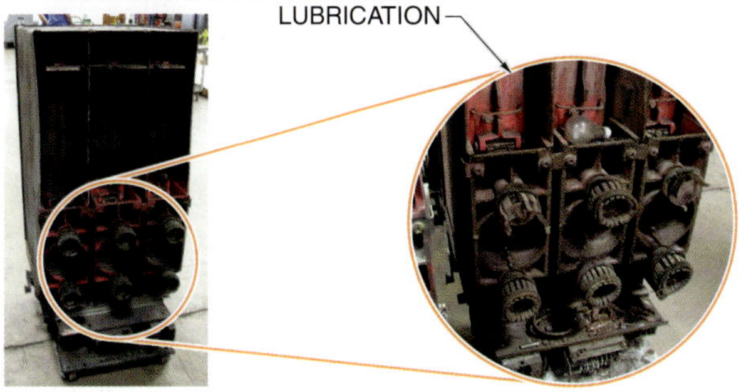

Vacuum circuit breakers often use specialized lifting trucks that attach directly to the circuit breaker cubicle. Extreme care must be taken in aligning these trucks and latching them to prevent damage to them or the breaker. **See Figure 3-12.**

CIRCUIT BREAKER LIFTING TRUCKS

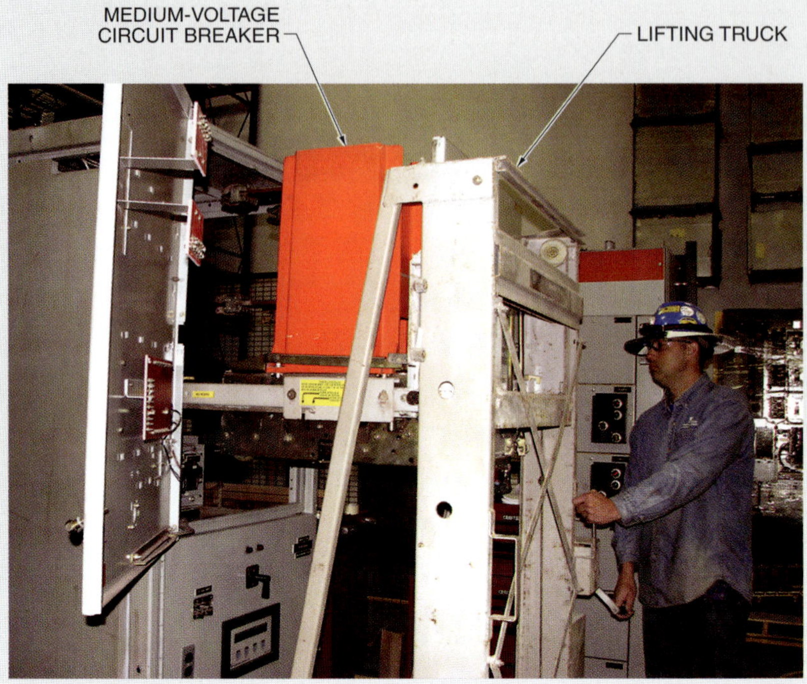

Shermco Industries

Figure 3-12. Medium-voltage circuit breakers often use specialized lifting trucks that attach directly to circuit breaker cubicles.

When removing a medium-voltage air-magnetic circuit breaker from its cubicle, it must be verified that the breaker is not positioned in such a manner that it could pull wiring from the switchgear or doors. **CAUTION:** When removing a circuit breaker from its cubicle, its barrier should never be pulled. Fingers can be pinched or the barrier can be damaged if it is used to move the breaker. Use the handles provided on the circuit breaker. They are designed for this purpose.

Once a circuit breaker is removed from its cubicle, a fifth wheel is used to manually move the circuit breaker along the floor to its desired location. A *fifth wheel* is a device with a swiveling, weight-bearing connection used to move heavy loads. A fifth wheel operates similar to a pallet jack by allowing the weight of heavy loads to be leveraged through the use of a wheel or casters with a swiveling mechanism placed under the load and manually controlled with a handle attachment. **See Figure 3-13.**

Care must be exercised when using a fifth wheel to remove a circuit breaker or insert it into a cubicle. The cubicle switch connects to the circuit breaker operating mechanism through an extension. If the circuit breaker is wheeled into the cubicle and carelessly inserted, the connection can be easily damaged. The fifth wheel should be used to position the circuit breaker in front of the switchgear. The circuit breaker should then be inserted by using the handles on its frame.

MOVING CIRCUIT BREAKERS

Figure 3-13. A fifth wheel is typically used to move a circuit breaker along the floor to its desired location within the facility where it is to be used.

CASE STUDY—ACCIDENTAL ELECTRIC EXPLOSION ANALYZED

This case concerns a service call for a 13.8 kV circuit breaker at an electrical utility generating station. In 1998, the federal Occupational Safety and Health Administration (OSHA) proposed penalties of $460,000 against the electric utility for failure to follow safety requirements at an energy center where three workers died in an electrical explosion.

OSHA cited the company for six alleged willful and eight alleged serious safety violations, which were mostly agency standards on electrical power generation. A *willful violation* is a safety violation committed with an intentional disregard of, or plain indifference to, the requirements of the Occupational Safety and Health Act and regulations. A *serious violation* is a safety violation in which there is a substantial probability that death or serious physical harm could result and the employer knew or should have known of the hazard. OSHA proposed fines of $70,000 for each of the willful violations, which was the maximum permitted under law.

When the accident occurred in the previous year, employees had been working within the minimum safe work distances of a high-voltage circuit breaker cubicle and removed safety guards from energized electrical conductors in the cubicle. Physical evidence indicates that it was likely that part of one of the safety guards touched the energized conductors, which caused an electrical explosion with a fireball estimated to be 26′ in diameter.

One employee was killed instantly and two other employees (a utility manager directing the work and an employee of a subcontractor) were fatally injured. Both died several days later as a result of their injuries. The energy center where the accident occurred is a power generation facility with 125 employees.

OSHA said that neither the two supervisors directing the work nor the employee who was killed instantly had been wearing PPE. When the work changed, exposing the employee to energized electrical equipment, the managers should have held a second job briefing to address the changed work situation and appropriate safety procedures but did not.

The violations of the power generation standard involve training, PPE, minimum safe work distances, work planning, hazardous energy control, and safe work practices.

continued on next page...

CASE STUDY (...continued from previous page)

The list of willful violations and resulting fines include the following:

- Protective equipment was not used when necessary whenever hazards capable of causing injury or impairment were encountered. **$70,000**
- Qualified employees were not trained and competent in the minimum approach distances specified in the standard for the voltages to which the qualified employee was exposed. **$70,000**
- Employees were not trained and competent in the proper use of special and precautionary techniques, PPE, insulating and shielding materials, and insulated tools for working on or near exposed energized parts of electrical equipment. **$70,000**
- The employer did not ensure that a second job briefing was held if significant changes, which might affect the safety of the employees, occurred during the course of the work. **$70,000**
- The employer did not ensure that no employee approached or took any conductive object closer to exposed energized parts than is permitted in the standard. **$70,000**
- The guarding of energized parts within a compartment was not maintained during operation and maintenance functions to prevent accidental contact with energized parts and to prevent tools or other equipment from being dropped on energized parts. **$70,000**

Total Penalties for Alleged Willful Violations: $420,000

OSHA also proposed penalties totaling $40,000 against the subcontractor for eight alleged serious violations. The list of serious violations and resulting fines include the following:

- The employer did not assess the workplace to determine whether hazards were present or likely to be present that necessitated the use of PPE. **$5,000**
- Electrical protective equipment subjected to periodic tests as required by the standard was not used. **$5,000**
- Employees were not trained and familiar with electrical safety-related work practices and procedures pertaining to their job assignments. **$5,000**
- The employer did not certify that employees had received required training. **$5,000**
- Existing conditions related to the safety of the work were not determined before work on or near electric lines or equipment was started. **$5,000**
- The employer did not require the use of a lockout on energy isolating devices that were capable of being locked out and did not demonstrate that the use of a tagout system would provide full employee protection. **$5,000**
- The utility maintained three violations of requirements for controlling hazardous energy and training employees in such controls. **$5,000** (grouped)
- The employer did not ensure that each employee exposed to hazards of flames or electric arcs did not wear clothing that would increase the extent of injury that might be sustained. **$5,000**

Total Penalties for Alleged Serious Violations: $40,000

GRAND TOTAL OF PROPOSED PENALTIES: $460,000

If proper safety procedures had been followed, the loss of life and equipment damage would not have occurred and financial penalties would not have been incurred.

SUMMARY

One of the most hazardous jobs involved with circuit breakers is installing and removing them from their cubicles. If interlocks fail, shutters do not operate properly, or a misplaced tool falls onto energized components, a short circuit can occur. Points of consideration when racking circuit breakers include the following:

- Obtain the proper clearances from the facility where work is to be performed.
- Follow the proper lockout/tagout procedures.
- Open the circuit breaker remotely or stand to one side when opening.
- Verify that the circuit breaker is open by manually operating the trip button.
- Remove the control fuses.
- Close the cabinet door and securely latch it.
- Rack out the circuit breaker while listening for arcing noises. If arcing noises are heard, rack the circuit breaker back in until it can be de-energized and repaired.
- If the door cannot be closed when racking, complete an energized electrical work permit (EEWP), and wear arc-flash protective clothing and PPE.
- Verify that the shutter/barrier mechanism is operational, if so equipped, and has covered the primary disconnects. If the shutter is not operational, do not enter the switchgear until it has been repaired.

The chances of an actual circuit breaker failure are minimal, but the consequences of such failures can be devastating. Since a circuit breaker is connected directly to the main power bus, the short-circuit current available is typically extremely high. Because there are no arc extinguishers, if there is an arc flash, it cannot be interrupted. The upstream overcurrent protective devices (OCPDs) would be required to clear the fault. Depending on how its OCPDs are set, it may process the fault as a large overcurrent condition rather than a short circuit. This could extend operating time, worker exposure to the incident energy, and cause more serious injuries than if it operated instantaneously. The appropriate PPE must always be worn.

DIGITAL RESOURCES
ATPeResources.com/QuickLinks
Access Code: 223436

chapter 3

CIRCUIT BREAKER RACKING

CHAPTER REVIEW

Name ______________________________ Date ______________

True-False

T F **1.** Circuit breakers should be racked with the door closed.

T F **2.** An arc flash boundary is required even if the door is closed.

T F **3.** Even if the indicator shows that the circuit breaker is open, it does not mean that it is.

T F **4.** Standard switchgear is designed to divert an arc flash upward through the top of the switchgear where the arc flash can be safely exhausted.

T F **5.** De-energizing the wrong circuit breaker can cause major, possibly fatal, accidents.

T F **6.** Operating time does not affect the arc flash boundary for low-voltage power circuit breakers.

T F **7.** The task of racking a circuit breaker is inherently hazardous.

T F **8.** Once the circuit breaker stabs are clear of the bus, the arc flash hazard is doubled.

T F **9.** Medium-voltage draw-out circuit breakers have a similar racking operation to low-voltage circuit breakers.

T F **10.** An overtorqued racking mechanism can fail mechanically.

T F **11.** When racking a circuit breaker, the load fed by the circuit breaker should be disconnected before operating.

T F **12.** Low-voltage draw-out circuit breakers are never equipped with shutter mechanisms.

T F **13.** A circuit breaker's primary disconnects are connected and disconnected to an energized bus.

T F **14.** An arc flash boundary on medium-voltage circuit breakers can be up to several feet.

T F **15.** Low-voltage circuit breakers are racked out of their cubicle differently than medium-voltage circuit breakers.

T F **16.** An EEWP should be reviewed at the end of a 30-day period.

T F **17.** Spreader bars are devices used to evenly lift long or wide loads.

T F **18.** Racking screws in circuit breakers do not require lubrication.

T F **19.** One or two seconds is considered a short time period by arc flash protection standards.

T F **20.** Class E hard hats are rated for high-voltage protection.

Multiple Choice

______ **1.** The effective arc rating of the clothing system does not include ___.
- A. the arc flash protective hood
- B. the arc flash protective jacket
- C. flammable underlayers
- D. arc-rated underlayers

______ **2.** The prohibited shock approach boundary for medium-voltage draw-out power circuit breakers (2.3 kV to 15 kV) is ___″.
- A. 7
- B. 12
- C. 26
- D. 60

______ **3.** The restricted shock approach boundary for 480 V low-voltage power circuit breakers (LVPCBs) is ___″.
- A. 1
- B. 7
- C. 12
- D. 21

______ **4.** A ___ is a device with a swiveling, weight-bearing connection used to move heavy loads.
- A. stab-finger cluster
- B. floor tripper
- C. spreader bar
- D. fifth wheel

______ **5.** Medium-voltage draw-out circuit breakers have ___ installed in the cubicle, which open the breaker and discharge the closing springs when racking.
- A. floor trippers
- B. safety interlock devices
- C. T-handles
- D. extension arms

______ **6.** Heavy-duty leather work shoes that meet the NFPA 70E minimum requirement of ___ oz/yd^2 weight should be worn when racking circuit breakers.
- A. 6
- B. 12
- C. 18
- D. 24

______ **7.** A ___ is a rigging device that is supported from a crane hook to support lifting lines at various locations.
- A. come-along
- B. block and tackle
- C. spreader bar
- D. web sling

_______________ **8.** The NFPA requirement for footwear to be worn when working with low- and medium-voltage circuit breakers is leather shoes or boots with a minimum ___ oz/yd^2 of weight.
- A. 10
- B. 12
- C. 16
- D. 20

_______________ **9.** HRC level ___ PPE is require for protection from arc flash hazards when racking 480 V circuit breakers.
- A. 1
- B. 2
- C. 3
- D. 4

_______________ **10.** The limited shock approach boundary for 208 V LVPCBs is ___″.
- A. 12
- B. 21
- C. 42
- D. 60

_______________ **11.** A(n) ___ violation is a safety violation committed with an intentional disregard of, or plain indifference to, the requirements of the Occupational Safety and Health Act and regulations.
- A. accidental
- B. deliberate
- C. serious
- D. willful

_______________ **12.** A(n) ___ violation is a safety violation in which there is a substantial probability that death or serious physical harm could result and that the employer knew or should have known of the hazard.
- A. accidental
- B. deliberate
- C. serious
- D. willful

Completion

_______________ **1.** A(n) ___ is a device used to ensure that a circuit breaker does not get racked while the circuit breaker is in the closed position.

_______________ **2.** A(n) ___ is a document issued by an employer to allow specific work to be performed on energized electrical equipment with various limitations.

_______________ **3.** The four positions LVPCBs must go through as they are being inserted or removed from their cubicles include connected, test, disconnected, and ___.

_______________ **4.** ___ is clothing, glasses, gloves, hard hats, respirators, or other safety devices designed to protect workers against safety hazards in the work area and from injury.

_______________ **5.** Medium-voltage circuit breakers use a(n) ___, or barrier, to cover the energized bus connections when the breaker is racked out.

_______________ **6.** A safety device that is used when racking in medium-voltage circuit breakers is the ___ assemblies, sometimes referred to as tulips, which connect the circuit breaker to the main bus.

_______________ **7.** In the test position, the ___ stabs are disconnected.

_______________ **8.** The total arc-rated clothing system worn when racking circuit breakers must have a minimum rating of ___ cal/cm^2.

_______________ **9.** When moving a circuit breaker with a cable reel, crossed cables can cause a circuit breaker to "___" when lowering and cause the cable to snap under the instantaneous weight/force increase.

_______________ **10.** A(n) ___ device is commonly installed on most circuit breakers to prevent them from being inserted or removed from a cubicle while they are closed.

_______________ **11.** The prohibited shock approach boundary for a 480 V LVPCB is ___″.

_______________ **12.** When racking medium-voltage circuit breakers, Class ___ rubber insulating gloves with leather protectors should be used.

_______________ **13.** Arc-resistant switchgear typically costs ___% to ___% more than standard switchgear.

_______________ **14.** The restricted shock approach boundary on medium-voltage draw-out circuit breakers (2.3 kV to 15 kV) is ___″.

_______________ **15.** When performing lockout/tagout procedures, the lock and tags used should be supplied by ___.

Short-Answer

1. Explain why an arc flash in an LVPCB is focused on the front opening of the circuit breaker cubicle.

2. Why must ropes and wire slings never be strung through the circuit breaker frame when moving a circuit breaker?

3. Explain how not cleaning and lubricating the racking screws on circuit breakers that use two racking screws (such as a GE PowerVac®) can affect the racking process.

4. What are two advantages of arc-resistant switchgear over standard switchgear?

5. Why do most facilities typically not allow work to be performed within the cubicle of medium-voltage circuit breakers, even if the shutter is covering the energized bus connections?

Matching

Arc Flash and Shock Approach Boundaries

______ 1. Extent of arc flash

______ 2. Limited approach

______ 3. Prohibited approach

______ 4. Restricted approach

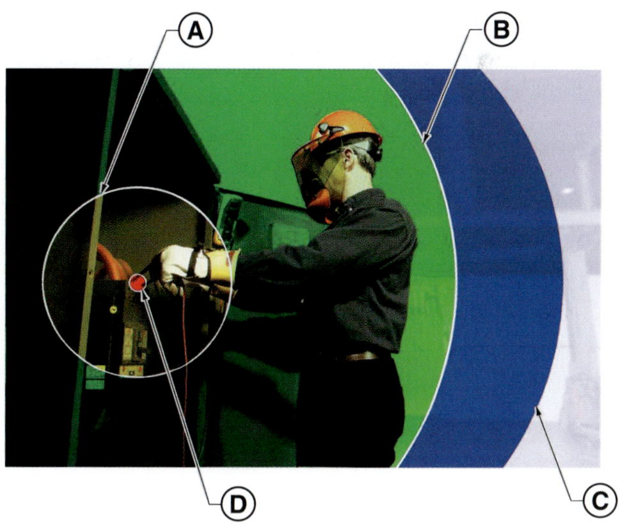

Low-Voltage Power Circuit Breaker (LVPCB) Racking Positions

______________ 1. Connected position

______________ 2. Disconnected position

______________ 3. Instrument module

______________ 4. Withdrawn position

______________ 5. Test position

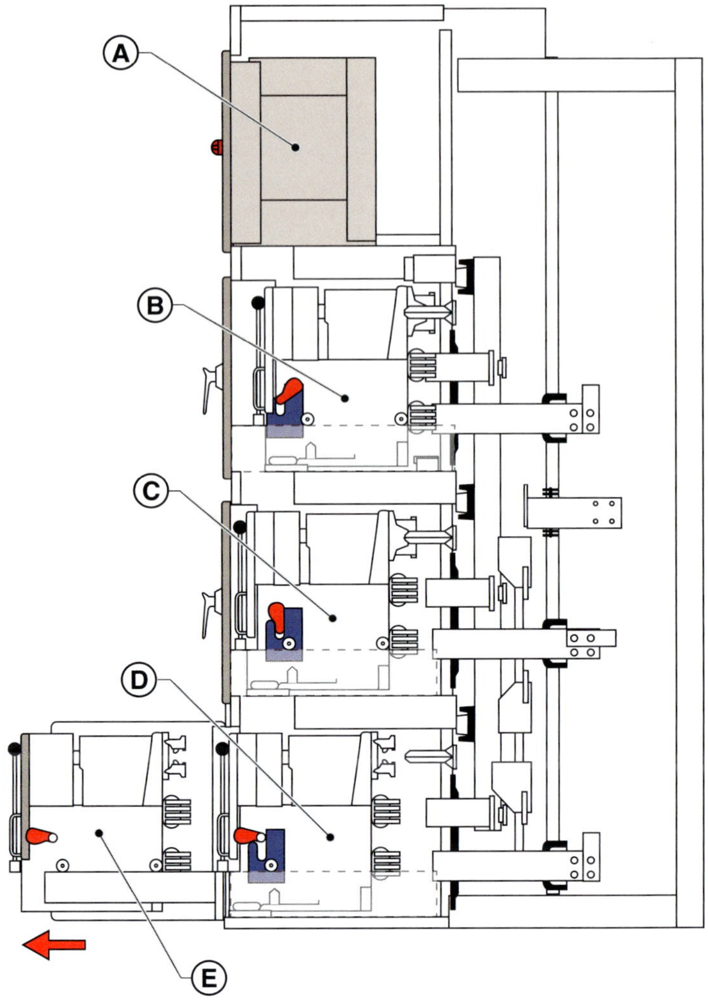

chapter 4

CIRCUIT BREAKER CUBICLE MAINTENANCE

Circuit breaker switchgear (enclosure) houses the circuit breaker and provides mounting for the main power bus, instrumentation, such as voltmeters and ammeters, and power monitoring. Draw-out circuit breakers also use switchgear to support the cradle, movable extension arms, and racking mechanism that the circuit breaker is inserted into and removed from. The switchgear that contains and supports the circuit breaker is the cubicle (cell). Switchgear is the entire enclosure including the circuit breaker, bus structure, control cabinet, metering, and protective relays.

A circuit breaker that cannot be safely removed or installed is a serious safety hazard. Servicing the cubicle ensures that it will be functional throughout the life of the circuit breaker. Cubicle service should be performed whenever the circuit breaker is removed for service.

OBJECTIVES

- Identify three sources for determining maintenance intervals.
- State the differences between routine maintenance and major maintenance.
- Identify the section in ANSI/NETA MTS-2011 that pertains to the maintenance and testing of switchgear.
- Use the maintenance frequency matrix in ANSI/NETA MTS-2011 Annex B to determine the recommended maintenance interval multiplier for low-voltage power circuit breaker switchgear.
- State the safety task that should be performed before physically working on any electrical system or device.
- List the four tasks that are usually performed during routine maintenance.
- State additional tasks that are performed during major maintenance.
- List the electrical tests that can be performed on switchgear during major maintenance.

MAINTENANCE INTERVALS

Many circuit breaker original equipment manufacturers (OEMs) recommend two different service intervals for the circuit breaker cubicle. These service intervals are the same as those for circuit breaker service. The first type of maintenance is routine maintenance, which is performed annually. This is minor maintenance performed to maintain the integrity of a cubicle. Typically, only the front side of the cabinet is accessed and no electrical tests are performed. Infrared thermography testing is often performed annually, which may involve removing switchgear covers, although infrared viewing ports can be used instead. The switchgear is normally not de-energized during routine maintenance.

The second type of maintenance is major maintenance, which is performed while the switchgear is de-energized and the bus is grounded. The rear covers of the switchgear are removed to access the bus and cables. Major maintenance is usually performed during the major maintenance interval, typically every three to five years.

KEY TERMS

- micro-ohmmeter
- thermal imager

The recommendations in ANSI/NETA MTS-2011, developed by the InterNational Electrical Testing Association (NETA), are used for the specific tasks to be performed. Section 7.1 covers switchgear and switchboard assemblies. Appendix B in the ANSI/NETA MTS-2011 contains a Maintenance Frequency Matrix that bases maintenance intervals on two factors: criticality of the system or device and device condition.

To accurately determine maintenance intervals, more factors must be considered, but the two listed in ANSI/NETA MTS-2011 are two of the more common critical factors. By choosing the appropriate reliability factor and condition, a multiplier is derived. That multiplier is then used with the recommended maintenance intervals for the specific type of device. With switchgear, mechanical and visual inspections are recommended once every 12 months, and electrical tests are recommended every 24 months. These intervals are then multiplied by the number derived from the Maintenance Frequency Matrix to determine the actual recommendations.

For example, if the equipment reliability need is high and the condition is poor, using data from the Maintenance Frequency Matrix would provide a recommendation of six months (multiplier of 0.25 × 24 months of electrical tests = 6 months). If the equipment reliability need is low and the equipment condition is good, using data from the Maintenance Frequency Matrix would provide a recommendation of 60 months (multiplier of 2.5 × 24 months of electrical tests = 60 months). **See Figure 4-1.**

RECOMMENDED MAINTENANCE INTERVALS—ANSI/NETA MTS-2007

Maintenance Frequency Matrix

Equipment Reliability Requirement	Equipment Condition: Poor	Average	Good
Low	1.0	2.0	2.5
Medium	0.50	1.0	1.5
High	0.25	0.50	0.75

EXAMPLE:
FOR ELECTRICAL TESTS TO BE PERFORMED ON EQUIPMENT WITH LOW RELIABILITY REQUIREMENT IN GOOD CONDITION, RECOMMENDED MAINTENANCE INTERVAL IS THE FOLLOWING:
2.5 × 24 = **60 MONTHS**

Inspection and Test Frequency in Months
(Multiply These Values by the Factor in the Maintenance Frequency Matrix)

Section	Description	Visual	Visual and Mechanical	Visual and Mechanical and Electrical
7.1	Switchgear and Switchboard Assemblies	12	12	24

Figure 4-1. Maintenance frequency intervals are multiplied by the number derived from the Maintenance Frequency Matrix to determine actual maintenance frequency recommendations.

Other factors, such as loading, operating environment, and equipment history, may modify these recommendations. However, the Maintenance Frequency Matrix provides reasonable guidance when more specific recommendations are not available. Additional information on recommended maintenance intervals can also be found in NFPA 70B, *Recommended Practice for Electrical Equipment Maintenance.* **See Figure 4-2.**

Unlike ANSI/NETA MTS-2011, NFPA 70B provides guidelines based on general industry-accepted maintenance time-based intervals, without regard to other factors. Table K.4(c), *Low-Voltage Equipment, Switchgear: Maintenance of Equipment Subject to Long Intervals Between Shutdowns—Electrical*

TECH TIP

NFPA 70B does not contain mandatory language, such as "shall" and "will" found in other standards and codes but uses terms such as "should" and "might."

Distribution from Annex K as well as Table L.1 from Annex L, *Maintenance Intervals,* of NFPA 70B contain recommendations for electrical equipment maintenance intervals. The introduction to these annexes takes into consideration that loading of equipment and the environment the equipment is operating in may change these recommendations. Specific recommendations are made for virtually every type of electrical device that may be installed in a system.

NFPA 70 B—MAINTENANCE RECOMMENDATIONS

ANNEX K 70B–227

Table K.2(e) Medium-Voltage Equipment, Circuit Breakers: Maintenance of Equipment Subject to Long Intervals Between Shutdowns — Electrical Distribution

Type	Inspections, Maintenance, and Tests	Typical Frequency and Remarks
Air-break, Drawout Type (15.4)	**Inspection and Maintenance (withdrawn from switchgear and deenergized) (7.1):**	Max. of 3 years or at manufacturer's maximum number of operations since previous maintenance, whichever occurs first. Immediately after breaker opens to interrupt a serious fault. Follow manufacturer's maintenance instructions. If breaker is stored-energy closing type, follow manufacturer's safety precautions, determine that closing springs are discharged, or mechanism is blocked to prevent personal injury. Keep hands away from contacts and mechanism while test operating breaker (15.4.1.1).
	Remove arc chutes. Inspect, adjust and clean where necessary:	
	Main contacts (15.4.3).	For pitting, spring pressure, overheating, alignment, overtravel, or wipe; adjust or replace accordingly.
	Arcing contacts (15.4.3.2).	For alignment, overtravel, or wipe and for arc erosion; adjust or replace accordingly.
	Moving parts and linkages (15.4.5.1 through 15.4.5.3).	For freedom of movement.
	Closing mechanism (15.4.5).	For quick and positive closing action.
	Tripping mechanism (15.4.5).	For freedom of movement and reliability to open breaker contacts.
	Interlocks and safety devices (15.4.6.2, 15.9.8).	Functionally test to prove proper operation.
	Primary disconnect finger clusters (15.4.3.8).	For proper adjustment and spring pressure; lubricate, unless manufacturer's instructions specify that they should not be lubricated.
	Secondary disconnect contacts (15.4.3.8).	For alignment and spring pressure. Lubricate.
	Closing and trip coils (15.4.6.1).	General condition and evidence of

Figure 4-2. Additional information on recommended maintenance intervals can be found in NFPA 70B, Recommended Practice for Electrical Equipment Maintenance.

OEM Instruction Manuals and Maintenance

OEM instruction manuals are possible sources of information regarding maintenance intervals for switchgear. However, when it comes to maintenance intervals, most OEMs use phrases such as, "*Breakers should be cared for through the implementation of a systematic periodic maintenance program. A periodic routine inspection is recommended. How frequently an inspection is made of an individual breaker will depend on the circumstances of its use. It would be well to inspect the breaker at least once per year*"(from General Electric maintenance manual) or "A periodic inspection and servicing, normally at intervals of one year, should be included in the maintenance routine" (from Siemens maintenance manual). Manufacturers often use somewhat vague language, since they do not know the environment, conditions, or applications under which the switchgear or circuit breakers may be used.

SAFETY

Safety must always be the primary concern and priority when working on or near electrical equipment. Even though circuit breakers and switchgear can be operated safely and have no issues 99.9% of the time, it is that 0.1% that leads to injury or death. Safe work procedures must always be used, and relevant OSHA regulations and NFPA 70E, *Standard for Electrical Safety in the Workplace*, work practices must be followed. Even when the circuit breaker is racked out of its switchgear and the shutter is closed, there may still be energized parts and components that can have voltages above 50 V. Some components that can be energized at the system voltage level include the following:

- exposed line-side primary disconnects on low-voltage power circuit breakers (LVPCBs)
- primary disconnects on medium-voltage air or vacuum circuit breakers with raised or defective shutter mechanisms
- line- and load-side primary disconnects on tie breakers
- possible backfeeds

Some components that can be energized at the control voltage level include the following:

- DC control circuits
- protective relays
- auxiliary or secondary disconnects
- indicating lamps
- space heaters

The presence of any potential voltage source must always be verified before performing work in a cubicle. Measuring to ground must not be done when taking measurements on DC control circuits, since battery-fed systems are not connected to ground. DC control circuits and other auxiliary DC sources should be verified for the presence of voltage by measuring from positive to negative. Control circuits and other auxiliary sources of voltage should be guarded by rubberized shielding so that inadvertent contact is not made while working inside the switchgear.

On medium-voltage metal-clad switchgear, maintenance on a cubicle should never be performed if the shutter mechanism does not cover the primary bus connections. The circuit must be either de-energized or an energized electrical work permit (EEWP) must be issued by the facility where the work is to be performed. In most cases, the incident energy will be too great to allow any work to be performed while the shutter mechanism is raised.

NFPA 70E provides guidance on what is required for the safety-related work practices referred to in OSHA standards. In general, the tasks are as follows:

- determine and document the incident energy exposure
- determine the Arc Flash Boundary
- determine the nominal voltage
- determine the shock approach boundaries (i.e., limited, restricted, or prohibited)
- perform a PPE assessment

OSHA 29 CFR 1910.333 — Selection and Use of Work Practices

OSHA 29 Code of Federal Regulations (CFR) 1910.333(a)(1) should be referenced prior to performing work on energized and de-energized electrical equipment. For example, excerpts from this standard that apply to working on or near electrical conductors and circuit parts include the following:

"(a) General. Safety-related work practices shall be employed to prevent electric shock or other injuries resulting from either direct or indirect electrical contacts, when work is performed near or on equipment or circuits which are or may be energized. The specific safety-related work practices shall be consistent with the nature and extent of the associated electrical hazards.

"(1) De-energized parts. Live parts to which an employee may be exposed shall be de-energized before the employee works on or near them, unless the employer can demonstrate that de-energizing introduces additional or increased hazards or is infeasible due to equipment design or operational limitations. Live parts that operate at less than 50 volts to ground need not be de-energized if there will be no increased exposure to electrical burns or to explosion due to electric arcs."

These steps are outlined in NFPA 70E, Annex J, *Energized Electrical Work Permit,* and should be completed prior to performing any high-risk energized work. Racking a circuit breaker in and out of its switchgear is high-risk and hazardous because electrical connections to the bus are being made and broken with no arc interrupters, even if the cubicle door is closed and secured. The spring-loaded fingers from a circuit breaker are forced onto the energized bus when racking the circuit breaker in. The circuit breaker must be in the open position in order to do this safely. **See Figure 4-3.**

RACKING CIRCUIT BREAKERS

Figure 4-3. The spring-loaded fingers from a circuit breaker are forced onto the energized bus when racking in the circuit breaker. The circuit breaker must be in the open position in order to do this safely.

A defective interlock or other malfunction could allow a closed circuit breaker to be inserted or removed from the energized bus and result in an electrical explosion. Defective interlocks are common with circuit breakers. Approximately 40% of the circuit breakers serviced by Shermco Industries are found to have defective interlocks. NFPA 70E Tables 130.7(C)(15)(a) and 130.7(C)(16) provide recommendations for this type of work activity. However, the available arc energy often exceeds the 40 cal/cm^2 limit recommended, which means that many low- and medium-voltage draw-out circuit breakers should not be routinely racked in or out while energized.

OSHA 29 CFR1910.269(a)(3)—Existing Conditions

Per OSHA 29 CFR 1910.269(a)(3), "*Existing conditions related to the safety of the work to be performed shall be determined before work on or near electric lines or equipment is started. Such conditions include, but are not limited to, the nominal voltages of lines and equipment, the maximum switching transient voltages, the presence of hazardous induced voltages, the presence and condition of protective grounds and equipment grounding conductors, the condition of poles, environmental conditions relative to safety, and the locations of circuits and equipment, including power and communication lines and fire protective signaling circuits.*"

In all situations, the qualified person performing the work is to determine whether it is safe to work given the conditions, choose the appropriate PPE or equipment required to safely perform the work, and determine the required safe work practices. Factors to consider when determining whether to work on equipment include the following:

- voltage level of the equipment
- short-circuit available current
- protective device settings
- equipment maintenance
- operation history
- equipment age
- operating environment
- loading
- visible deterioration or defects
- proper clearances to remove the equipment from service

One additional factor that is often overlooked until it is too late to make a change is the other equipment or processes that could be affected when the circuit breaker is opened. De-energizing one circuit could affect several other processes and lead to a dangerous situation, especially if it affects control systems, emergency systems, or hazardous materials processes.

SAFETY FACT

Qualified persons working with circuit breakers must receive training. They must demonstrate the skills and knowledge received related to the construction, operation, and hazards associated with the equipment they are working on as well as the power system.

CASE STUDY—TESTING EQUIPMENT FOR THE PRESENCE OF VOLTAGE

During shutdown at a primary metals processing facility, a technician was performing maintenance on a 480 V motor starter. As he was tightening the connection, his torque wrench contacted a grounded part of the support structure and drew an arc. The resulting arc flash did not injure the technician, even though he was not wearing his arc-rated PPE or rubber insulating gloves. He thought the equipment was de-energized.

Upon investigation it was discovered that the A-phase contact inside the main circuit breaker had welded itself shut and could not be opened. The other phases opened when the circuit breaker handle was operated, but the A-phase contact was welded in the closed position.

Conclusion: The primary cause of the accident was not the welded contacts inside the circuit breaker but that the technician did not follow OSHA and NFPA 70E work practices and check for the absence of voltage before he began maintenance. He assumed that the circuit was de-energized, since the circuit breaker was in the OFF position. Electrical equipment may appear to be de-energized but cannot be considered electrically safe until all steps have been taken to secure it.

ROUTINE MAINTENANCE

Customers of InterNational Electrical Testing Association (NETA) member companies usually perform a combination of the recommended routine and major maintenance tasks during regular facility shutdown and turnaround periods. The tasks are performed during one maintenance interval in which the equipment is locked, tagged, and grounded.

Routine maintenance is typically performed by the equipment owner on an annual basis, although the trend in industry is to extend maintenance intervals. During routine maintenance, all work performed is limited to the front section of the cabinet, with any exposed energized conductors covered with a suitable rubber insulating blanket or shielding. The equipment is not de-energized and maintenance is not performed in the energized compartments. However, the circuit breaker is usually removed from its cubicle while routine maintenance is being performed. When performing routine maintenance on switchgear, the four steps usually performed include inspection, cleaning, lubrication, and tightening.

Inspecting Switchgear Cubicles

For all items visually inspected, the condition and any corrective action required must be documented in accordance with NFPA 70E-2015, Chapter 2, Section 205.3 and 205.4. **See Figure 4-4.**

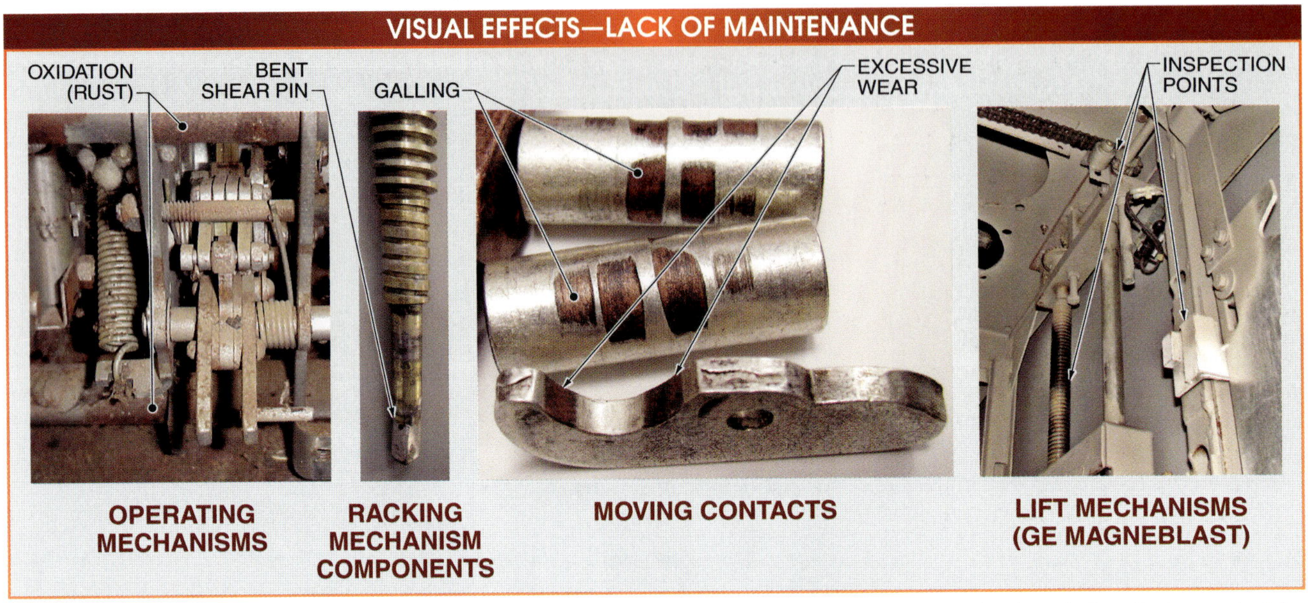

Figure 4-4. Before any work can be performed on circuit breakers, they must be de-energized and visually inspected for any effects from a lack of maintenance.

Items that must be visually inspected on a cubicle include the following:

- disconnected or loose components that may have fallen off the circuit breaker and that may be resting on the bottom of the cubicle; components that have been dislodged from either the circuit breaker or the switchgear must be replaced; all loose parts and debris must be removed from the cubicle
- spring-charging motors and operating mechanisms for excess lubricant
- primary, auxiliary, and ground stabs for damage, deformity, discoloration, and signs of overheating
- ground stabs for galling, tightness, and dried lubricant
- protective relay cases and wiring on medium-voltage switchgear for damage and tightness of the connections (*Note:* Testing and calibration of protective relays is often performed at this time.)
- all parts for signs of moisture or rusting (*Note:* On outdoor switchgear, this could indicate defective seals on cabinet doors or inoperative heaters. The presence of moisture on indoor equipment could be caused by moisture dripping from above or seeping in from the floor. Space heaters can help with condensation problems but cannot stop damage from leaks or other excessive moisture sources.)
- conductors that are scraped, cut, or bare, especially around the door hinge area and in wire ducts
- wiring that is secure and not exposed or is not in a position to be pulled loose when removing the circuit breaker from its cubicle
- terminal connections for signs of overheating

- racking mechanisms for smooth operation, broken shear pins, and misaligned or bent parts
- GE MagneBlast circuit breaker mechanisms for wear on the chain and bent, broken, or misaligned parts on the lift mechanism
- pressure (Mercoid®) switches, cell switches, and other mechanisms as required
- burned-out indicator light bulbs (caution must be used when replacing light bulbs because the circuits they are connected to can be energized)

Caution: Shutter mechanisms on medium-voltage switchgear are not typically inspected during initial inspection, since to do so would require exposing the energized bus. If a shutter mechanism is operating erratically or malfunctioning, the circuit must be de-energized so that it can be repaired safely.

Cleaning Circuit Breakers

After the control cabinet, cubicle, and switchgear have been thoroughly inspected, the equipment must be cleaned. Cleaning a control cabinet, cubicle, and switchgear is performed by applying the following procedure:

1. Vacuum the cabinet interior. Vacuuming removes dirt and excess debris, whereas low-pressure air moves it around. **CAUTION:** Use only vacuum attachments approved for use near electrical equipment and without metal or rubber-covered metal components.
2. Wipe or brush loose dirt from wiring, relay cases, and other surfaces not accessible to a vacuum. Do not use brushes with metal ferrules. Only brushes with plastic or stiff fabric ferrules should be used. Use only clean, dry cotton cloths when wiping surfaces.
3. Use a stiff-bristle brush or clean, dry cloth and clean any dried lubricant from auxiliary and ground stabs. Use caution by testing for the absence of voltage with a voltmeter or multimeter to prevent inadvertent contact with energized stabs.
4. Remove any excess lubricant with an approved safety solvent and clean cloth. **CAUTION:** Use only an amount of solvent required and only work in well-ventilated areas. Solvent containers should be plastic. Metal solvent containers must be bonded and grounded. Although safety solvents are considered nonflammable, they can have a flashpoint as low as 140°F. Do not attempt to clean energized bus stabs on low-voltage switchgear.

SAFETY TIP

Do not raise any shutter or barrier within the cubicle unless the entire switchgear bus has been de-energized, locked out, tagged out, and grounded. If it is necessary to open the shutter while the bus is energized, obtain the required clearances and energized electrical work permits (EEWPs) from a facility safety manager or appropriate person.

Lubricating Circuit Breakers

After the control cabinet, cubicle, and switchgear have been thoroughly inspected and cleaned, the equipment must be lubricated. Lubrication of the control cabinet, cubicle, and switchgear is performed by applying the following procedure:

1. Apply a light coat of nonconductive grease on the racking mechanisms, chains, and racking screws. Nonconductive lubricants are typically petroleum-based and designed for use on parts that do not carry electrical current.

2. Apply a light coating of conductive lubricant on the ground stab, auxiliary stabs, and, if de-energized, main stabs. Conductive lubricants are not actually conductive but are designed for use on components that carry electrical current.
3. Apply a thin coat of light oil on hinges, linkage mechanisms, and bearings. One or two drops of oil placed on a bearing with dried grease reactivates the grease.

Note: All lubricants must be applied lightly. Any excess lubricant must be wiped off. Lubricant that runs, drips, or squeezes out of a pivot point is considered excessive. Circuit breaker OEMs typically provide a list of recommended lubricants for specific equipment. **See Figure 4-5.** Synthetic greases, such as Mobilgrease 28 HT (high temperature) can be used in most instances for conductive and nonconductive locations, for both the switchgear and circuit breaker and can be used in place of most of the OEM-recommended lubricants.

RECOMMENDED LUBRICANTS BY OEM		
OEM	Grease for Stabs & Pivot Points	Grease for Mechanical Parts & Bearings
General Electric	D50H47, D50H38	D50H15, D50H38 (Mobilgrease 28)
Westinghouse	53701AN10J	693C500G04
ABB/ITE	NO-OX-ID, Special Grade A	Anderol, L757
Seimens-Allis Allis-Chalmers	S-A Contact Lube, 15-171-370-002	Beacon, P-290
Square-D	PJC-7201	Lubriplate 630-2

Figure 4-5. Circuit breaker OEMs typically provide a list of recommended lubricants for specific equipment.

Tightening Circuit Breakers

After the control cabinet, cubicle, and switchgear have been thoroughly inspected, cleaned, and lubricated, the equipment can be tightened. When tightening fasteners within a control cabinet, a cubicle, or switchgear, the following considerations are taken:

- Fasteners must be torqued to OEM-recommended torque values (where specified) and not just tightened. A calibrated torque wrench or, for screws and small fasteners, a torquing screwdriver should be used.
- Steel fasteners are torqued to the fastener head marking, usually SAE grade 5.
- If a steel bolt or screw is to be fastened to a different material (bronze, aluminum, etc.), it should be torqued to the recommended specification of the softer of the two materials.
- If a screw or bolt is to be fastened into a plastic-encapsulated nut, such as a standoff insulator, it should be tightened until the lock nut flattens, then an additional ⅛ turn. These types of fasteners should not be torqued because overtorquing can break or permanently damage the plastic or fiberglass housing the nut.

Other sources for torque values are available in ANSI/NETA MTS-2011, *Maintenance Testing Specifications,* Table 100.12.1 and per ANSI/NFPA Standard 70B, *Recommended Practice for Electrical Equipment Maintenance*. The torque chart from ANSI/NETA MTS-2011 is included in Appendix A. **See Figure 4-6.**

TIGHTENING FASTENERS—ANSI RECOMMENDED TORQUE VALUES

PART 1—HEAT-TREATED STEEL, CADMIUM OR ZINC PLATED

Grade	SAE 1&2	SAE 5	SAE 7	SAE 8
Head Marking				
Minimum Tensile Strength‡	64K	105K	133K	150K
Bolt Diameter*	Torque†			
1/4	4	6	8	8
5/16	7	11	15	18
3/8	12	20	27	30
7/16	19	31	44	48
1/2	30	48	68	74
9/16	42	70	96	105
5/8	59	96	135	145
3/4	96	160	225	235
7/8	150	240	350	380
1.0	225	370	530	570

PART 2—SILICON BRONZE FASTENERS TORQUE†

Bolt Diameter*	Nonlubricated	Lubricated
5/16	15	10
3/8	20	14
1/2	40	25
5/8	55	40
3/4	70	60

PART 3—ALUMINUM ALLOY FASTENERS TORQUE†

Bolt Diameter*	Lubricated
5/16	8.0
3/8	11.2
1/2	20.0
5/8	32.0
3/4	48.0

PART 4—STAINLESS STEEL FASTENERS TORQUE†

Bolt Diameter*	Uncoated
5/16	14
3/8	25
1/2	45
5/8	60
3/4	90

* in in.
† in ft-lb.
‡ in ft-lb./psi

Figure 4-6. ANSI/NETA MTS-07, Maintenance Testing Specifications, Table 100.12.1 contains recommended torque values for various types of fasteners.

MAJOR MAINTENANCE

Major maintenance is normally performed by NETA-member companies on their customers' electrical power systems during regularly scheduled facility maintenance shutdowns and turnarounds. During major maintenance, the electrical system is de-energized, rear covers are removed, and the bus is grounded. Proper clearances must be obtained, and safety measures must be taken in accordance with OSHA 29 CFR 1910.331, 1910.335, and 1910.269 as well as the current edition of NFPA 70E. Systems must be placed in an electrically safe work condition (i.e., locked out, tagged out, verified as de-energized by testing, and grounded before being considered de-energized).

All items performed during routine maintenance are also to be performed during major maintenance. In addition to routine maintenance procedures, the following tasks are performed:

- An infrared scan with a thermal imager may be performed before the electrical power system is de-energized. Since the rear covers must be removed, there are additional and/or increased hazards involved. Therefore, proper arc flash protection must be used. Infrared scans can indicate loose connections on bus joints, cable lugs, and other areas of overheating. For the scan to be most effective, the system must be under load.
- Prior to cleaning, any as-found tests that may be required must be performed. These tests represent the condition of the switchgear before maintenance is performed and can aid in determining the frequency of maintenance. If problems are found whenever maintenance is performed, it may be advisable to decrease the length of time between service intervals. If a small number of minor problems are discovered during maintenance, usually the length of time between service intervals can be increased.
- The de-energized bus, cables, standoff insulators, and flooring must be thoroughly cleaned. Dirt is conductive. As it builds up, the possibility of tracking, arcing, and subsequent problems increases. An approved safety solvent must be used when wiping down components, ventilating as needed to reduce fumes. Vacuuming the floor and bus is a good method of cleaning those areas. Low-pressure air should not be used for cleaning, because it only moves the dirt around and does not remove it.
- With the system de-energized and grounded, all primary stabs, including auxiliary stabs, should be cleaned of dried greases and relubricated with conductive lubricant.
- Fuse and/or circuit breaker sizes, types, and settings must be verified to correspond to drawing specifications and the coordination study. The circuit breaker addresses must be verified for those circuit breakers that include microprocessor communication packages.
- Current and voltage transformer ratios must be verified to correspond to drawing specifications.
- Bus connections must be inspected for any signs of overheating, loose standoff insulators, and cracked standoff insulators.
- Cables and conductors must be inspected for tracking as well as cracked, broken, or scraped insulation.
- The tightness of connections must be verified using one of the following methods:
  - Bus joints and cable connections may be torqued.

  Note: Cable connections and the hex-head set screws used to secure them should be inspected, since they have a tendency to loosen over time, even when securely tightened previously.
  - A low-resistance ohmmeter may be used to take resistance measurements. A *micro-ohmmeter* is a device used to measure the resistance of an electrical conductor down to the micro-ohm range.
  - A thermographic survey may be performed with a thermal imager. A *thermal imager* is a device that detects heat patterns in the infrared wavelength spectrum without making direct contact with the workpiece. **See Figure 4-7.**

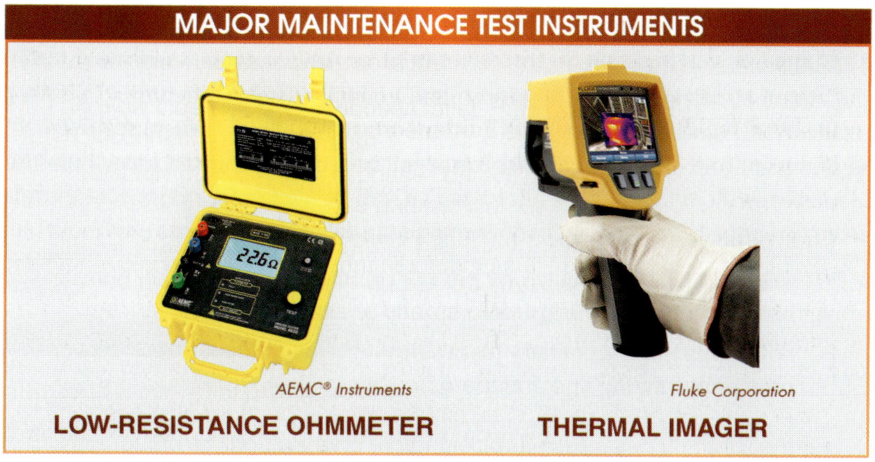

Figure 4-7. Test instruments that can be used to perform major maintenance tasks include low-resistance ohmmeters and thermal imagers.

- The correct operation and sequencing of electrical and mechanical interlock systems must be confirmed as follows:
  - The closing of locked-open devices and the opening of locked closed devices should be attempted.
  - Key change sequences with all devices included in the interlock scheme may be made as applicable (Kirk® Key systems). If these Systems are not tested during maintenance, failure is more likely to occur during an emergency. The failure of keyed interlock systems can be time consuming to correct, since they are tamper-resistant.
- Shutter and barrier installation and operation must be verified to be correct. The shutter mechanism must be inspected for smooth operation, ensuring that it covers the stabs completely.

Most mechanical interlock systems installed on circuit breakers are the Kirk® Key or similar type locking systems.

TECH TIP

On GE MagneBlast® switchgear, be certain to inspect the lift mechanism and shutter assembly since the primary stabs can penetrate the shutter if it does not lift or close properly. This type of malfunction can cause damage to the epoxy-covered bushings and shutter which can result in a short circuit.

- All active components, such as lift or racking mechanisms, should be exercised.
- It must be verified that all filters are in place and/or vents are clear. Animal and insect nests, debris, and other items that can block the flow of cooling air must be removed. Existing filters must be replaced with new filters.
- A visual and mechanical inspection of instrument transformers must be performed.
- Control power transformers must be inspected as follows:
  - Wiring must be inspected for physical damage, cracked insulation, broken leads, tight connections, defects, and general condition.
  - All primary and secondary fuse ratings or circuit breakers must be verified to match drawing specifications.
  - Draw-out disconnecting and grounding contacts and interlocks must be verified to function correctly.

Electrical Tests

Electrical tests must also be performed on the switchgear during major maintenance since it is de-energized. The correct functioning of control transfer relays located in switchgear with multiple power sources must be verified. In addition, the operation of switchgear heaters and controllers must be verified. Thermostats and heaters must turn on and off properly. Furthermore, the accuracy of all test instruments must be verified. Watt-hour meters must be calibrated in accordance with ANSI/NETA MTS-2011 Section 7.11.

Low-Resistance Testing of Bus Joints and Other Connections. Low-resistance tests of bus joints are performed with a micro-ohmmeter, which injects pulses of DC, usually 10 A, across a connection. The resistance of the voltage drop is measured and read in micro-ohms (μΩ). Clean, tight connections must have no more than 20 μΩ across them. However, this reading can vary. Another method of grading connections recommended by NETA involving the use of micro-ohmmeters requires that there be no more than a 50% difference between the highest and lowest reading taken across connections of the same type and size.

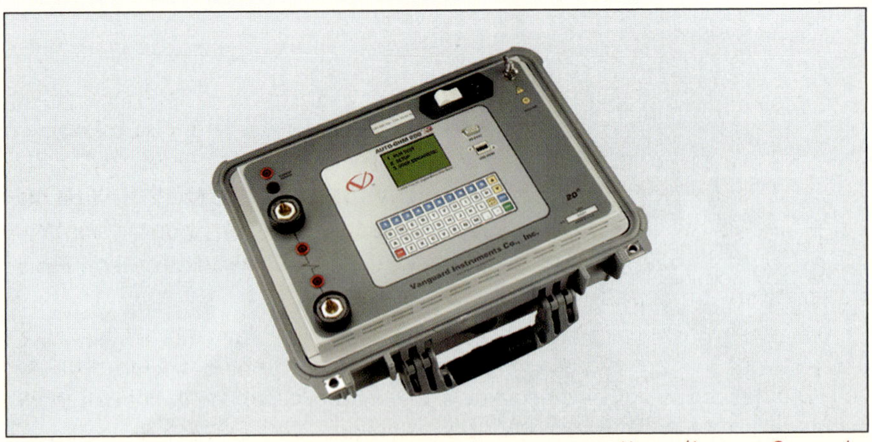

Vanguard Instruments Company, Inc.

Resistance testing of bus joints and other connections are performed using a micro-ohmmeter, which injects pulses of DC, usually about 10 A, across a connection.

Insulation Resistance Testing. Insulation resistance tests are performed with a megohmmeter, which is used to apply DC voltage across an insulation system. Busbars can be tested phase-to-phase and phase-to-ground. At the same time, cable can be tested phase-to-phase and phase-to-ground and evaluated. In order for insulation to be tested, it must be ungrounded and isolated. CAUTION: When a bus or conductor is ungrounded, it must be treated as an energized conductor. Wear electrical insulating gloves with leather protectors when connecting test leads. Be aware that if control-power transformers are connected to ground, they can appear as short circuits unless they are ungrounded. Table 100.1 of ANSI/NETA MTS-2011 shows allowable insulation resistance values for switchgear. **See Figure 4-8.**

INSULATION RESISTANCE TEST VOLTAGES—ANSI/NETA MTS 2011

Nominal Rating of Equipment*	Minimum Test Voltage, DC	Recommended Minimum Insulation Resistance†
250	500	25
600	1000	100
1000	1000	100
2500	1000	500
5000	2500	1000
8000	2500	2000
15,000	2500	5000
25,000	5000	20,000
34,500 and above	15,000	100,000

* in V
† in MΩ

Figure 4-8. Table 100.1 in ANSI/NETA MTS-2011 shows allowable insulation resistance values for switchgear.

TECH FACT

Insulation resistance testing has proven worth for detecting critical current leakage, poor workmanship, and equipment defects and for diagnosing otherwise unknown equipment failures. The key point to remember with insulation resistance testing is that the single test value often does not provide enough information to evaluate the insulation. By trending the test results, a sharp decrease in the insulation resistance can be detected. This sharp decrease would indicate imminent failure, even if the insulation resistance value is above the minimum.

Insulation resistance tests should be performed on control wiring with respect to ground. For cable rated 300 V and lower, 500 VDC should be applied. For cable rated 301 V to 600 V, 1000 VDC should be applied for 60 sec. It is important to be cautious of any solid-state components that may be in the system, because they could be damaged unless properly isolated. OEM recommendations must be followed for the treatment of solid-state components when performing insulation resistance tests.

Electrical tests on instrument transformers must be performed per ANSI/NETA MTS-2011 Section 7.10. Insulation resistance tests from winding-to-winding and winding-to-ground should be performed on control power transformers. Test voltages are to be in accordance with OEM specifications. If no values are available, voltages from Table 100.1 of ANSI/NETA MTS-2011 must be used. Table 100.5 provides the recommended insulation resistance values for control power transformers. **See Figure 4-9.**

TRANSFORMER INSULATION RESISTANCE TEST VOLTAGES—ANSI/NETA MTS-2011			
Transformer Coil Rating Type*	Minimum Test Voltage, DC	Recommended Minimum Insulation Resistance†	
		Liquid Filled	Dry
0–600	1000	100	500
601–5000	2500	1000	5000
Greater than 500	5000	5000	25,000

* in V
† in MΩ

Figure 4-9. Table 100.5 in ANSI/NETA MTS-2011 provides the recommended insulation resistance values for control power transformers.

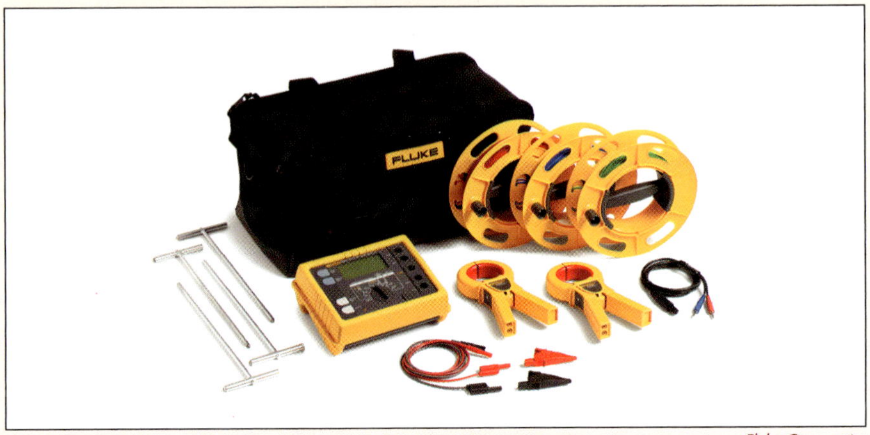

Fluke Corporation

Specialized ground resistance testers are used for taking ground resistance tests.

Ground Resistance Tests. Ground resistance tests must be performed in accordance with ANSI/NETA MTS-2011 Section 7.13. Ground resistance tests are used to verify the integrity of the grounding system in order to properly protect buildings, electrical equipment, and personnel. Typically, the fall-of-potential (three-terminal) test method is used, although clamp-on ground testers can provide equally accurate measurements in most cases.

Overpotential Testing. While a megohmmeter can only detect gross defects in insulation, an overpotential (hipot) test stresses the insulation and detects such problems as tracking, moisture intrusion, weaknesses, and damage. Overpotential tests are performed on each bus section phase-to-phase and phase-to-ground for 60 sec. Allowable voltage for overpotential tests on switchgear are listed in ANSI/NETA MTS-2011 Table 100.2. If no evidence of distress or insulation failure is observed by the end of the 60-sec test, the switchgear is considered to have passed the test.

When performing an overpotential test on bus systems, either an AC or DC hipot tester can be used. However, often a DC hipot tester is used due to the smaller size of the equipment and lower cost compared to an AC hipot tester. Another consideration is that AC hipot testers are "go, no-go" testers, in which voltage is applied to the devices for 60 sec, and no measurement is recorded. If the devices fail within that time period, they are considered bad. If they do not fail, are considered good. Overpotential test voltages for bus systems are listed in ANSI/NETA MTS-2011 Table 100.2. **See Figure 4-10.**

OVERPOTENTIAL TEST VOLTAGES FOR BUS SYSTEMS—ANSI/NETA MTS-2011			
Type of Switchgear	Rated Maximum Voltage*	Maximum Test Voltage*	
		AC	DC
Low-Voltage Power Circuit Breaker Switchgear	0.254/0.508/0.635	1.6	2.3
Metal-Clad Switchgear	4.76	14.0	30.0
	8.25	27.0	37.0
	15.0	27.0	37.0
	27.0	45.0	a
	38.0	60.0	a
Station-Type Cubicle Switchgear	15.5	37.0	a
	38.0	60.0	a
	72.5	120.0	a
Metal-Enclosed Interrupter Switchgear	4.76	14.0	20.0
	8.25	19.0	27.0
	15.0	27.0	37.0
	27.0	45.0	a
	38.0	60.0	a

* in kV

Figure 4-10. Table 100.2 of ANSI/NETA MTS-2011 lists overpotential test voltages for bus systems.

SAFETY GUIDELINES FOR ELECTRICAL TESTING

The first requirement for performing electrical testing is to always follow safety guidelines. OSHA mandates that electrical tests using high-voltage or high-power test equipment follow OSHA 29 CFR 1910.269(o), *Testing and Test Facilities—(3) Guarding of Test Areas,* and OSHA 29 CFR 1910.269(o), *Testing and Test Facilities—(4) Grounding Practices.*

According to OSHA 29 CFR 1910.269(o), *Testing and Test Facilities—(3) Guarding of Test Areas,* "*(i) Permanent test areas shall be guarded by walls, fences, or barriers designed to keep employees out of the test areas. (ii) In field testing, or at a temporary test site where permanent fences and gates are not provided, one of the following means shall be used to prevent unauthorized employees from entering: (A) The test area shall be guarded by the use of distinctively colored safety tape that is supported approximately waist high and to which safety signs are attached, (B) The test area shall be guarded by a barrier or barricade that limits access to the test area to a degree equivalent, physically and visually, to the barricade specified in paragraph (o)(3)(ii)(A) of this section, or (C) The test area shall be guarded by one or more test observers stationed so that the entire area can be monitored. (iii) The barriers required by paragraph (o)(3)(ii) of this section shall be removed when the protection they provide is no longer needed. (iv) Guarding shall be provided within test areas to control access to test equipment or to apparatus under test that may become energized as part of the testing by either direct or inductive coupling, in order to prevent accidental employee contact with energized parts.*"

continued on next page...

SAFETY GUIDELINES FOR ELECTRICAL TESTING (...continued from previous page)

According to OSHA 29 CFR 1910.269(o), *Testing and Test Facilities—(4) Grounding Practices*, "*(i) The employer shall establish and implement safe grounding practices for the test facility. (A) All conductive parts accessible to the test operator during the time the equipment is operating at high voltage shall be maintained at ground potential except for portions of the equipment that are isolated from the test operator by guarding. (B) Wherever ungrounded terminals of test equipment or apparatus under test may be present, they shall be treated as energized until determined by tests to be de-energized. (ii) Visible grounds shall be applied, either automatically or manually with properly insulated tools, to the high-voltage circuits after they are de-energized and before work is performed on the circuit or item or apparatus under test. Common ground connections shall be solidly connected to the test equipment and the apparatus under test.*" These standards can be used as a guide to compose and implement specific facility testing procedures.

SUMMARY

Depending on the type of maintenance required, maintenance can be performed annually or every three to five years. Some form of maintenance is required for all circuit breakers, cubicles, and related components and devices. Routine maintenance is intended to keep the equipment in good working order until the next scheduled major maintenance period. Major maintenance is investigative, and problems that could cause unscheduled outages are corrected. It is normally performed on circuit breakers, switchgear, and circuit breaker cubicles at the same time. Proper safety procedures must always be followed when performing any type of maintenance.

DIGITAL RESOURCES
ATPeResources.com/QuickLinks
Access Code: 223436

chapter 4

CIRCUIT BREAKER CUBICLE MAINTENANCE

CHAPTER REVIEW

Name ______________________________ Date ______________

True-False

T F **1.** NFPA 70B provides guidelines based on general industry-accepted maintenance time-based intervals, without regard to other factors.

T F **2.** Routine maintenance is typically performed every three to five years.

T F **3.** Even when the circuit breaker is racked out of its switchgear and the shutter is closed, there may still be energized parts and components.

T F **4.** For major maintenance, systems must be placed in an electrically safe work condition.

T F **5.** All lubricants must be applied in excess.

T F **6.** Defective interlocks are common with circuit breakers.

T F **7.** If a steel bolt or screw is to be fastened to a different material, it should be torqued to the recommended specification of the harder of the two materials.

T F **8.** If a shutter mechanism is operating erratically or malfunctioning, the circuit must be de-energized so that it can be repaired safely.

T F **9.** The circuit breaker is never removed from its cubicle while routine maintenance is being performed.

T F **10.** Circuit breaker cubicle service intervals are the same as those for circuit breaker service.

T F **11.** On medium-voltage metal-clad switchgear, maintenance on a cubicle should never be performed if the shutter mechanism does not cover the primary bus connections.

T F **12.** All items performed during routine maintenance are also to be performed during major maintenance.

T F **13.** During routine maintenance, all parts should be inspected for signs of moisture or rusting.

T F **14.** Low-pressure air should be used for cleaning.

T F **15.** Do not attempt to clean energized bus stabs on low-voltage switchgear.

T F **16.** Fasteners must be tightened and not torqued.

T F **17.** A method of grading connections recommended by NETA requires that there be no more than a 50% difference between the highest and lowest reading taken across connections of the same type and size.

T F **18.** When a bus or conductor is ungrounded, it must be treated as an energized conductor.

T F **19.** During major maintenance, existing filters must be replaced with new filters.

T F **20.** Insulation resistance tests are used to verify the integrity of the grounding system.

Multiple Choice

______ **1.** Appendix B in the ANSI/NETA MTS-2011 contains a Maintenance Frequency Matrix that bases maintenance intervals on two factors: criticality of the system or device and ___.
A. loading
B. operation environment
C. equipment history
D. device condition

______ **2.** Clean, tight connections of bus joints must have no more than ___ μΩ across them.
A. 20
B. 30
C. 40
D. 50

______ **3.** Routine maintenance is performed ___.
A. monthly
B. annually
C. every two years
D. every three years

______ **4.** A(n) ___ may be used to take resistance measurements to verify the tightness of connections.
A. AC hipot tester
B. DC hipot tester
C. thermal imager
D. low-resistance ohmmeter

______ **5.** During routine maintenance, a light coat of nonconductive grease should be applied on the racking mechanisms, chains, and ___.
A. auxiliary stabs
B. main stabs
C. racking screws
D. hinges

______ **6.** For cable rated 300 V and lower, ___ VDC should be applied during an insulation resistance test.
A. 250
B. 500
C. 750
D. 1000

________ **7.** If a screw or bolt is to be fastened into a plastic-encapsulated nut, such as a standoff insulator, it should be tightened until the lock nut flattens, then an additional ___ turn.
A. ⅛
B. ¼
C. ½
D. ¾

________ **8.** Overpotential tests are performed on each bus section phase-to-phase and phase-to-ground for ___ sec.
A. 30
B. 60
C. 90
D. 120

________ **9.** DC control circuits and other auxiliary DC sources should be verified for the presence of voltage by measuring from ___.
A. ground to positive
B. ground to negative
C. positive to negative
D. positive to phase

________ **10.** Use only ___ when wiping surfaces during the cleaning of a circuit breaker.
A. brushes with metal ferrules
B. abrasive cloths
C. clean, dry cotton cloths
D. clean, wet cotton cloths

Completion

________ **1.** ___ maintenance is minor maintenance performed to maintain the integrity of a cubicle.

________ **2.** A(n) ___ is a device that detects heat patterns in the infrared-wavelength spectrum without making direct contact with the workpiece.

________ **3.** With switchgear, mechanical and visual inspections are recommended once every ___ months.

________ **4.** The circuit breaker must be in the ___ position in order to safely rack the circuit breaker in.

________ **5.** A(n) ___ is a device used to measure the resistance of an electrical conductor down to the micro-ohm range.

________ **6.** When performing an overpotential test on bus systems, either an AC or DC ___ can be used.

________ **7.** To perform maintenance on medium-voltage metal-clad switchgear with a shutter mechanism that does not cover the primary bus connections, the circuit must be either de-energized or a(n) ___ permit must be issued by the facility where the work is to be performed.

______________________ **8.** When performing routine maintenance on switchgear, the four steps usually performed include inspection, cleaning, lubrication, and ___.

______________________ **9.** During ___ maintenance, the electrical system is de-energized, rear covers are removed, and the bus is grounded.

______________________ **10.** Insulation resistance tests are performed with a(n) ___, which is used to apply DC voltage across an insulation system.

______________________ **11.** During routine maintenance, all work performed is limited to the ___ section of the cabinet.

______________________ **12.** For cable rated 301 V to 600 V, ___ VDC should be applied for 60 sec during an insulation resistance test.

______________________ **13.** Typically, the ___ test method is used for ground resistance testing, although clamp-on ground testers can provide equally accurate measurements in most cases.

______________________ **14.** During routine maintenance, a light coating of ___ lubricant should be applied on the ground stab, auxiliary stabs, and, if de-energized, main stabs.

______________________ **15.** Electrical tests must also be performed on the switchgear during ___ maintenance since it is de-energized.

Short Answer

1. List five factors to consider when determining whether to work on equipment.

2. List five tasks outlined in NFPA 70E, Annex J, that should be completed prior to performing any high-risk energized work.

3. What insulation problems can be detected by an overpotential (hipot) test?

4. What are three factors that must be checked during the inspection of a control power transformer?

5. What is the purpose of an as-found test that is performed prior to cleaning during major maintenance?

Matching

Circuit Breaker Racking Components

______________ **1.** Primary disconnects

______________ **2.** Power bus

______________ **3.** Springs

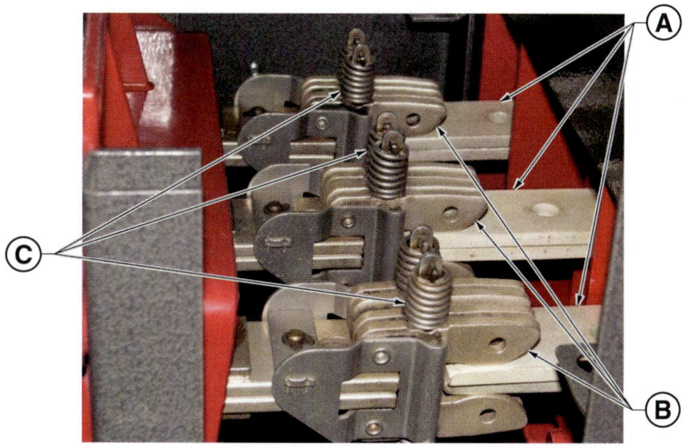

Circuit Breaker Inspection

________________ **1.** Excessive wear

________________ **2.** Oxidation (rust)

________________ **3.** Galling

________________ **4.** Bent shear pin

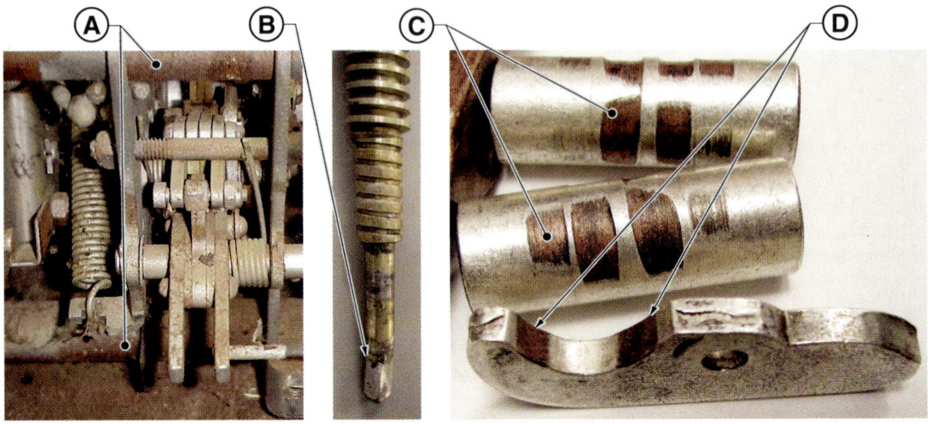

chapter 5

CIRCUIT BREAKER SERVICING

Circuit breakers require regular maintenance. Many original equipment manufacturer (OEM) instruction manuals have clear, concise instructions with photographs of working components. Some instruction manuals may only have line drawings of circuit breakers or only show exterior circuit breaker components. Incomplete illustrations can cause circuit breaker service to be difficult and time consuming. OEM manuals include images and illustrations of components as installed in a specific type of circuit breaker. Common models of low- and medium-voltage circuit breakers include the following: General Electric (GE) MagneBlast, GE Power/Vac®, Cutler-Hammer VCP-W, ABB/ITE HK, Westinghouse/Cutler-Hammer Square D DS, GE AK50, GE AKR/AKRD-30 and 50, Siemens/Allis-Chalmers LA, ABB/ITE K series.

OBJECTIVES

- List five factors that affect the reliability of a circuit breaker.
- Identify the type of circuit breaker that cannot be opened for maintenance.
- Describe the four conditions that indicate a circuit breaker is in need of major maintenance (teardown).
- List the five tasks often required for routine (annual) maintenance of a circuit breaker.
- List the eight tasks required when performing major maintenance.
- List the seven areas of inspection for arc chutes (extinguishers).
- State the most common cause of circuit breaker misoperation.
- State the components that are the most frequent causes of high resistance readings.
- State the three electrical tests that can be performed during routine maintenance.
- State the lubricant authorized to be used by all circuit breaker manufacturers.
- Identify the mechanical procedure that must be performed in order to properly adjust contacts in a circuit breaker.

CIRCUIT BREAKER SERVICE SAFETY

Electrical workers and technicians must follow safe work practices as required by OSHA regulations and NFPA 70E. Circuit breaker operating mechanisms and contacts are extremely fast-acting. Tools and fingers can be caught and crushed. Prior to inspecting a circuit breaker, the circuit breaker should be tripped open and an attempt to reclose it should be made. Blocking devices should be installed as recommended by the OEM to prevent accidental operation.

Proper precautions must be observed during testing, including the following:

- grounding of test equipment and components to be tested
- securing the test area
- using proper PPE

CIRCUIT BREAKER MAINTENANCE

There are two maintenance intervals typically recommended by original equipment manufacturers (OEMs): routine maintenance, which is performed on an annual basis, and major maintenance, which is recommended every three to five years. Actual maintenance practice in industry is often a combination of the

KEY TERMS

- slow closing (maintenance closing)

two maintenance intervals. Unless there is a problem, circuit breakers stay in service with no maintenance until a scheduled shut down or turn-around. At that time, the system is deenergized and servicing is performed in accordance with NFPA 70B, *Recommended Practice for Electrical Equipment Maintenance,* or ANSI/NETA Section 7.6, *Standard for Maintenance Testing Specifications for Electrical Power Distribution Equipment and Systems*.

OEMs recommend tearing down the current-carrying components so that they can be cleaned and relubricated on a periodic basis, based either on the number of operations or the time in service. Because this practice is expensive and disruptive to manufacturing, the circuit breaker is left in service until the next maintenance cycle if it exhibits no outward appearance of problems. Industry often sets maintenance intervals based on experience as well as the costs involved in a system-wide shutdown. If circuit breakers are not maintained properly, not only will it increase downtime and the costs of operation, but it can increase the risks to employees maintaining and operating the electrical system.

CASE STUDY—FAULTY MAINTENANCE EFFECTS

This circuit breaker was in service for several years without having maintenance performed on it. When a 13.8kV feeder cable failed, the circuit breaker's trip coil burned open as a result of a lack of lubrication. This incident resulted in the total loss of six circuit breakers and their associated switchgear. The facility where these circuit breakers were located was completely out of production for seven days while emergency repairs were performed. Because this facility did not perform maintenance as required by the manufacturer, its insurance claim was denied. Total cost of the repairs exceeded $250,000, not including loss of production.

There are several factors to be considered when determining maintenance intervals. The operating environment the circuit breaker is in is one such factor. As an example, circuit breakers that are located indoors in a positive-pressure switchgear room that has filtered and conditioned air will have fewer problems than one that is sitting in outdoor switchgear where the filters are probably rarely (if ever) changed and the circuit breaker is subjected to constant temperature extremes, moisture, and dirt.

The circuit breaker application and its duty cycle are other major factors to be considered. A circuit breaker that is operated infrequently, such as one supplying a feeder circuit, may require much less maintenance than one that is used as an across-the-line starter and is operated dozens of times each day.

Maintenance intervals are also determined by the amount of heat created by current flow through a conductor. As current flows through a conductor, it creates heat. For example, for a 12 AWG wire, the heat generated by 20 A of current warms the conductor. If 30 A of current were to flow through that same conductor, the conductor heats considerably more, causing the insulation to deteriorate and fail over a shorter period of time. Circuit breakers require lubrication in the operating mechanism and in the pivot point of the contact assembly. **See Figure 5-1.**

Because of the heat created by current flow through the circuit breaker, the lubricant in the contact pivots dries out over time. It eventually flakes off the contacts and results in metal-to-metal wear. Heavily loaded circuit breakers operate at much higher temperatures than lightly loaded circuit breakers and cause the lubricant to dry out at a much faster rate. The contact pivot must be disassembled, cleaned, and relubricated at regular intervals in order to function properly.

In addition to environment, reliability requirements, condition, and loading, other factors to consider when determining maintenance intervals include the age of the equipment, OEM quality, number of cycles, and results of maintenance tests. Equipment that is old may have components that are unreliable due to wear, heat, and aging. Older plastic components found in electrical components are especially at risk. A large number of operating cycles increases wear, but circuit breakers that do not operate at all are also at risk, since their lubricants never get mixed and freshened by operating them. Circuit breakers require regular operation (cycling) to maintain functionality. As electrical or mechanical components fail, test results may indicate a requirement for more frequent testing and maintenance.

Low-voltage, molded-case circuit breakers can come in sealed-case or insulated-case designs. Sealed-case, molded-case circuit breakers cannot be opened because opening them voids their UL certification. Larger, interchangeable-trip molded-case circuit breakers can be opened and serviced, but often parts are not readily available, which makes servicing them at the facility difficult. The large numbers of these circuit breakers also makes it difficult to maintain and service each molded-case circuit breaker since there can be hundreds or possibly thousands at large industrial facilities. Many companies try to make this more manageable by limiting maintenance on molded-case circuit breakers to a certain size and above, such as 200 A or 400 A and greater, or only those molded-case circuit breakers that may feed more critical loads.

Insulated-case circuit breakers are typically used for more critical loads and have higher interrupting and continuous current ratings. They can be opened and serviced, and parts can be secured to repair them, although the parts can often take several weeks to be delivered.

Circuit breakers require a certain amount of maintenance each year to maintain their suitability for safe, reliable service. Improperly maintained circuit breakers may not open. Indications that a circuit breaker is in need of major maintenance include failure to reset after tripping, failure to open, slow or sluggish operation, and high contact resistance (microohmmeter) readings.

Failure to Reset after Tripping. The application of spray-type penetrating liquids into the contact pivot or operating mechanism can damage the circuit breaker. Improperly maintained circuit breakers may not open in accordance with OEM specifications because of seized or sluggish contact assemblies or operating mechanisms. Spray-type penetrating liquids are not lubricants. Penetrating liquids remove any remaining grease from the contact pivot. The circuit breaker may initially appear to operate normally and is placed back into service. The issue is not corrected, just delayed, probably until an emergency situation occurs.

CIRCUIT BREAKER SERVICE—PIVOT POINTS

Figure 5-1. Circuit breakers require lubrication in the operating mechanism and in the pivot point of the contact assembly.

Penetrating liquids are typically water-based and are not actual lubricants. Their effects are short-lived and will promote wear, not prevent it. When sprayed into the contact pivot, they not only remove any grease, but also spray other surfaces. The pivot is still not lubricated properly and the overspray collects dirt and abrasive grit that can make its way into the pivot over time, increasing wear substantially.

Failure to Open. Contacts may fail to open or open completely. If the contacts do not open fully, the circuit breaker will not be able to extinguish an arc. **See Figure 5-2.**

CIRCUIT BREAKER SERVICE—CONTACTS

Figure 5-2. Circuit breaker contacts must be serviced at regular intervals in order to open fully and prevent overheating and arcing.

Slow or Sluggish Operation. Sometimes there can be a delay from a few seconds to several minutes from when the trip button is depressed to when the circuit breaker actually opens. The time required for a circuit breaker to open from when the trip button is depressed varies depending on the circuit breaker OEM.

High Contact Resistance Readings. Higher than normal contact resistance readings indicate that the lubricant in the contact pivot is hardening and losing its ability to function. Hard, dry lubricant can flake away from the mating parts and cause metal-to-metal contact, which results in friction, heating, and permanent damage to the mating parts.

Routine Maintenance

As with switchgear cubicles, routine maintenance on circuit breakers consists of tasks and tests that are aimed to maintain the reliability of the circuit breaker until a major maintenance period. Many of the steps in routine maintenance are the same as in major maintenance, but they are not done to the same degree.

The primary tasks associated with routine maintenance on low-voltage power circuit breakers (LVPCBs) include inspection, cleaning, lubricating, adjusting, and testing. During inspection, arc chutes must be removed, but the circuit breaker should not be disassembled. The arc chutes should be inspected for damage. The circuit breaker should be inspected inside for any misaligned or missing components or keepers. Any evidence of overlubrication, tracking, or contamination must be addressed. Also, any evidence of tracking or contamination that could lead to other issues should be addressed. Light lubrication of latch surfaces and linkages is required but not disassembly of the contact pivot point or mechanism.

The performance of the circuit breaker should be checked and it should be adjusted as needed. Normally, no adjustments are necessary unless there is a problem with wear or misaligned components. Contact resistance and secondary injection testing of the overcurrent protective device (OCPD) low-voltage circuit breakers are often performed during routine maintenance. A thermographic survey can also be performed at this time.

Molded-Case and Insulated-Case Circuit Breakers. A molded-case circuit breaker (MCCB) is a circuit breaker that uses the housing as part of its insulating structure. An insulated-case circuit breaker (ICCB) is similar in construction to an MCCB but has higher interrupting ratings and uses digital trip units. Sealed-case MCCBs can only have routine maintenance performed on them. Larger-frame MCCBs may have their cases opened for limited inspection and maintenance.

Routine maintenance for MCCBs and ICCBs consists of cleaning, inspection, and tightening. The outside case should be cleaned of dirt and dust by wiping with a rag. The case should be visually inspected for cracks and the handle for damage that could cause it to fail. The terminal lugs and wiring connected to the circuit breaker should be checked for signs of overheating such as discoloration. All terminal lugs must be tightened to proper torque values, as specified by the OEM. It should be noted that if the set screw in the terminal lug has loosened and the lug shows signs of overheating, there may have been arcing between the set screw and the barrel of the lug. Once arcing has taken place, although the set screw appears tight, it cannot tighten any further on the conductor than the point of arcing. **See Figure 5-3.**

An additional task required in the routine maintenance of ICCBs is to check the battery in the electronic overcurrent device (trip unit). Although the electronic overcurrent trip device uses line power to operate, it also uses replaceable batteries to provide back-up power to preserve settings. Most manufacturers of electronic overcurrent devices state that the battery can be changed while the unit is in operation. However, safe work practice is to verify this by consulting the OEM instruction manual of the specific overcurrent device.

WARNING

Cleaning while energized and tightening terminal connections is considered energized work and requires proper safe work practices and procedures. This may include using an energized electrical work permit, performing a hazard/risk analysis, and using appropriate PPE. Refer to NFPA 70E, Standard for Electrical Safety in Employee Workplaces, *for additional information regarding electrical safe work practices.*

CIRCUIT BREAKER SERVICE—TERMINAL LUGS

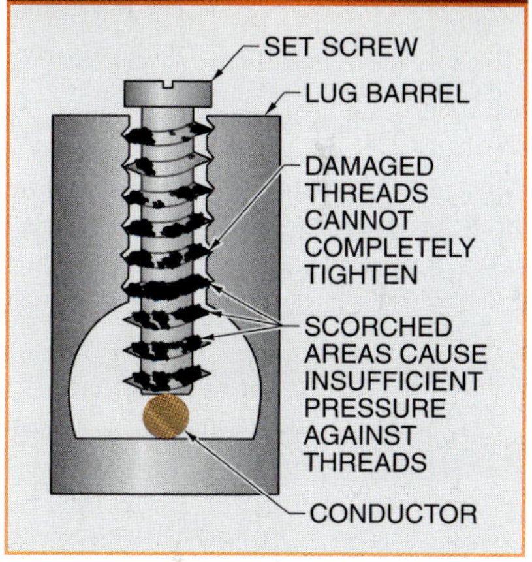

Figure 5-3. Terminal lugs must be inspected for arcing because once arcing has occurred, the terminal lug set screw cannot tighten any further than the point of arcing.

Major Maintenance

Major maintenance covers the same tasks associated with routine maintenance, however, at a higher degree. In addition to the tasks performed during routine maintenance, during major maintenance, the circuit breaker current path is disassembled, thoroughly cleaned, and inspected; the contact pivots are relubricated; and the circuit breaker is reassembled, adjusted, and tested. All accessible rollers, sliding surfaces, and latches in the operating mechanism are also cleaned and lubricated at this time.

Due to time constraints and difficulty in scheduling plant shutdowns, maintenance is typically performed in two to three-year intervals and disassembly is delayed until the breaker exhibits distress. A regularly scheduled maintenance interval, as specified by either NFPA 70B, *Recommended Practice for Electrical Equipment Maintenance,* or ANSI/NETA MTS-2011, *Standard for Maintenance Testing Specifications for Electrical Power Distribution Equipment and Systems,* must be followed to ensure a safe working environment for electrical workers. While routine maintenance can require 1 hour to 2 hours per breaker to complete, major maintenance can require 4 hours to 8 hours per circuit breaker to complete.

On low-voltage power circuit breakers (LVPCBs) the backboard assembly is disconnected from the operating mechanism frame by removing the bolts securing the two frame halves together. The operating mechanism is disconnected by removing the link pins holding the moving contacts to the backboard assembly. **See Figure 5-4.** Medium-voltage air circuit breakers have components within the current path that can be disassembled while installed on the circuit breaker. Only one phase unit should be disassembled at any time.

Large MCCBs and ICCBs have cases that can be opened for a thorough inspection. Maintenance tasks performed for ICCBs are the same as for LVPCBs and include removing the front cover of the circuit breaker and then removing the contact mechanism. **See Figure 5-5.** Large MCCBs do not usually have major maintenance of this type performed on them. For MCCBs, damaged or worn parts are usually replaced or the circuit breaker is replaced.

LOW-VOLTAGE POWER CIRCUIT BREAKERS (LVPCBs)

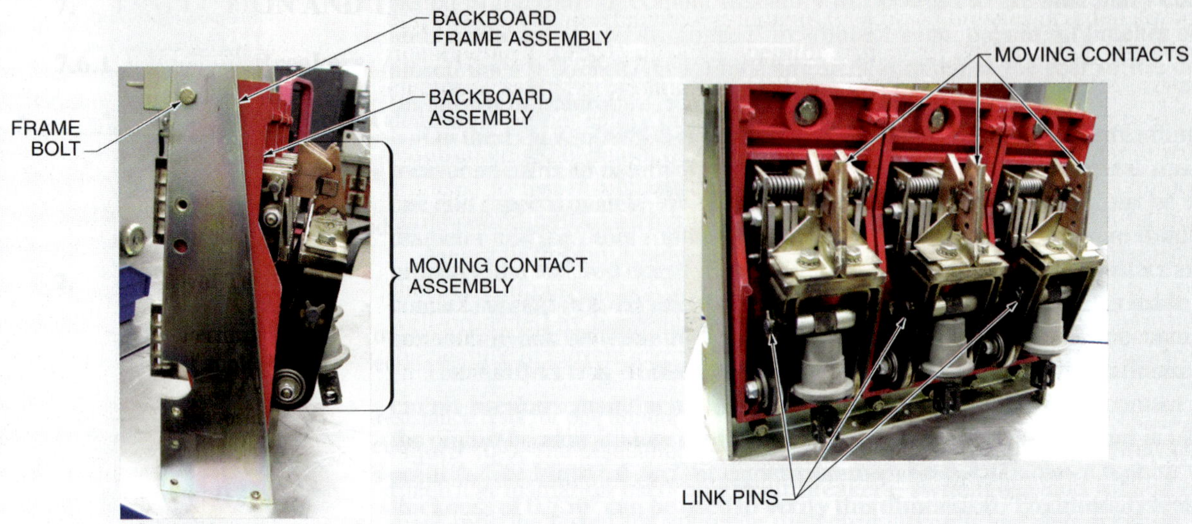

Figure 5-4. With LVPCBs, the backboard assembly is disconnected from the operating mechanism frame by removing the bolts securing the two frame sections together. The operating mechanism is disconnected by removing the link pins that hold the moving contacts to the backboard assembly.

CIRCUIT BREAKER SERVICE—INSULATED-CASE CIRCUIT BREAKERS

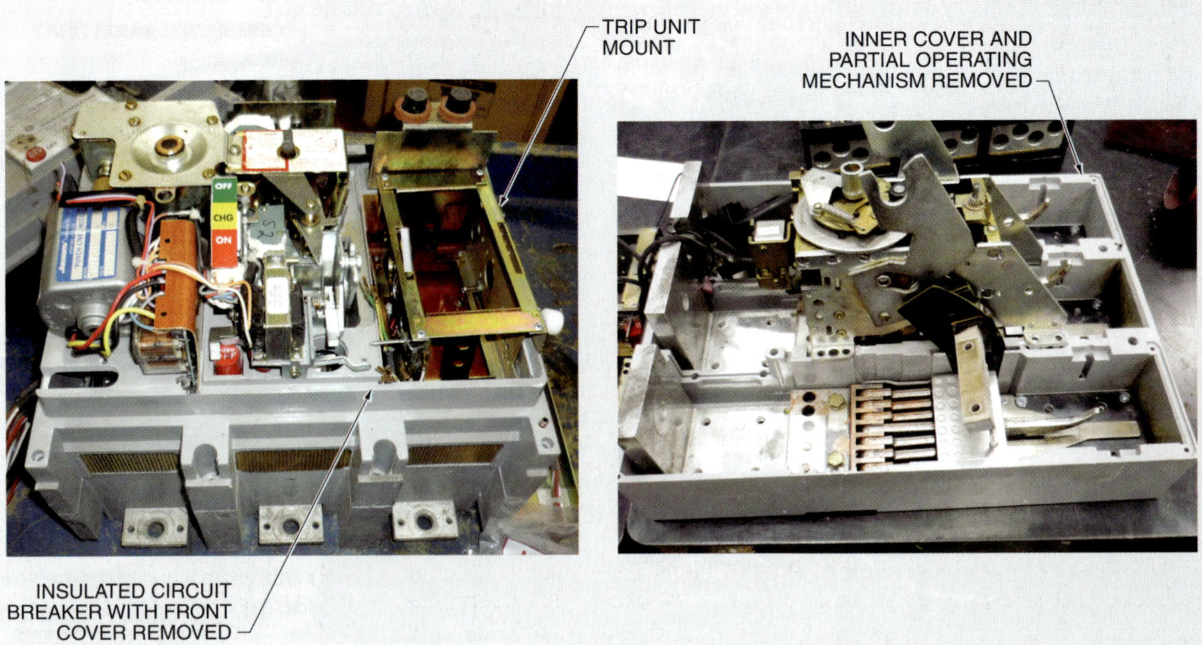

Figure 5-5. Maintenance tasks performed on insulated-case circuit breakers include removing the front cover of the circuit breaker and then servicing the contact mechanism.

Vacuum circuit breakers do not have externally-accessible contacts or arc chutes and do not require that area of maintenance. However, the operating mechanism still requires attention, such as cleaning and lubrication. The connections between the bottles and primary disconnects also must be inspected and checked for tightness. Other maintenance tasks include the following:

- inspection—A thorough inspection of all components, especially contact assemblies and their pivot points, is required.
- cleaning—All components must be cleaned thoroughly, including contact pivots, linkages, and other accessible parts, including the operating mechanism.
- lubrication—Contact pivots, latch surfaces, and anything else that rubs or slides must be lubricated with the proper lubricant. Care must be taken not to overlubricate, especially when using synthetic greases, which melt and run when heated. Lubrication is to be just enough to get the job done, no more.
- reassembly—Once all three phases have been cleaned and renewed, the backboard to the operating mechanism can be reassembled.
- torquing—Overtorquing is as harmful as undertorquing. The breaker must be properly torqued to withstand the mechanical stresses that result when a fault occurs.
- adjustment—The OEM instruction book (IB) must be consulted for the specific circuit breaker being worked on. As circuit breaker frame size varies, the design varies. Also, it must be ensured that the correct IB is being used. The latest IB must be used for that specific circuit breaker, but older versions of the same IB should not be discarded. Often, the manufacturer will shorten the IB from one version to the next.
- testing—Contact resistance, insulation resistance, vacuum bottle integrity, overpotential or insulation power factor (depending on the voltage rating of the circuit breaker) and mechanical timing must be tested. Also, this is an opportunity to verify the operation of control devices such as open and close coils.

The backboard assembly (phase or pole units on medium-voltage breakers) requires periodic disassembly in order to service the current-carrying components and to clean any carbon or other contaminants. This step is not performed during routine maintenance, since it can be time consuming. Usually, it is not needed on vacuum circuit breakers because the vacuum bottle contains the current-carrying components and is sealed. However, the connection points between the vacuum bottle and the primary disconnect should be cleaned and relubricated.

Low-voltage and medium-voltage power circuit breakers can be separated into two main sections: the operating mechanism and the backboard assembly (low-voltage) or pole unit (medium-voltage). Typically, the operating mechanism is not disassembled, due to its complexity and the presence of heavy springs inside. However, if a problem is discovered during the inspection, it should be corrected when a circuit breaker is being serviced.

The backboard assembly consists of the primary disconnect assemblies, moving and stationary contact structures, and the backboard/frame/pole units. All components are removed from the backboard, cleaned, inspected, lubricated, and replaced as needed. Medium-voltage pole units are usually maintained without being removed from the circuit breaker.

WARNING

Circuit breakers operate extremely fast and can cause severe injury. Prior to performing work on a circuit breaker, discharge the stored energy mechanism by manually opening the breaker and then attempting to reclose it.

CAUTION: Only disassemble one phase unit at a time. If problems arise, such as remembering the order of assembly, the other two phase units can be used as guides. Also, never mix parts from one phase unit with another.

Arc Chute Inspection. For arc chute inspection on medium-voltage circuit breakers, the barrier is removed around and/or between the arc chutes. Arc chute barriers prevent contact with the arc chute when the circuit breaker is energized and also prevents spark scatter, where molten or vaporized metal or debris can be ejected from one phase to another and cause a cross-phase arc. They must be visually inspected for damage or missing parts, which can occur when the circuit breaker is moved or handled. Nylon hardware is often used as fasteners with arc chutes. Nylon hardware is often used as fasteners for arc chute barriers and can break easily. **See Figure 5-6.**

ARC CHUTE INSPECTION

Figure 5-6. Arc chute inspection on medium-voltage circuit breakers requires removal of the barrier around and/or between the arc chutes and inspection of the arc chutes and nylon fasteners.

This hardware must not be replaced with other makes or types of hardware. Only replacement parts as specified by the OEM should be used. On LVPCBs, the interphase barriers must be removed from between the arc chutes. Interphase barriers are used to prevent arc scatter. **See Figure 5-7.**

One of the main devices requiring inspection on a circuit breaker is the arc chute. When an arc chute fails to clear the arc properly, it can easily lead to restrike. A restrike is a resumption of current between two contacts of a switching device during or after an opening operation after an interval of zero current of a quarter cycle at normal frequency. Restrike can occur whenever an arc is not fully extinguished and then re-establishes itself, which often leads to catastrophic failure.

Arc chutes must be removed from the circuit breaker using a lift or hoist if needed on medium-voltage circuit breakers. Medium-voltage breakers may have arc chutes weighing hundreds of pounds. On circuit breakers that have their arc chutes mounted on the top, only one arc chute can be tilted back at a time. Tilting two or more arc chutes will cause the circuit breaker to become imbalanced and tip over. **See Figure 5-8.**

Once the arc chute is removed from the circuit breaker, the arc chute can be visually inspected. Small cracks in asbestos or ceramic are common and can be disregarded if they do not extend completely through the part. Corners that are broken can be repaired using epoxy to glue them back on temporarily. Any missing hardware or components must be replaced using parts specified by the OEM.

Any spatter should be removed from the arc divider plates. Scraping a screwdriver or putty knife across the face of the plate can often remove arc spatter. If there is any carbon buildup on metal divider plates or the throat of the arc chute, it should be cleaned with solvent and a nonconductive, plastic scouring pad. Also, the vent/muffler assembly must be kept clear to allow combustible gasses to escape. Blocked vents are one possible cause of restrike because the gasses can ignite.

The most common signs of damage caused by heat are discoloration, brittleness of parts, and warped or melted components. **See Figure 5-9.** In addition, arc chutes deteriorate as they interrupt faults. The higher the value of short-circuit current, the faster the arc chute deteriorates. LVPCBs often indicate erosion on the arc divider plates. When erosion reaches a level of ⅛″, the arc chute must be replaced.

On medium-voltage air circuit breakers, uniform glass beading of the ceramic material in the throat of the arc chute indicates the end of life. Glass beading is caused by the ceramic being overheated and drawing silica out of it. A small amount of glass beading is typically not a problem, but large areas of uniform glass beading indicate the need for replacement. The magnetic pole pieces on GE Magne-Blast circuit breakers have an insulating cover on them, since they become energized during fault interruption. Cracks or gouges in this type of insulation can be repaired by using an insulating coating or similar product.

TECH FACT

Vacuum circuit breakers have vacuum bottles rather than arc chutes. A vacuum bottle consists of a hardened ceramic or glass container. Other than obvious signs of damage, such as cracks, overheating, or discoloration, there are no additional inspection points.

Contact Assembly Inspection. Once the arc chutes are removed, the interior of the circuit breaker must be visually inspected for bent, broken, or missing parts and misaligned parts that rub against or catch on each other. Due to the powerful springs in the operating mechanism, any misalignment will cause damage. Also, it should be ensured that springs or keepers are not missing.

In addition, the circuit breaker should be inspected for parts showing excessive wear, especially lubricated parts. As the old lubricant dries, metal-to-metal wear can occur in areas other than the contact assemblies.

It should be ensured that there is no excessive lubrication, burning or pitting of contacts, or contact pressure. Overlubrication, especially in the contact pivot, leads to failure. Dirt and moisture are drawn in by the excessive lubricant and hastens wear. Another area to inspect is the spring charging motor. If the seals on it deteriorate, lubricant will leak out of the motor and onto the cubicle floor.

Some burning on the arcing contacts is normal or closes. Arcing contacts are designed to interrupt the arc when the circuit breaker opens. However, defects on the contact face or adjustment problems can cause severe damage to the contacts as well as the circuit breaker. Main contacts should show little arc damage.

INTERPHASE BARRIERS

Figure 5-7. On LVPCBs, interphase barriers are used to prevent arc scatter.

ARC CHUTE SERVICING

Figure 5-8. When servicing medium-voltage air circuit breakers, arc chutes can only be tilted out of the breaker assembly one at a time because the circuit breaker can become unbalanced and tip over.

Figure 5-9. Once the arc chute is removed from the circuit breaker, it must be inspected for conditions such as broken, cracked, or missing components; arc spatter; carbon buildup; heat/arcing damage; erosion of arc divider plates; and damaged insulation on insulated magnetic poles.

The face of the contact should also be inspected for cracking. This type of damage is usually caused by excessive contact pressure or improper silver-soldering of the contact face to the contact mounting point.

If the breaker interrupts repeated fault-level currents, or if the contacts are out of adjustment, the arcing (and main) contacts may show excessive pitting. This can be visually verified or, on breakers with butt-style contacts, a contact print test can be performed. Arcing contacts are replaced when the circuit breaker can no longer be properly adjusted or if significant erosion is evident. **See Figure 5-10.**

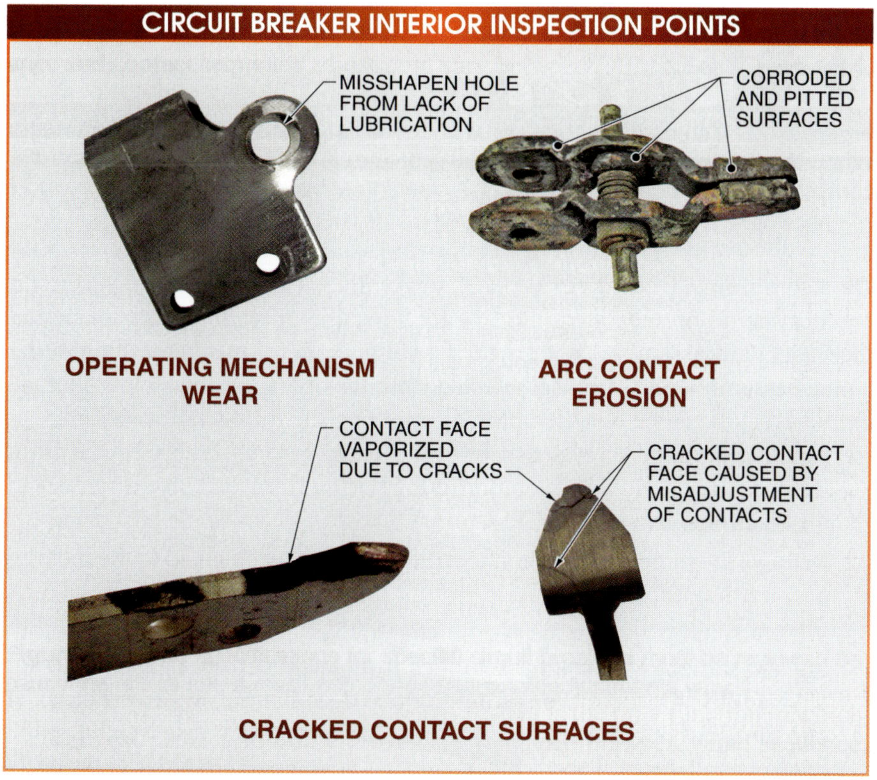

Figure 5-10. Once the arc chutes are removed, the interior of the circuit breaker must be visually inspected for conditions such as operating mechanism wear, arc contact erosion, and cracked contact surfaces.

Contact Print Tests. To perform a contact print test, a piece of carbon paper is sandwiched between two pieces of white (bond) paper and is placed between the open contacts of a phase unit. The breaker is closed at normal operating speed and opened. A print of the contact area is left on the white paper. Contacts must have approximately 80% make. **See Figure 5-11.** Newer breakers with multiple contacts or breakers that use bayonet-style contacts cannot be tested with this method. One other indicator that the arcing contacts need to be replaced is if the main contacts show evidence of arcing damage.

Contacts can be filed with a fine file, down to their original contour to remove high spots from arcing. However, one should never attempt to remove surface pitting. Doing so could change the contour of the contact and possibly interfere with the normal "wiping" action of the contacts. **See Figure 5-12.**

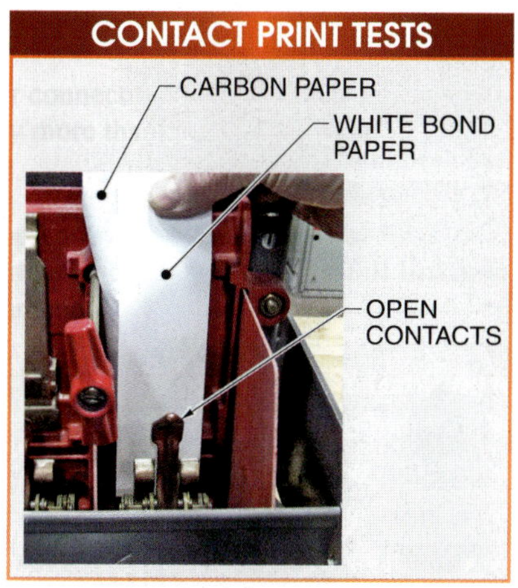

Figure 5-11. To perform a contact print test, a piece of carbon paper is sandwiched between two pieces of white (bond) paper and is placed between the open contacts of a phase unit.

CONTACT WIPE

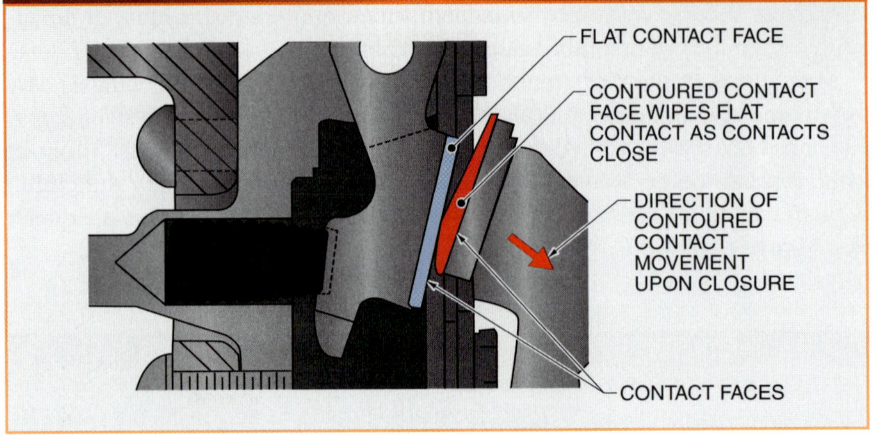

Figure 5-12. With contacts, one contact face is flat, while the other is contoured. When the breaker opens or closes, the contoured contact rubs against the flat contact, which provides a wiping or self-cleaning action.

Disassembling Circuit Breakers. On low-voltage circuit breakers, the circuit breaker should be separated into two sections: the operating mechanism and the backboard assembly. With medium-voltage circuit breakers, the moving contact assembly must be disconnected from the operating rod. A low-voltage circuit breaker can be disassembled by applying the following procedure:

1. Remove bolts or screws that hold the backboard assembly to the front frame member and any interconnecting wiring from the current sensors to the electronic trip unit (OCPD).
2. Remove the three pins that connect the operating mechanism to the moving contact assembly.
3. With the circuit breaker disassembled into two sections, disassemble moving and stationary contact assemblies from backboard or phase/pole unit. **CAUTION:** Observe the warning not to disassemble all three phases at one time, but only one phase at a time. Use care when disassembling the contact assemblies. The springs that must be removed in order to properly clean and relubricate the circuit breaker are strong.

Additional inspections are performed after disassembly of the circuit breaker. **See Figure 5-13.** Disassembled circuit breakers should be visually inspected as follows:

- Contact pivot points should be inspected for galling, wear, or flaking of any silver areas. Sometimes silver is bonded to the pivot area to increase conductivity. If the copper is not cleaned properly, the silver will not adhere and begins to flake.
- Contact components should be inspected for wear and damage from lack of lubrication. It is important to disassemble all components of the moving and stationary contact assemblies. Although rare, it is also important to inspect for components that were improperly machined or manufactured.
- If inspecting vacuum breakers, the wear indicators should be checked. Vacuum breakers usually have wear (erosion) indicators, since the contacts themselves cannot be visually checked. Most circuit breaker OEMs recommend against resetting wear indicators.

Figure 5-13. After disassembly of a circuit breaker, inspections are performed on contacts, erosion wear indicators, pole insulators, backboard assemblies, operating and racking mechanisms, and electrical components.

- Phase unit pole insulators should be checked for tracking, cracking, and deterioration. Often, these types of circuit breakers have an epoxy-based insulation that can be galled or gouged by rough handling.
- The backboard assembly on LVPCBs and the phase/pole units on medium-voltage air circuit breakers should be inspected for mechanical damage and for damage caused by misalignment of components. The primary disconnect fingers should be checked for signs of overheating or arcing. This could be an indication of poor spring tension.
- Operating and racking mechanisms should be checked for corrosion, damage, wear, and misalignment. They should also be checked for loose or missing keepers, hardware, and springs. Often circuit breakers are subjected to high humidity conditions and receive very little maintenance. This leads to seized mechanisms and the galling of latch surfaces.
- Electrical components on the interior of circuit breakers such as control wiring, coils, solenoids, and spring-charging motor should be inspected. Frayed wiring, corroded solenoids, worn brushes in the motor, and loose or broken connectors are the most commonly seen wiring problems. However, high resistance on switch contacts is the most common electrical problem overall. Switches should have resistance of no greater than 0.1 Ω.
- The circuit breaker should be slow (maintenance) closed.
- The ground contact or shoe must be inspected for tightness, straightness, and alignment. It must be ensured that there are no signs of galling or misalignment.
- The primary disconnect fingers should be inspected for signs of overheating or arcing. This could be an indication of poor spring tension.

Cleaning Circuit Breakers. In order to remove rust, corrosion, and dried or contaminated lubricants, the circuit breaker must be cleaned thoroughly. Inadequate cleaning could lead to tracking and arc-over phase-to-phase or phase-to-ground. Cleaning is also required in order to inspect components that have old, dried lubricant on them.

Using low-pressure, dry air is the best method of cleaning a circuit breaker because it can reach into areas inaccessible with a vacuum or brush. Arc chutes should be blown through with low-pressure (≤30 psi), dry air to remove dust and ensure that their vents are clear. An approved safety nozzle should be used. All required PPE, such as safety glasses or goggles, should be worn.

Another method of cleaning is to wipe affected areas with an approved safety solvent. Solvents must only be used in well-ventilated areas. Even though all facilities require the use of "safety solvents," this label does not mean they are safe to use in all conditions. After cleaning with a solvent, residue must be removed by wiping the component with a clean, dry cloth. Alcohol is an effective cleaner but is highly flammable and should only be used where there is adequate ventilation. **CAUTION:** Never use alcohol or a safety solvent in a confined area with low movement of air.

Even though a facility specifies the type of solvent used at a particular location, they may have some negative effects. Certain types of solvents can soften and degrade plastic materials. A solvent with a high dielectric strength is necessary, along with a fast evaporation rate. However, a fast evaporation rate can cause condensation. This can cause problems when performing electrical tests soon after applying cleaning solvents. On medium-voltage air circuit breakers that have acrylic puffer tubes, applying cleaner/solvent to puffer tubes could cause them to shatter. This is due to the rapid cooling caused by the propellant of the solvent. The solvent should be applied to a small test area and its reaction should be checked with the material to ensure it does not damage the component.

Vacuuming and wiping circuit breakers with a clean, dry cloth (without using an approved safety solvent) are not recommended. Vacuuming, while effective with switchgear, is ineffective with circuit breakers due to the difficulty of reaching tight areas inside the circuit breaker. Furthermore, the suction from most vacuums is inadequate. Wiping with a clean, dry cloth and not using an approved safety solvent is also ineffective since dirt is simply moved around and not removed.

Although low-pressure, dry air is the best method for the initial cleaning of a circuit breaker, cleaning individual components must be performed with different methods. Nonconductive abrasive cleaning pads are ideal for polishing conductive surfaces, latch surfaces, and arc chutes that do not contain asbestos. A fine grade cleaning pad should be used. Abrasive cleaning pads should never be used on plastic or fiberglass components. There are additional areas that require careful cleaning such as arc chutes, contact pivot points, and primary disconnect fingers.

Older types of arc chutes may contain asbestos. Asbestos is not hazardous as long as it is bonded together. Arc chutes containing asbestos may not be marked as containing asbestos, especially older circuit breakers or spare arc chutes.

Circuit breakers interrupting large fault currents or that operate frequently often have a buildup of carbon on them. This buildup is not a problem if it is light, but a heavy covering of carbon can cause a flashover or restrike. Carbon can be cleaned from arc chutes that do not contain asbestos with safety solvent and a nonconductive abrasive pad. Care should be taken not to disturb asbestos if present. Arc chutes containing asbestos cannot be sanded or cleaned in any manner because it would create asbestos dust. Arc chutes that are severely carbonized must be replaced.

As current flows through the contact pivot point, the heat produced from current flowing through the contacts and other conductors dries the lubricant. As the lubricant dries, it thickens, which slows the operating speed of the circuit breaker. A slow-operating circuit breaker can cause an increase in incident energy. Increasing the incident energy can cause properly selected arc flash protective equipment to fail, resulting in serious injury or death. This can occur even if circuit breakers do not operate on a regular basis.

As the lubricant continues to dry, it begins to flake off. Eventually, metal-to-metal contact occurs and the pin wears into the pivot. When this occurs, the contacts cannot be adjusted and the breaker can become jammed. **See Figure 5-14.**

Bolted and pivoting surfaces should be polished with a nonconductive abrasive cleaning pad. It is not necessary to polish the entire contact structure. Only the contact surfaces, pivot, and any part that either contains silver or makes contact with another component should be polished. When cleaning, they should be lightly polished to remove tarnish and corrosion, but the silver plating must not be removed.

TECH FACT

Shell Oil Company did a study of circuit breaker failure rates after they have been in service with no maintenance. They found that after 3 to 5 years in service, approximately 30% of the circuit breakers failed to function normally. After 7 to 10 years, 50% of the circuit breakers failed to function normally. After 17 to 20 years in service, more than 90% failed to function normally.

CONTACT PIVOT LUBRICATION

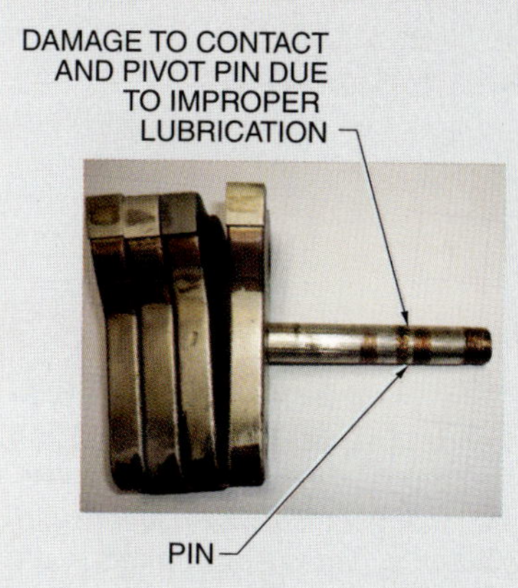

Figure 5-14. Contact pivots that are not properly lubricated are subject to metal-to-metal contact, which causes the pin to wear into the pivot.

The backboard and phase/pole units should be cleaned with a nonabrasive cleaner. Usually, a heavy-duty multisurface cleaner is adequate. Carbon tracking and similar types of contaminants can be removed using stronger solvents, but the solvent should be tested on a test area first to ensure it does not damage the insulation.

Accessible latches, rollers, and sliding surfaces in the operating mechanism should be cleaned using a nonconductive abrasive cleaning pad. The operating mechanism should be cleaned and lubricated as much as possible without disassembly. Also, old and dried lubricant must be removed from primary disconnect fingers, the ground stab, and the shoe and racking mechanism.

Lubricating Circuit Breakers. Once circuit breaker components have been cleaned, the lubricant must be replaced. Suitable lubricants are listed in the OEM manuals. Only those lubricants recommended by the OEM should be used. Lubricants should not be mixed because the additives in the lubricants may react with each other and cause corrosion. **See Figure 5-15.**

RECOMMENDED CIRCUIT BREAKER LUBRICANTS		
Original Equipment Manufacturer	**Grease and Stabs and Pivot Points**	**Grease for Mechanical Parts and Bearings**
General Electric	D50H47 D50H38 (Mobilgrease 28)	D50H15 D50H38 (Mobilgrease 28)
Cutler-Hammer/ Westinghouse	53701AN10J	693C500G04
ABB/BBC/ITE/ ITE Imperial	NO-OX-ID Special Grade A	Anderol L757
Siemens/Siemens-Allis Allis-Chalmers	S-A Contact Lube 15-171-370-002	Beacon P-290
Square-D	PJC-7201	Lubriplate 630-2

NOTE: Most OEMs also recommend Mobilgrease 28 HT Lubricant.

Figure 5-15. Only those lubricants recommended by the original equipment manufacturer are recommended for use with specific circuit breakers.

The three primary types of lubricants that can be used on circuit breakers are nonconductive, conductive, and oil. Nonconductive lubricants are used for mechanical components not in the current path. This includes latches, latch surfaces in the operating mechanism, racking screws and gears, and any mechanical non-current-carrying component. Conductive lubricants are formulated for use in conductive areas, such as primary disconnect fingers, ground stabs, and contact pivot points. These lubricants are not actually conductive but are made for use in conductive areas. Usually 20W or 30W oil is used for linkages, pivot pins, and other light-duty areas. One or two drops of oil in rollers and other open bearings can reactivate applied lubricating grease and allow use in the breaker.

Lubricants must always be applied sparingly and any excess should be wiped off. This is especially true for synthetic lubricants because they "melt" or thin when heated, which causes them to drip or run. Dripping or running lubricant must be wiped off so that it does not collect dirt, which will be transferred into the contact area and cause wear.

Occasionally, in industries such as oil drilling, petroleum processing, or pulp and paper manufacturing, silver is exposed to chemicals such as hydrogen sulfide (H_2S) or sulfur dioxide (SO_2). Silver exposed to these chemicals can develop silver "whiskers," which are conductive, or corrosive. **See Figure 5-16.** In cases such as these, silver plating can be replaced with tin or zinc plating in order to prevent

TECH FACT

Circuit breaker OEMs do not recommend the application of lubricant between bolted joints. However, there are circumstances when lubricating contact surfaces is warranted, such as when a circuit breaker is installed in a highly corrosive atmosphere, for instance where hydrogen sulfide (H_2S), chlorine (Cl), or fluorine (F) are present.

this type of chemical reaction. Even though these platings are not as conductive as silver and they do not last as long, they are effective in solving the problem.

Reassembling Circuit Breakers. Contact assemblies are serviced one phase at a time, using the other phases as guides for proper reassembly. Contact assemblies and components should not be mixed from phase to phase because they develop wear patterns with use over time and mixing parts could cause misalignment and damage.

The moving and stationary contacts should be replaced together rather than trying to replace just one. If the main contacts are worn, the arcing contacts will also be worn. For example, replacing just the main moving contact or the main stationary contact can result in the contacts not mating properly once the circuit breaker is assembled, due to wear and erosion. The same applies to when main contacts and arcing contacts are replaced. If the arcing contacts are eroded or worn, then the main contacts from that phase should be replaced as well.

The pivot joint must be assembled correctly in order for the circuit breaker to function properly. Adjusting the pivot too snuggly results in slower circuit breaker operation, whereas a pivot that is too loose results in high resistance and overheating. The pivot joint must be tight enough to conduct electrical current well and maintain its position but loose enough so that the circuit breaker operates at the correct speed. This balance is obtained through a variety of methods, which varies among OEMs.

For example, most pivots on medium-voltage air circuit breakers are adjusted so that they can maintain their position when placed at a 45° angle from horizontal. If this method does not seem to give satisfactory results, the OEM manual should be consulted. **See Figure 5-17.** Additional recommendations to observe during reassembly include the following:

- Old keepers and locks should not be reused when reassembling a circuit breaker. Most keepers are considered one-time-use fasteners and can fail in service if used again.
- Rusty or corroded fasteners must be replaced. These fasteners cannot be torqued properly and could fail during the stress of a fault interruption. Oil should never be applied to a fastener before tightening the fastener. The oil will cause overtorquing. Fasteners are designed to be tightened when clean and dry. If the threads seem to be tight or rusty, a tap and die can be used to clean the threads so it can be torqued to the proper setting.
- Fasteners that go through plastic should never be overtightened. Overtightening can cause the plastic to crack. A flat washer is used to back up the nut or bolt and should be placed next to the plastic surface so it can spread the pressure.
- Plastic-encapsulated nuts should not be torqued. The lock washer should be tightened until it is flattened and then tightened an additional ⅛ of a turn.
- Bolts and other fasteners used on circuit breakers are usually SAE Grade 5. Fasteners should usually be tightened to the torque value indicted by the head mark of the bolt. Fasteners of different materials have torque applied to the softer material specification. For example, while not common, some manufacturers can install a steel bolt threaded into a brass or aluminum nut or retainer. The amount of torque applied to fasteners should be based on the brass or aluminum specification rather than steel. Recommended torque values for bolts are available from ANSI/NETA tables. **See Figure 5-18.**

Figure 5-16. Silver-plated surfaces in circuit breakers can become corroded or develop "whiskers" when exposed to chemicals such as hydrogen sulfide (H_2S) or sulfur dioxide (SO_2).

Figure 5-17. The pivot joint of pole unit assemblies must be assembled correctly in order for the circuit breaker to function properly.

RECOMMENDED TORQUE VALUES FOR BOLTS Heat-Treated Steel—Cadmium or Zinc Plated				
Grade	SAE 1&2	SAE 5	SAE 7	SAE 8
Head Marking				
Minimum Tensile Strength*	64,000	105,000	133,000	150,000
Bolt Diameter†	Torque‡			
¼	4	6	8	8
5/16	7	11	15	18
⅜	12	20	27	30
7/16	19	32	44	48
½	30	48	68	74
9/16	42	70	96	105
⅝	59	96	135	145
¾	96	160	225	235
⅞	150	240	350	380
1	225	370	530	570

Silicon Bronze Fasteners Torque‡		
Bolt Diameter†	Nonlubricated	Lubricated
5/16	15	10
⅜	20	14
½	40	25
⅝	55	40
¾	70	60

Aluminum Alloy Fasteners Torque‡	
Bolt Diameter†	Lubricated
5/16	8.0
⅜	11.2
½	20.0
⅝	32.0
¾	48.0

Stainless Steel Fasteners Torque‡	
Bolt Diameter†	Uncoated
5/16	14
⅜	25
½	45
⅝	60
¾	90

* in ft lb/sq in.
† in in.
‡ in ft lb

Figure 5-18. Recommended torque values for bolts are provided in ANSI/NETA tables.

Slow Closing Circuit Breakers. *Slow closing,* also known as maintenance closing, is a method of mechanically blocking closing springs from operating when a circuit breaker is manually closed. Slow closing is used for inspections and when contact adjustments are needed. Depending on the type of circuit breaker, a separate blocking tool may be required. Most circuit breakers must be cycled completely through the slow closing procedure in order to reset them into normal operation. Some must have blocking mechanisms installed before attempting a slow closing procedure. The OEM manual must always be consulted prior to slow closing a circuit breaker.

DIFFERENCES IN SLOW CLOSING COMMON MAKES OF CIRCUIT BREAKERS

The slow closing of circuit breakers can differ from one OEM to another. Common makes of circuit breakers include General Electric (GE), Westinghouse/Cutler-Hammer, ABB/ITE, and Siemens.

GE MagneBlast. The GE MagneBlast circuit breaker is prepared for slow closing by charging the closing springs and inserting a safety lockout pin in each closing spring guide so that the springs cannot discharge, then attaching a manual spring charging/maintenance closing tool (½″ drive ratchet) to the mechanism at the eccentric cam. The front sheet metal cover is often removed since it can cause bruising of the hand when the eccentric rotates. *Note*: When slow closing a GE MagneBlast circuit breaker, it may be difficult at first and then much easier to rotate the eccentric. This is normal since the closing springs are only compressed during part of the rotation of the eccentric.

GE Power/Vac®. The GE Power/Vac vacuum circuit breaker uses a "gag" plate, which is inserted into the front of the circuit breaker to slow charge the circuit breaker.

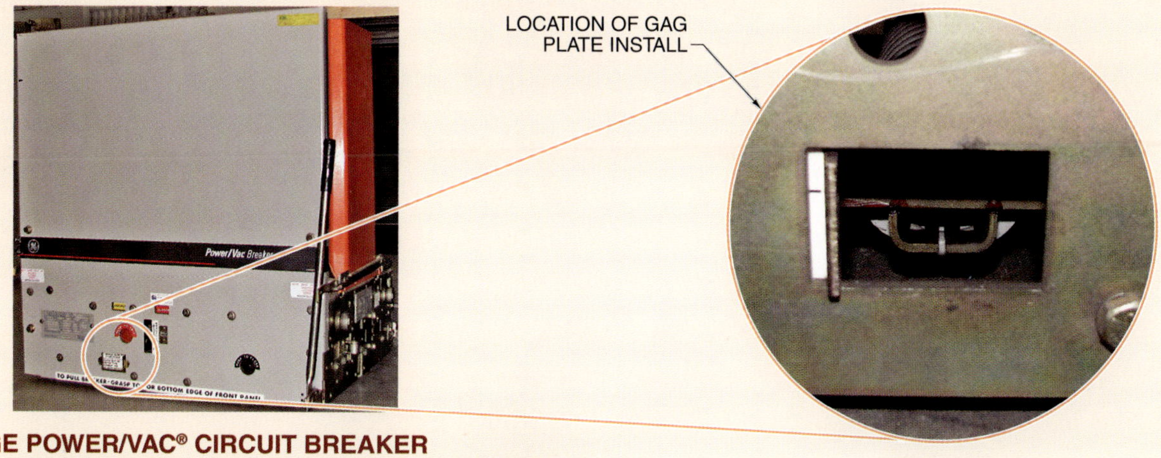

GE POWER/VAC® CIRCUIT BREAKER

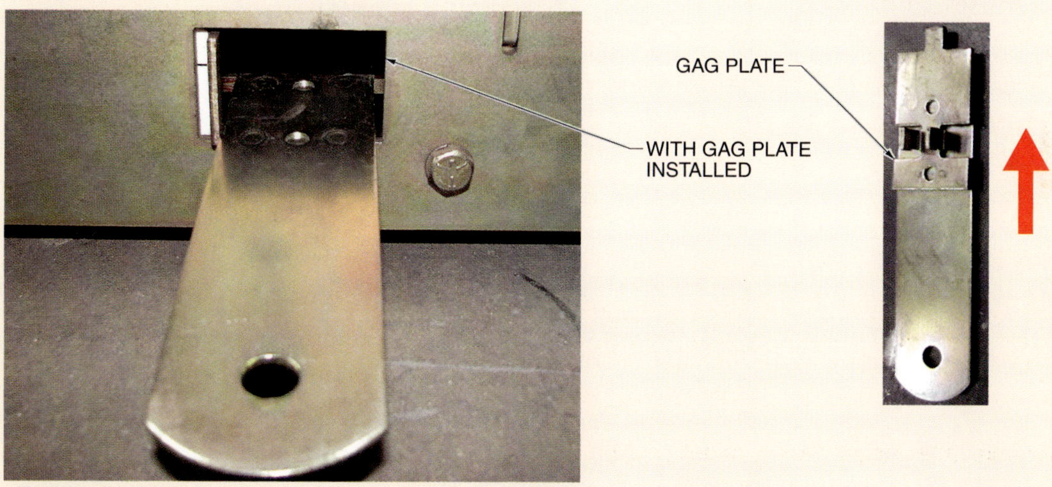

To slow close a GE Power/Vac circuit breaker, apply the following procedure:

1. Use manual charging handle to charge the opening springs.
2. Install gag plate.
3. Depress close button to lock out closing springs.
4. Use manual charging handle to close contacts.

To release the circuit breaker from the slow close set up, the circuit breaker must be completely closed. The manual charging handle can be used to close the circuit breaker. The circuit breaker should be tripped by depressing the trip button, and the gag plate should be removed.

continued on next page...

DIFFERENCES IN SLOW CLOSING COMMON MAKES OF CIRCUIT BREAKERS

(...continued from previous page...)

Cutler-Hammer VCP-W. A Cutler-Hammer VCP-W circuit breaker does not have a provision for slow closing. The only conditions to check are erosion and wipe, and they are performed in a manner that does not require slow closing.

ABB/ITE HK. An ABB/ITE HK circuit breaker is slow closed by using blocking plates, which are similar in function to the gag plate used with a GE Power/Vac.

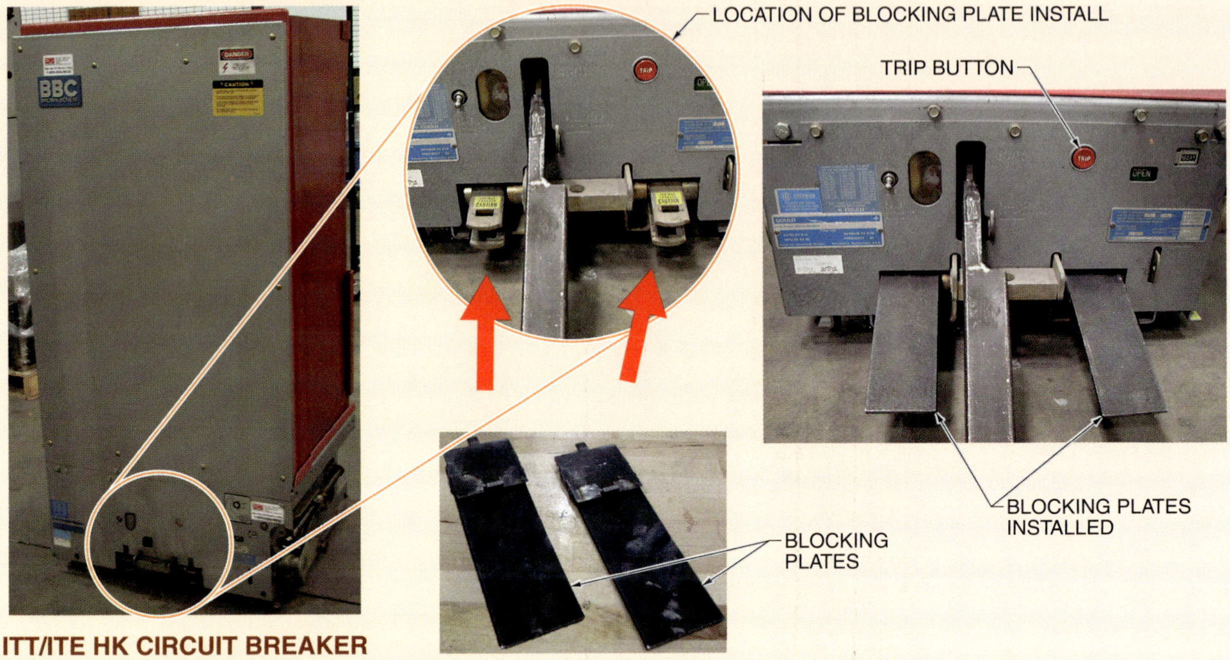

ITT/ITE HK CIRCUIT BREAKER

To slow close an ABB/ITE HK circuit breaker, apply the following procedure:

1. Insert manual spring charging handle.
2. Charge closing springs.
3. Insert blocking plates.
4. Depress close button.

To release the circuit breaker from the slow close set up, the contacts must be completely closed using the manual spring charging handle. The circuit breaker should be opened using the trip button, and the blocking plates should be removed.

Westinghouse/Cutler-Hammer Square-D DS. Unlike other types of circuit breakers, no special tools are required to slow close a Westinghouse/Cutler-Hammer Square-D DS circuit breaker.

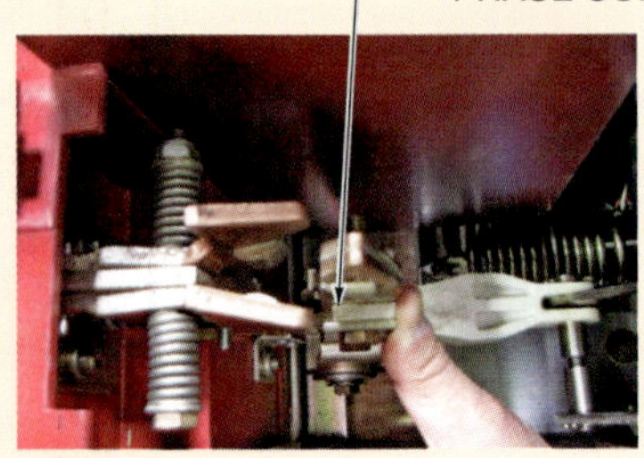

WESTINGHOUSE/CUTLER-HAMMER SQUARE-D TYPE DS CIRCUIT BREAKER

continued on next page...

DIFFERENCES IN SLOW CLOSING COMMON MAKES OF CIRCUIT BREAKERS
(...continued from previous page...)

To slow close a Westinghouse/Cutler-Hammer/Square-D DS circuit breaker, apply the following procedure:

1. Disconnect opening spring.
2. Manually push the center phase contact, and slow close the circuit breaker making sure to push as close to the connecting rod as possible and not high up on contact assembly.

General Electric AK50. General Electric AK50 circuit breakers can be electrically operated (EO) or manually operated (MO). To slow close an EO GE AK50 circuit breaker, apply the following procedure:

1. Use maintenance handle to charge closing springs.
2. Insert restraining pin into restraining pin hole.
3. Depress close button to discharge springs against restraint pin. This blocks the closing springs. Contacts are closed using a maintenance handle (or a ½″ drive socket and ratchet wrench) and by rotating the cam on the end of the spring-charging motor clockwise.

To disable the slow close feature, the circuit breaker is tripped, the springs are recharged, and the restraining pin is removed. The circuit breaker is closed and then tripped to complete the disabling process.

The MO GE AK50 circuit breaker is slow closed in a similar manner. The spring charging handle, attached to the front of the circuit breaker, is rotated 3½ turns until the restraining pin can be inserted. Turning the maintenance handle an additional ½ to 1 turn places the circuit breaker into the maintenance mode. The spring charging handle is rotated several times to slowly close the contacts.

GE AK 50 CIRCUIT BREAKER

GE AKR/AKD-30/50. The GE AKR/AKD-30 and GE AKR/AKD-50 circuit breakers are slow closed in a different manner than the GE AK50.

To slow close a GE AKR/AKD-30 or GE AKR/AKD-50 circuit breaker, apply the following procedure:

1. Remove bolt connecting the closing spring to the frame.
2. Connect maintenance handle to triangular-shaped extension located on right-hand side of circuit breaker operating mechanism.
3. Charge closing springs by moving the maintenance handle in a back-and-forth motion.

GE AKR/AKD-30/50 CIRCUIT BREAKER

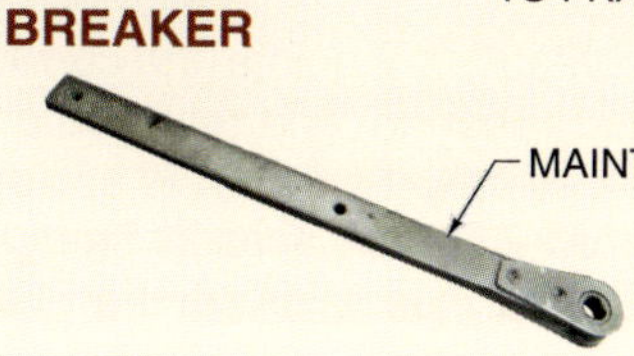

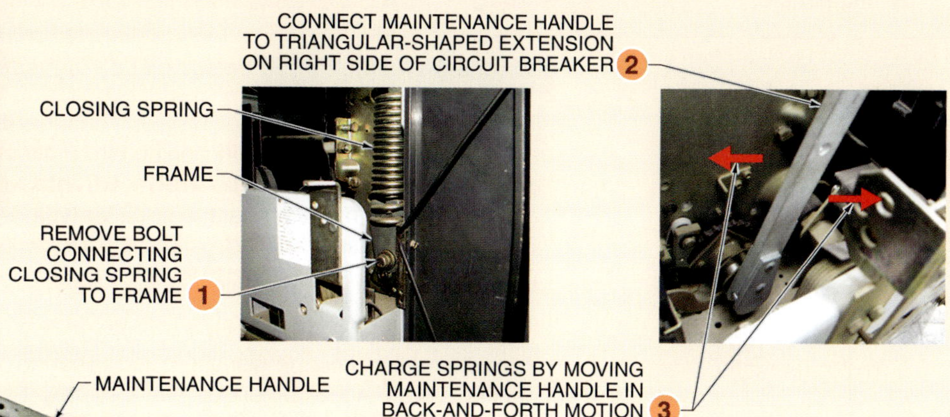

continued on next page...

DIFFERENCES IN SLOW CLOSING COMMON MAKES OF CIRCUIT BREAKERS

(...continued from previous page)

Siemens/Allis-Chalmers LA. Siemens/Allis-Chalmers LA circuit breakers are slow-closed differently than Westinghouse, Cutler-Hammer, and GE circuit breakers. A Siemens/Allis-Chalmers LA circuit breaker is slow closed by applying the following procedure:

1. Remove set screw that attaches manual charging handle to circuit breaker.
2. Insert flathead screwdriver into spring disengagement area with the manual charging handle in the full down position.
3. While pressing down on screwdriver, slowly control the manual charging handle to release to the up position. As handle travels upward, the contacts close. Use caution when releasing the manual charging handle. It is under strong tension and can return to the up position with great force if uncontrolled.

SIEMENS/ALLIS-CHALMERS TYPE LA CIRCUIT BREAKER

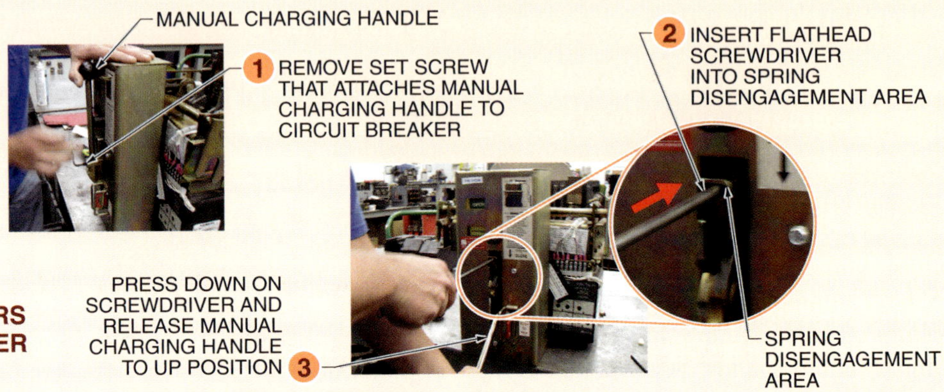

ABB/ITE K-Line. The ABB/ITE K-Line circuit breaker has two different procedures for slow closing, depending on whether the circuit breaker is EO or MO. To slow close an EO circuit breaker, apply the following procedure:

1. Charge closing springs, either electrically or manually. The manual spring charging handle is inserted into a pivoting plate that is slotted for the handle.
2. Insert flathead screwdriver into opening on right side of escutcheon (extension) on the front of circuit breaker and pull lever down.
3. While holding lever down, press the close button.
4. Use manual charging handle and slow close circuit breaker. *Note*: Charging handle is inserted from bottom of circuit breaker.

MO circuit breakers are simpler to slow close than EO circuit breakers. To slow close an MO circuit breaker, apply the following procedure:

1. Insert flathead screwdriver into right-side escutcheon and press the C-shaped linkage down.
2. Slowly operate manual closing handle while holding linkage down. Contacts will move at same speed as manual closing handle.

SUMMARY

Proper maintenance of circuit breakers ensures they will operate quickly and safely when needed. Malfunctioning circuit breakers can not only damage switchgear and cause unscheduled down time but are also a safety hazard to personnel working on or near them. OEMs typically recommend maintenance annually, for cleaning and inspection, and a more thorough major maintenance interval, where the breaker is disassembled, cleaned, inspected, lubricated, reassembled, and tested.

In order to properly service the contact pivot points, the current path of a circuit breaker must be disassembled, cleaned, and relubricated. Spray lubricants are not recommended because they cause dirt to accumulate from overspray and are ineffective at lubricating the surfaces that require it. Circuit breakers also require slow closing for final inspections and adjustments.

DIGITAL RESOURCES
ATPeResources.com/QuickLinks
Access Code: 223436

chapter 5

CIRCUIT BREAKER SERVICING

CHAPTER REVIEW

Name ______________________________ Date ______________

True-False

T F 1. The operating environment the circuit breaker is in does not affect the maintenance interval.

T F 2. Surface pitting should be removed from contact surfaces.

T F 3. Circuit breakers that do not operate at all are not at risk of failure.

T F 4. Abrasive cleaning pads should never be used on plastic or fiberglass components.

T F 5. Unless there is a problem, circuit breakers stay in service with no maintenance until a scheduled shut down or turn-around.

T F 6. Overtorquing is as harmful as undertorquing.

T F 7. Any misalignment of components will cause damage to the contact assembly.

T F 8. Lower than normal contact resistance readings indicate that the lubricant in the contact pivot is hardening and losing its ability to function.

T F 9. Different lubricants should not be mixed.

T F 10. When reassembling a circuit breaker, contact assemblies and components should not be mixed from phase to phase.

T F 11. Once arcing has taken place, although the set screw on a MCCB or ICCB terminal appears tight, it cannot tighten any further on the conductor than the point of arcing.

T F 12. For MCCBs, damaged or worn parts are usually replaced or the circuit breaker is replaced.

T F 13. Spray-type penetrating liquids are considered lubricants.

T F 14. Arc chutes on medium-voltage circuit breakers can weigh up to several hundred pounds.

T F 15. Circuit breaker maintenance intervals can be determined by the amount of heat created by current flow through a conductor.

T F 16. Only one arc chute of medium-voltage air circuit breakers (that have arc chutes mounted on top of the pole units) should be tilted at a time because the circuit breaker can become unbalanced and tip over.

T F 17. All large MCCBs and ICCBs have cases that cannot ever be opened.

T F 18. Arcing contacts are replaced when the circuit breaker can no longer be properly adjusted or if significant erosion is evident.

T F **19.** Routine maintenance on MCCBs and ICCBs typically requires over 8 hr to complete.

T F **20.** An advantage of using nylon fasteners is that nylon is not easily broken or damaged.

T F **21.** "Safety solvents" are safe to use in all locations.

T F **22.** Newer circuit breakers with multiple contacts or breakers that use bayonet-style contacts cannot be tested with the contact print test method.

T F **23.** A heavy covering of carbon in an arc chute can prevent a flashover or restrike.

T F **24.** Older types of arc chutes can contain asbestos.

T F **25.** Often circuit breakers are subjected to high humidity conditions and receive very little maintenance.

Multiple Choice

______________ **1.** Due to time constraints and difficulty in scheduling plant shutdowns, major maintenance is typically performed in ___- to___-year intervals and disassembly is delayed until the breaker exhibits distress.

A. two; three
B. four; five
C. six; eight
D. eight; ten

______________ **2.** When erosion reaches a level of ___″, the arc chute must be replaced.

A. 1⁄64
B. 1⁄32
C. 1⁄16
D. 1⁄8

______________ **3.** Switches should have resistance of no greater than ___ Ω.

A. 0.1
B. 0.5
C. 1.0
D. 1.5

______________ **4.** ___ circuit breakers cannot be opened because opening them voids their UL certification.

A. Gas
B. Air-magnetic
C. Sealed-case, molded-case
D. Low-voltage power

______________ **5.** ___ circuit breakers usually have wear (erosion) indicators, since the contacts themselves cannot be visually checked.

A. Oil
B. Vacuum
C. Low-voltage
D. Medium-voltage

_______________ **6.** Using ___ is the best method of cleaning a circuit breaker.
A. an alcohol solvent
B. abrasive cleaning pads
C. a clean, dry cloth
D. low-pressure, dry air

_______________ **7.** The two types of maintenance typically recommended by circuit breaker OEMs are routine maintenance and ___ maintenance.
A. major
B. predictive
C. preventive
D. specialized

_______________ **8.** Arc spatter is best removed from arc divider plates by ___.
A. disassembling components and soaking in a chemical bath
B. heating with a torch
C. scraping with a screwdriver or putty knife
D. using abrasive cleaning agents

_______________ **9.** Most pivots on medium-voltage air circuit breakers are adjusted so that they can maintain their position when placed at a ___° from horizontal.
A. 30
B. 45
C. 60
D. 120

_______________ **10.** The outside case of an MCCB or an ICCB should be cleaned of dirt and dust by ___.
A. using low-pressure air
B. spraying with chemical cleaners
C. using a pressure washer
D. wiping with a rag

Completion

_______________ **1.** Low-voltage and medium-voltage power circuit breakers can be separated into two main sections: the ___ and the backboard assembly (low-voltage) or pole unit (medium-voltage).

_______________ **2.** The three primary types of lubricants that can be used on circuit breakers are nonconductive, conductive, and ___.

_______________ **3.** When a contact print test is performed, contacts must have approximately ___% make.

_______________ **4.** Penetrating liquids are typically ___-based and are not actual lubricants.

_______________ **5.** A(n) ___ circuit breaker is similar in construction to a molded-case circuit breaker but has higher interrupting ratings and uses digital trip units.

_______________ **6.** ___ is a method of mechanically blocking closing springs from operating when a circuit breaker is manually closed.

______ 7. Electrical workers and technicians must practice all OSHA regulations and requirements as stated in ___.

______ 8. A ___ is a resumption of current between the contacts of a switching device during an opening operation after an interval of zero current of quarter cycle at normal frequency or longer.

______ 9. The time required for a circuit breaker to open from when the trip button is depressed varies depending on the circuit breaker ___.

______ 10. As current flows through a conductor, it creates ___.

______ 11. Sealed-case MCCBs can only have ___ maintenance performed on them.

______ 12. On low-voltage power circuit breakers (LVPCBs), the backboard assembly is disconnected from the operating mechanism frame by ___.

______ 13. On medium-voltage air circuit breakers, uniform glass beading of the ceramic material in the throat of the arc chute indicates the ___.

______ 14. Interphase barriers are used on medium-voltage circuit breakers to prevent arc ___.

______ 15. ___ of the contact face is usually caused by excessive contact pressure or improper silver-soldering of the contact face to the contact mounting point.

______ 16. Circuit breakers require lubrication in the operating mechanism and in the ___.

______ 17. Hardware made from ___ is often used as fasteners with arc chutes on medium-voltage circuit breakers.

______ 18. To prevent conductive silver "whiskers" that can develop after exposure to certain chemicals, silver plating in circuit breakers can be replaced with ___ plating.

______ 19. Arc chutes should be blown through with low-pressure (≤ ___ psi), dry air to remove dust and ensure that their vents are clear.

______ 20. Arcing contacts are designed to interrupt the arc when the circuit breaker ___.

Short Answer

1. List five variables that can be used to determine circuit breaker maintenance intervals.

2. List three indicators that a circuit breaker is in need of major maintenance.

3. In addition to inspecting, cleaning, lubricating, reassembling, torquing, adjusting, and testing, what are five primary tasks associated with the major maintenance of LVPCBs?

4. Briefly explain why on medium-voltage air circuit breakers that have acrylic puffer tubes, cleaning solvent should be applied to a small test area and its reaction should be checked with the material to ensure it does not damage the component.

5. Briefly explain why lubricants must always be applied sparingly and any excess should be wiped off.

Matching

Low-Voltage Power Circuit Breakers (LVPCBs)

______________	1. Backboard assembly	______________	4. Moving contacts
______________	2. Frame bolt	______________	5. Operating mechanism frame
______________	3. Link pins	______________	6. Operating mechanism

Arc Chute Inspection Points

____________________ **1.** Arc spatter

____________________ **2.** Blocked vents

____________________ **3.** Broken components

____________________ **4.** Carbon buildup

____________________ **5.** Cracks

____________________ **6.** Erosion on arc divider plates

____________________ **7.** Erosion indicators on arc divider plates

____________________ **8.** Heat/arc damage

____________________ **9.** Insulated magnetic poles

____________________ **10.** Erosion

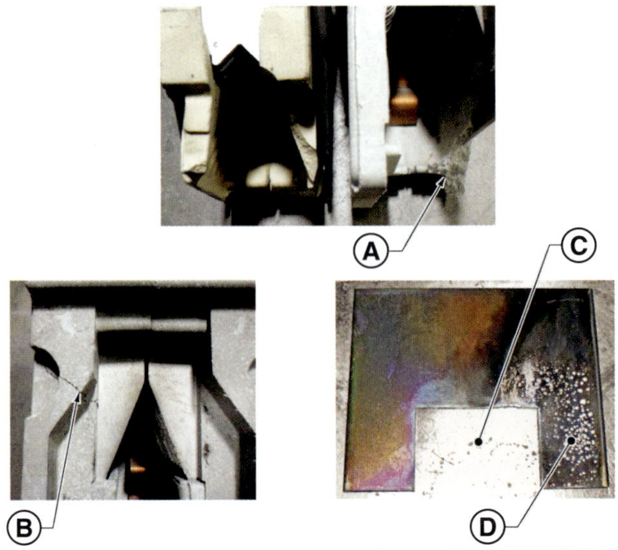

Circuit Breaker Interior Inspection Points – Common Problems

____________________ **1.** Corroded and pitted surfaces

____________________ **2.** Contact damage from cracked face

____________________ **3.** Cracked contact face caused by misadjustment of contacts

____________________ **4.** Misshapen hole

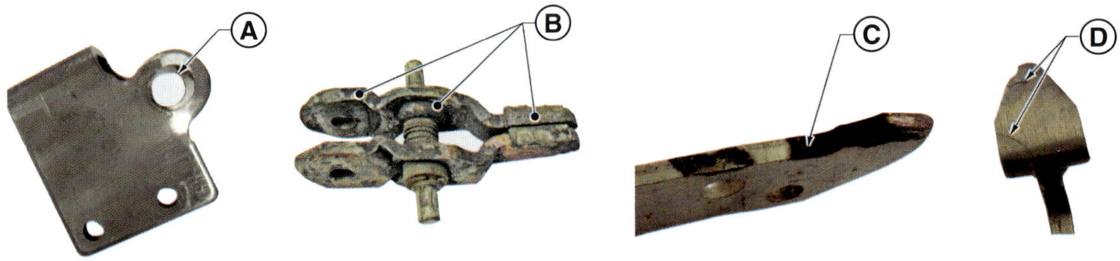

chapter 6

CIRCUIT BREAKER ADJUSTMENT

Disassembly, replacement of components, and reassembly of circuit breakers often require that a circuit breaker have its adjustments reset. The main types of adjustments performed on low- and medium-voltage circuit breakers are contact adjustments and operating mechanism adjustments. The most common makes of circuit breakers include GE, ABB/ITE, Westinghouse, Cutler-Hammer, and Siemens LA.

OBJECTIVES

- Identify the mechanical procedures that must be performed in order to properly adjust contacts in a circuit breaker.
- List the common contact adjustments that are made to low- and medium-voltage air circuit breakers.
- Determine the appropriate adjustments to make on various makes and models of circuit breakers.

KEY TERMS

- overtravel
- contact wipe
- contact gap
- contact sequence
- diving pawl
- latching pawl
- buffer assembly

CONTACT ADJUSTMENTS

Through normal use, circuit breaker contacts can become worn or misaligned over time. The circuit breaker contacts most commonly used include bayonet-style and butt-style contacts. These contacts are adjusted differently, and the manufacturer's instructions must be consulted to verify procedures and specifications for the specific circuit breaker.

Contact Centering and Penetration Adjustments—Bayonet Contacts

Bayonet contacts can have two adjustments performed on them: penetration and centering. Penetration adjustment determines how far the moving contact penetrates the stationary contacts. Centering ensures that the moving contact is centered between the stationary contacts. **See Figure 6-1.**

TECH FACT

A Westinghouse DHP circuit breaker uses bayonet contacts for both main and arcing contacts. Other original equipment manufacturers (OEMs) may use bayonet-style contacts on the arcing contacts and butt-style on the main contacts or vise-versa.

Contact penetration is measured by slow-closing the breaker and measuring how far the moving contact penetrates the stationary. With Westinghouse DHP circuit breakers, the back of the moving arcing contact faces must be in line with the front of the stationary arcing contact faces, $\pm 1/32''$. The amount of contact wear can be measured by calculating the difference between the contact gap with the breaker both open and closed.

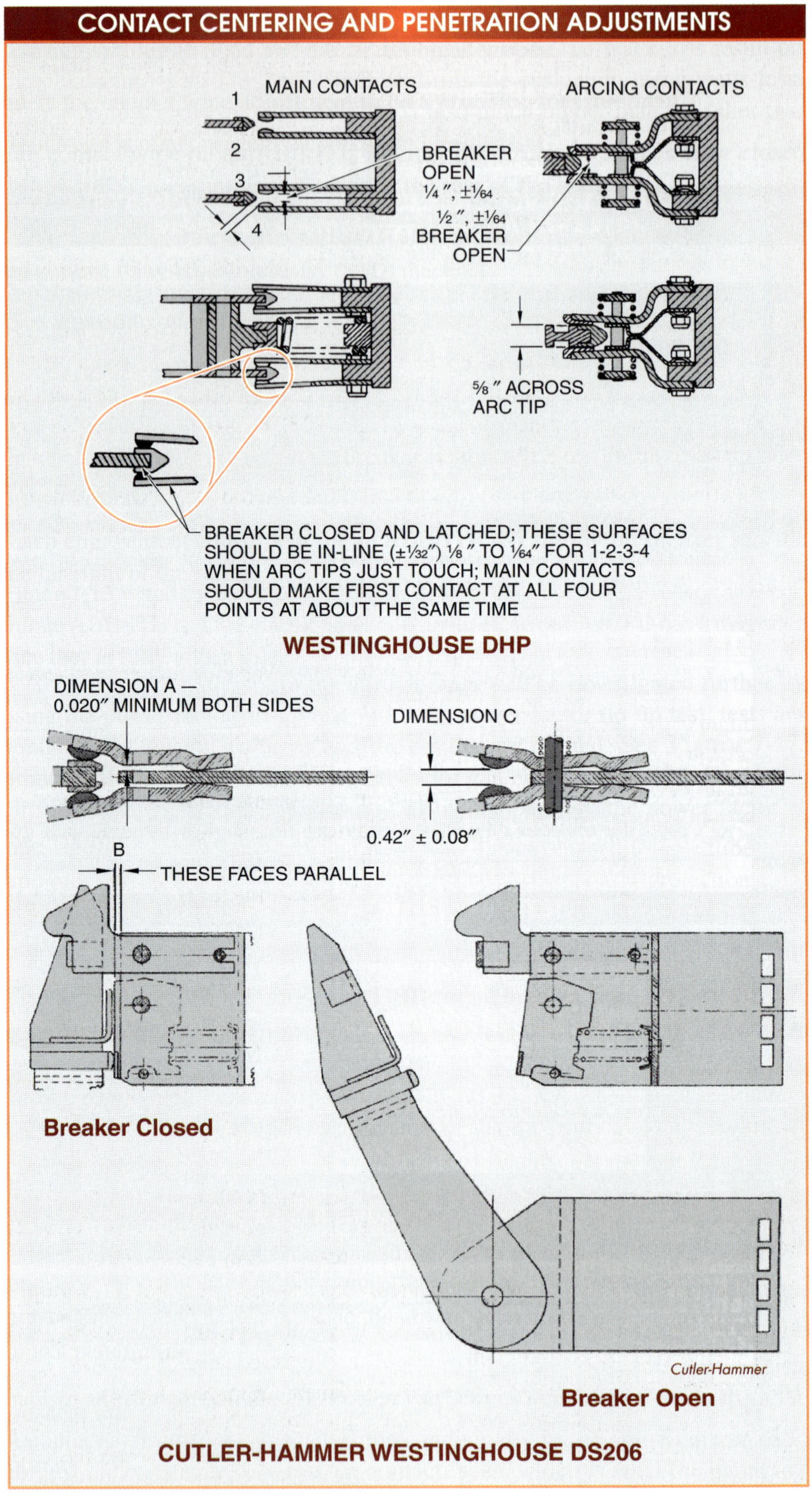

Figure 6-1. Low-voltage power and medium-voltage air-magnetic circuit breakers with bayonet contacts have a centering adjustment and a penetration adjustment that is dependent on critical dimensions.

Although somewhat similar in design to the Westinghouse DHP medium-voltage air circuit breaker, the low-voltage DS style has different specifications. Contact adjustments for the DS416 and DS532 are similar to the DS206 model circuit breaker with the following exceptions:

- With the circuit breaker in the closed position, dimension A (the gap between the center piece of the stationary arcing contact assembly and the arcing contact) must be 0.020″ or greater. If this dimension cannot be maintained, the arcing contacts must be replaced.
- The X and Y dimensions given for the DS206 do not apply on the DS416 or DS532.
- Dimension B must be able to accept a ¼″ round rod. If a ¼″ rod does not fit in this gap, the main contacts must be inspected to ensure that they have a minimum thickness of 1⁄32″. If their thickness is less than 1⁄32″ they require replacement. If the gap at B is less than 3⁄16″, the main contact assembly must be replaced.
- When the circuit breaker is closed, the face of the moving contact assembly is parallel to the center section of the stationary contact assembly.

Overtravel and Wipe Adjustments—Butt Contacts

Overtravel is the distance a circuit breaker mechanism travels once the contacts touch and the mechanism latches. Overtravel is the tendency for the mechanism/contacts to continue to move after the mechanism has opened or closed. When a circuit breaker is closed, the mechanism does not just stop, it continues to travel due to inertia.

Contact wipe is the amount of overtravel a contact assembly has once it initially makes contact. Contact wipe provides a self-cleaning action of the contact surfaces. When the moving contact strikes the stationary contact, the spring-loaded contact is compressed, which causes the faces of the contacts to rub together. This wiping action keeps the contact faces clean and ensures that they do not "bounce" when closing. The spring-loaded contact can be either the stationary or moving contact assembly. *Note:* To ensure consistent results, all adjustments are made by closing the breaker using slow closing rather than closing at normal speed, except as noted by the OEM.

Contact Wipe Adjustments—Medium-Voltage Air Circuit Breakers

Contact wipe adjustments are performed on the most commonly used medium-voltage air circuit breaker makes including GE MagneBlast and ABB/ITE HK. Medium-voltage air circuit breakers have arcing and main contacts.

GE MagneBlast. With GE MagneBlast circuit breakers, primary (main) contact wipe is measured from the primary contacts fully extended (open position) to when the contacts are closed and latched. The dimension is 5⁄16″ (+0″, –1⁄16″). The easiest method to take this measurement is to measure from the top contact support down to the tip of the primary stationary contact with the circuit breaker open. The circuit breaker should be slow closed to the tip of the primary stationary contact and remeasured from the top contact support to the primary moving contact. The two measurements must be subtracted to determine wipe. **See Figure 6-2.**

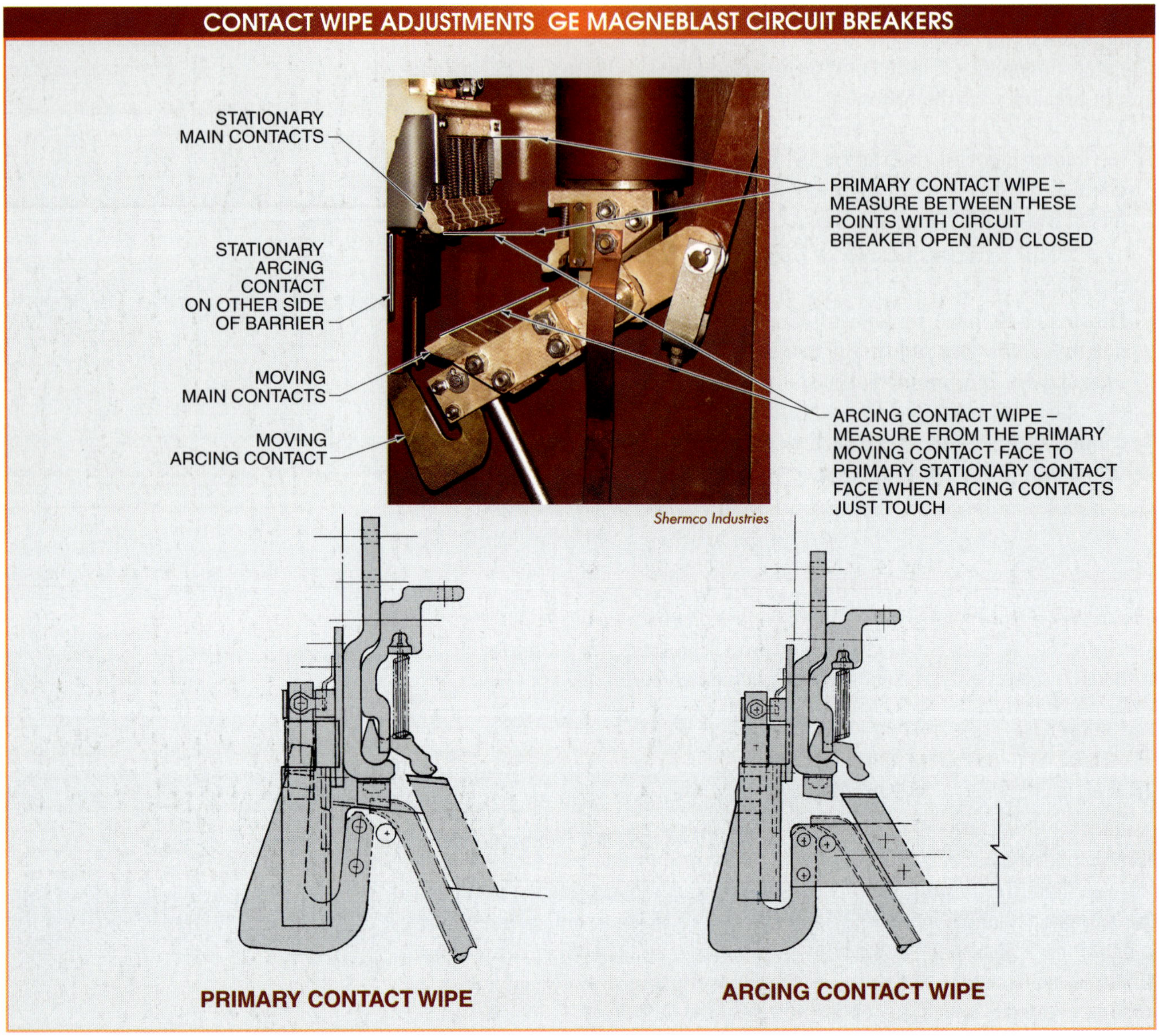

Figure 6-2. Typical GE MagneBlast circuit breaker primary and arcing contact wipe adjustments.

Primary contact wipe is adjusted at the lower bushing. Some medium-voltage circuit breakers have intermediate contacts, which function as a second set of arcing contacts. They are sometimes used in circuit breakers that have higher interrupting and continuous current ratings as well as wipe for the arcing contact.

To check the arcing contact wipe, the circuit breaker should be slow closed until the arcing contacts just touch, as indicated by an ohmmeter or light box. The distance should be measured from the main moving contact to the tip of the main stationary contact. There must be a minimum clearance of 5⁄16″. There is no adjustment for arcing contact wipe. A smaller distance than specified indicates the arcing contacts have eroded sufficiently to require replacement. To center the arcing contact, the side of the contact should be struck using a rubber mallet while supporting the contact near its base.

ABB/ITE HK. Nearly all medium-voltage air circuit breakers use an operating rod for adjusting contact wipe, sequence, and gap. The operating rod between the operating mechanism and the contact assembly has an adjusting bolt and nut that move the contact assembly closer to or farther from the operating mechanism. This is a simple solution that reduces the complexity of adjustments but also involves compromises when adjusting circuit breakers.

The typical sequence of adjustments is to set the contact wipe, then the contact sequence, and to check the contact gap last. If the contacts have not eroded or worn too badly, all three measurements will be within OEM specifications. If all three adjustments cannot be obtained within the OEM specifications, it is an indication that the pivots or contacts are worn or contact faces have eroded. Functionally, this adjustment is the same as the adjustment made on low-voltage power circuit breakers, just on a much larger scale.

The contact wipe of an ABB/ITE HK series circuit breaker is measured by slow closing the circuit breaker completely and measuring the gap from the edge of the plate to the front edge of the main contact. **See Figure 6-3.**

CONTACT WIPE ADJUSTMENTS — ABB/ITE HK CIRCUIT BREAKERS

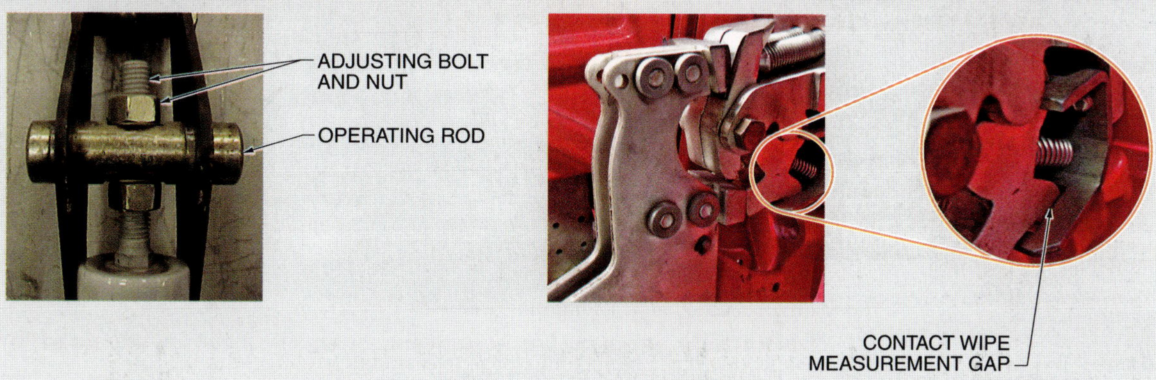

Shermco Industries

Figure 6-3. Nearly all medium-voltage air circuit breakers use an operating rod for adjusting contact wipe, gap and sequence. Contact wipe of ABB/ITE HK circuit breakers is measured by slow closing the circuit breaker completely and measuring the gap from the edge of the plate to the front edge of the main contact.

Contact Wipe Adjustments—Vacuum Circuit Breakers

Wipe is the additional compression of a preloaded spring that is used to apply force to the vacuum interrupter contacts and to provide opening kick-off force. Because OEMs change the design of their products over time, the appropriate operator's manual for the specific circuit breaker model should always be referenced.

GE Power/Vac®. Contact wipe on vacuum circuit breakers is measured differently than on air circuit breakers since there are no arcing contacts or main contacts, just contacts inside a bottle. The procedure given in this section differs somewhat from the one given by GE but has been shown to provide adequate results. On a GE Power/Vac circuit breaker with ML-18 mechanism, contact wipe is measured by making a mark on the wipe indicator then slow closing the circuit breaker completely. The distance is measured from the mark to the top of the spring cap. **See Figure 6-4.** GE requires a minimum contact wipe distance of 0.075″. Any less than this amount requires adjustment of the circuit breaker. To adjust contact wipe, apply the following procedure:

1. Close the circuit breaker at normal speed and loosen the nuts securing the interrupter clamp. *Note:* The interrupter clamp must be loose. If it is not, lightly pry it using a flat-head screwdriver to break it free.
2. Using a 1″ wrench, hold the adjusting nut, and loosen the ¾″ locknut. Adjust each phase by rotating the operating rod insulator (rotating insulator toward the vacuum bottle increases the contact wipe).
3. Once all three phases are adjusted, torque the locknut to 40 lb-ft to 50 lb-ft while holding the adjusting nut secure. The clamp screws are torqued to 8 lb-ft to 10 lb-ft.

Note: GE does not consider this adjustment to be a precision measurement, so allow some latitude when performing it. When performing adjustments or maintenance on any circuit breaker, always reference the OEM instruction manual. Specifics about each type and model of circuit breaker are provided, along with the required clearances and dimensions.

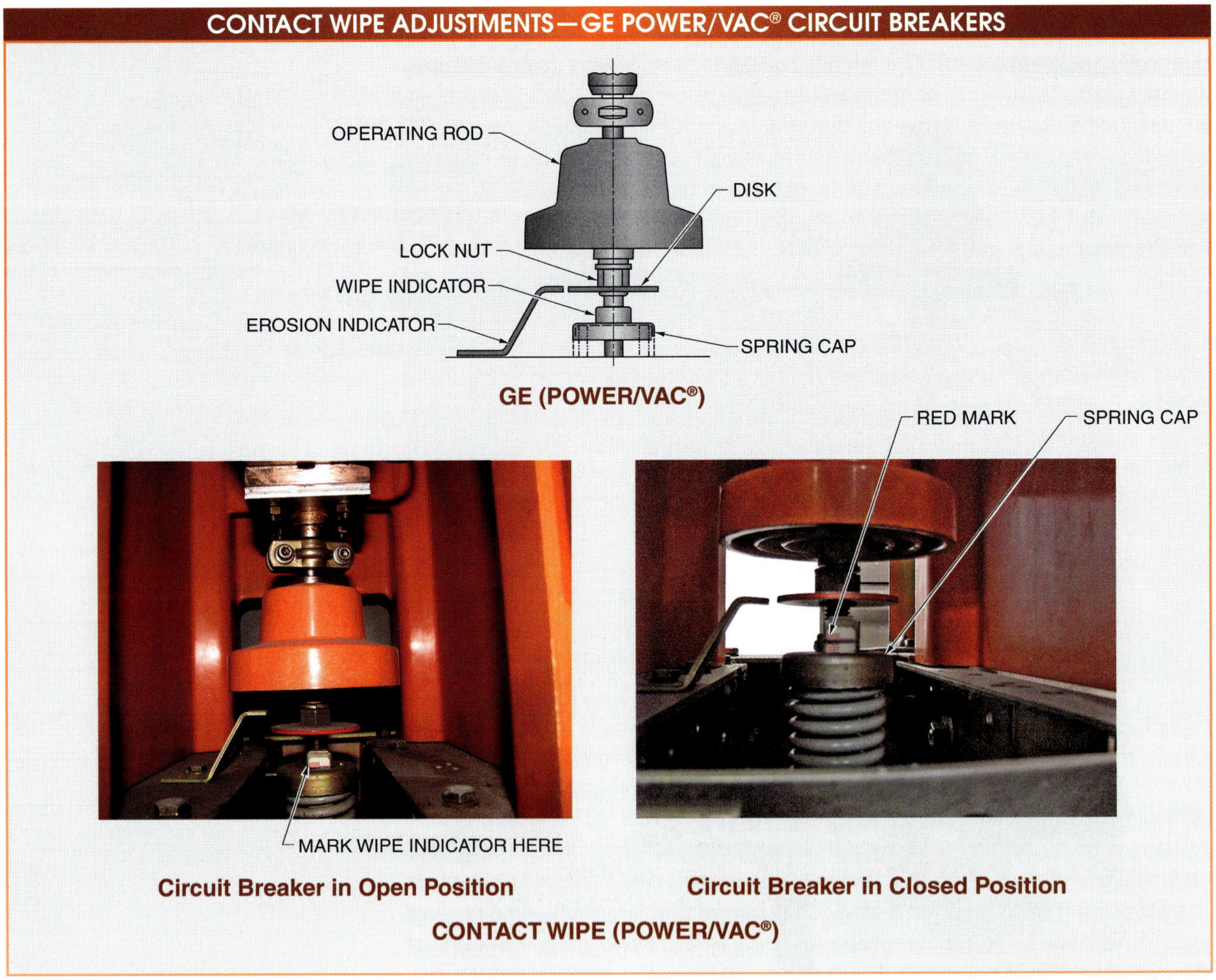

Shermco Industries

Figure 6-4. With the circuit breaker open, a mark should be made on the wipe indicator just above the top of the spring cap. The circuit breaker should be slow closed and measured from the mark to the top of the spring cap.

Cutler-Hammer VCP-W. The contact wipe of Cutler-Hammer VCP-W vacuum circuit breakers is less difficult to measure than that of other types of circuit breakers. To check contact wipe, the circuit breaker must be closed at normal speed and the indicator observed. The type of indicator depends on the color of the closing springs: red, blue, white, orange, or black. There is no adjustment for contact wipe. **See Figure 6-5.**

CONTACT WIPE INDICATORS — CUTLER-HAMMER VCP-W CIRCUIT BREAKERS

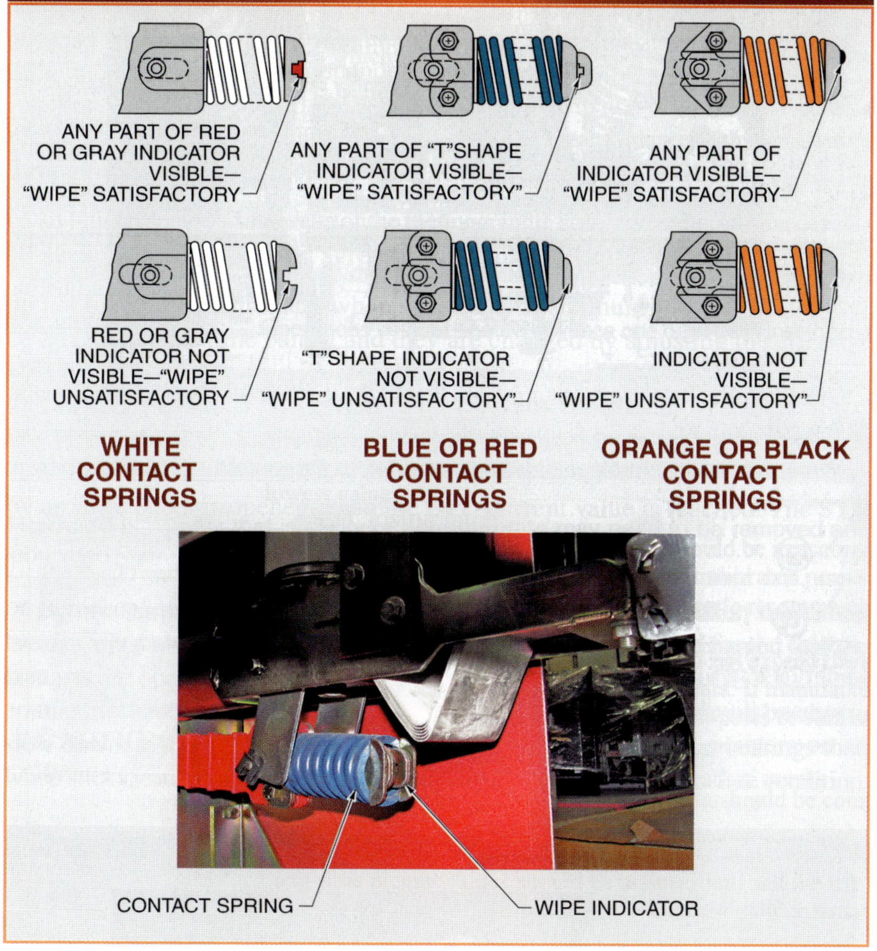

Shermco Industries

Figure 6-5. Contact wipe on Cutler-Hammer VCP-W vacuum circuit breakers is checked by closing the circuit breaker at normal speed and observing the indicator.

Contact Wipe Adjustments— Low-Voltage Power Circuit Breakers

The most common types of low-voltage power circuit breakers are GE AKR/AKRD-30/50, GE AK 50, ABB/ITE K-Line, and Siemens LA. Low-voltage circuit breakers are similar to medium-voltage air circuit breakers in construction, although adjustments can vary considerably from OEM to OEM.

GE AKR/AKRD-30/50. To measure the contact wipe on the GE AKR/AKRD-30/50 circuit breaker, the circuit breaker is slow closed, and the gap between the back of the stationary contact and the contact pivot is measured. The contact gap must be 0.060″ ±0.020″.

Shermco Industries

Contact wipe adjustment can be checked with a continuity tester or lighting circuit.

A GE AKR/AKRD-50 circuit breaker uses a different contact structure than the AKR-30, so the adjustment procedure also differs. Since this circuit breaker model uses intermediate contacts, it is necessary to place a thin piece of 2″ wide insulation over the stationary arcing and intermediate contacts. **See Figure 6-6.** To adjust the contact wipe on a GE AKR/AKRD-50 circuit breaker, apply the following procedure:

1. Slow close the circuit breaker while holding the insulation in place. Verify that the insulation does not come between the main contacts as the circuit breaker is slow closed.
2. On the phase to be adjusted, connect a continuity tester or ohmmeter between the line and load primary disconnects. The continuity tester/ohmmeter must show continuity (indicating that the contacts are closed).
3. Adjust the connecting link adjusting stud to open the contacts (a continuity tester indicates no continuity or ohmmeter indicates an open circuit).
4. Rotate the adjusting stud until the contacts make light contact, as indicated by the continuity tester/ohmmeter. Rotate the adjusting stud an additional 4½ turns. Per GE, it should take more than 40 lb-ft of torque to rotate the adjusting stud once this adjustment is made. If not, add a thread-locking compound to threads of the adjusting stud to ensure that it does not come loose in service.
5. Remove the insulating material and repeat the same procedure for the remaining two phases.

Note: A lighting circuit can also be used in lieu of a continuity tester to perform this adjustment.

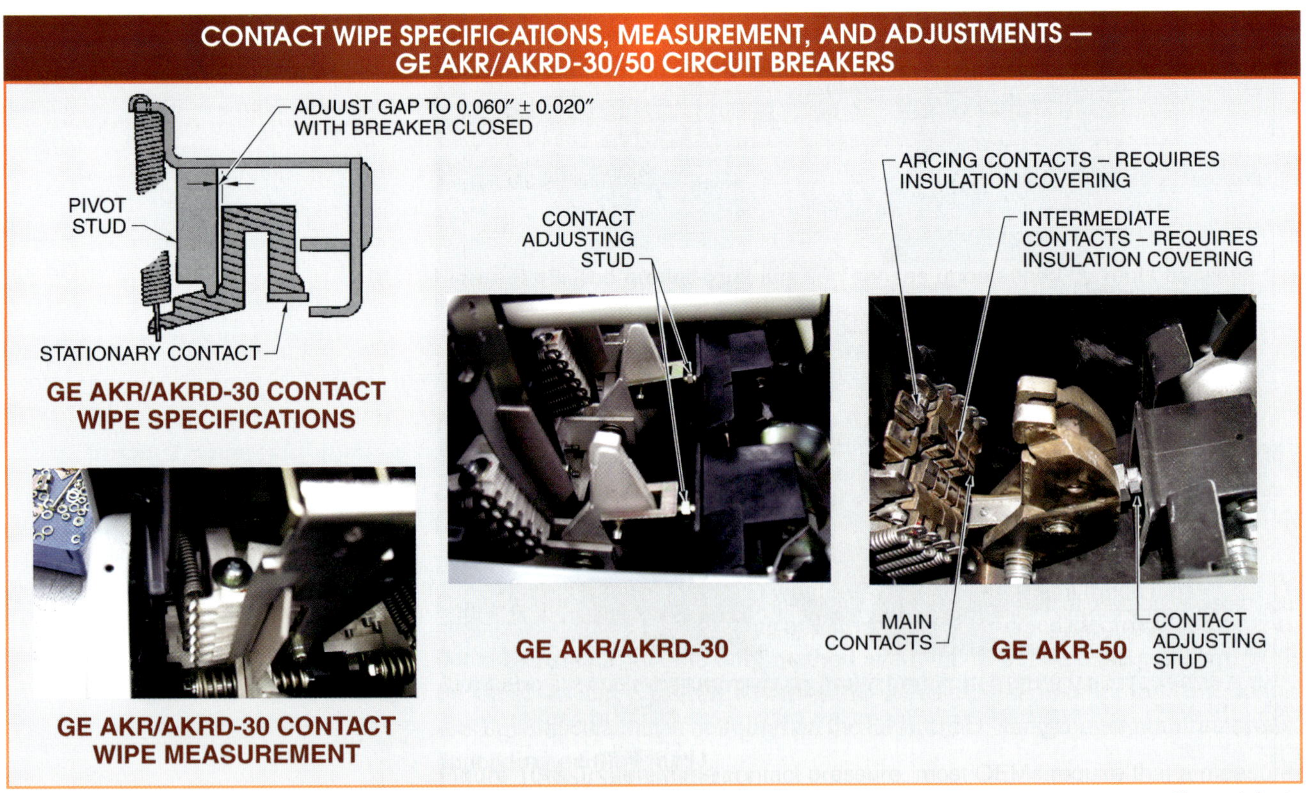

Shermco Industries

Figure 6-6. The AKR/AKRD-50 requires a thin piece of 2″ wide insulation over the stationary arcing and intermediate contacts. The circuit breaker contacts are monitored using a light circuit or ohmmeter.

GE AK 50. With the circuit breaker in the open position, measure from the arcing contact's face to the backboard and the main contact's face to the backboard. Close the circuit breaker and remeasure both dimensions. The difference between the two measurements (open to close) is the contact wipe. The arcing contacts require a contact wipe between 3/16″ to 9/32″ and the main contacts require a contact wipe between 1/16″ to 7/64″. **See Figure 6-7.**

Making contact wipe adjustments on a GE AK 50 circuit breaker is different than other models of GE low-voltage power circuit breakers in that the clevis pin on the center phase moving contact assembly must be removed and the link rotated. The distance between the connecting clevis (link) and the contact assembly must be no more than 3/16″ and filled to within 0.005″ of being solid by using shims. This prevents the adjusting threads on the clevis from being overstressed. Once the center phase is adjusted properly, the outside contact wipe can be set by using the crossbar adjusting plates.

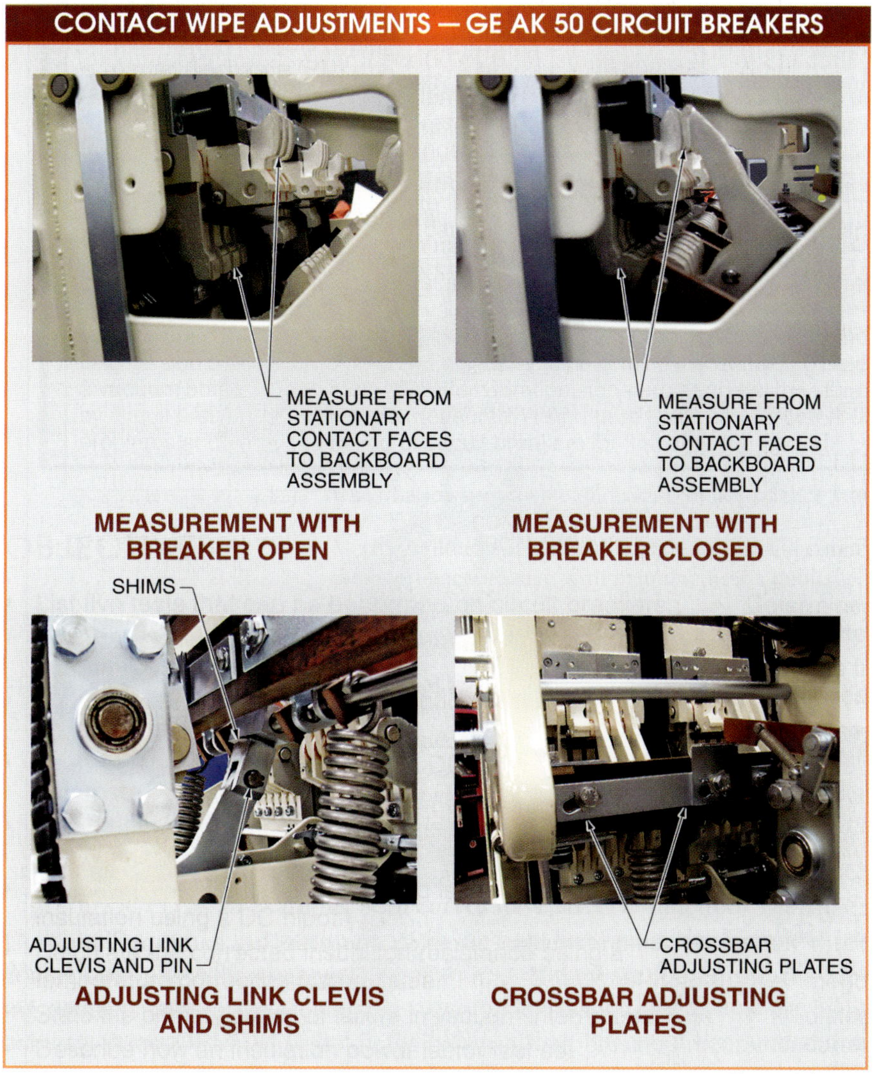

Shermco Industries

Figure 6-7. Making contact wipe adjustments on a GE AK 50 requires that the clevis pin on the center phase moving contact assembly be removed and the link rotated.

ABB/ITE K-Line. The contact assemblies in ABB/ITE K-line circuit breakers are different than most other makes of circuit breakers. The springs are contained in the moving contact assembly as opposed to the stationary contacts and backboard assembly. To measure contact wipe, the circuit breaker is slow closed until it latches, then a measurement is taken at the rear of the contact (as shown in point A).

On the ITE K-600/800 circuit breakers, the best method for performing this measurement is to use two pieces of welding rod, or similar metal rod, with one end (approximately ½″ to ¾″) bent at a 90° angle. One rod must be 5/64″ in diameter and the other rod 3/32″ in diameter. When contact pressure is adjusted properly, the 5/64″ rod does not slip through the back side of the contact and the contact support (the 3/32″ diameter rod slips through). Adjustment is made at the connecting link between the contact assembly and the operating mechanism.

The ABB/ITE K-1600/2000 circuit breaker illustrates the differences in circuit breakers manufactured by the same OEM. To measure contact wipe, the circuit breaker is slow closed until it latches and a measurement is taken at point A. **See Figure 6-8.** This dimension must be 0.750″. *Note:* A shim with a thickness of 0.750″ can be used to verify this dimension. Traditional measuring instruments can be difficult to use on installed circuit breakers.

CONTACT WIPE ADJUSTMENTS — ABB/ITE K-LINE CIRCUIT BREAKRES

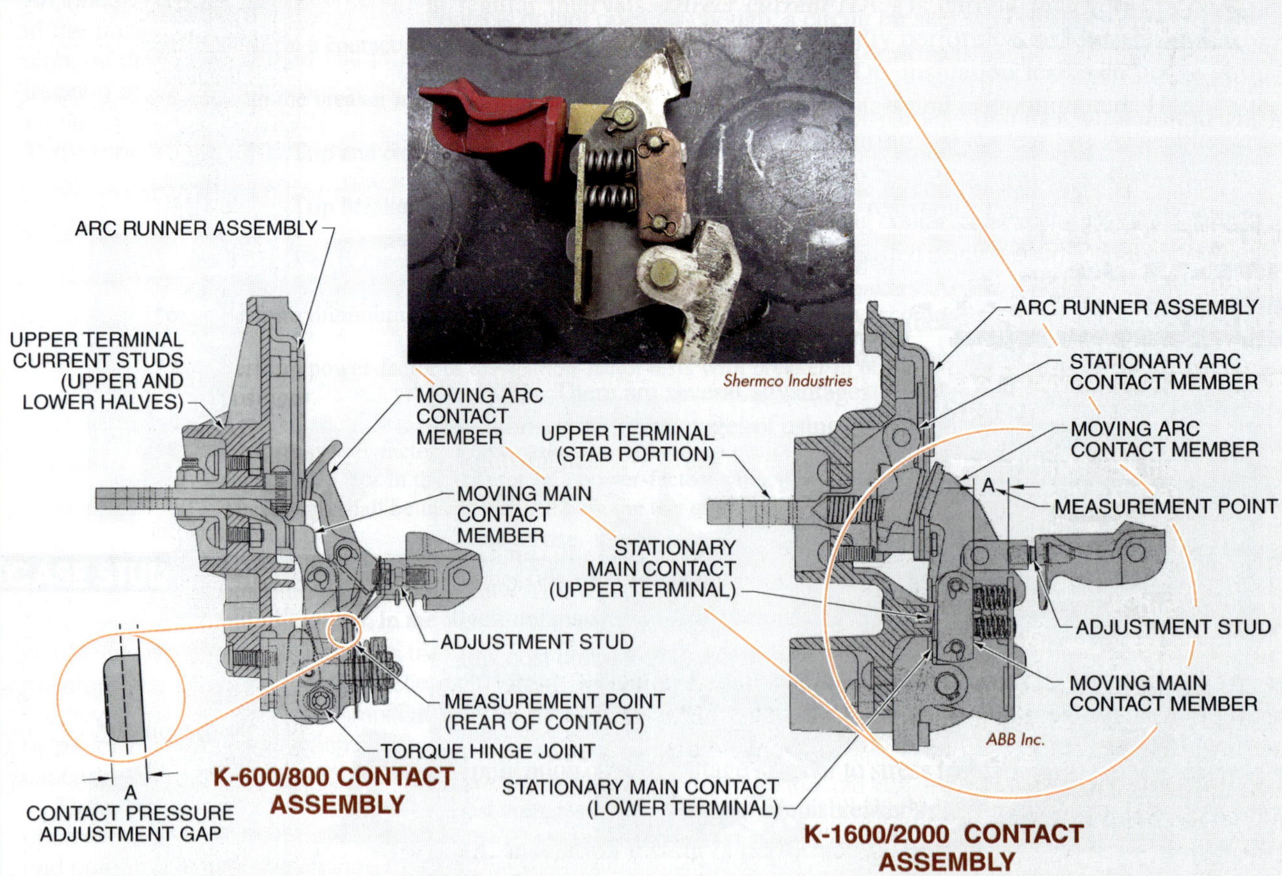

Figure 6-8. ABB/ITE K-Line contact pressure is measured on the moving contact assemblies at point A. Most other low-voltage power circuit breakers measure contact pressure at the stationary contact assemblies.

Siemens LA-600/800. The contact wipe on a Siemens LA-600/800 circuit breaker is measured from the backboard to the edge of the stationary contact with the circuit breaker in the open position. The circuit breaker is closed at normal speed, and the contact wipe measurement is retaken. **See Figure 6-9.** Contact wipe (the difference between the open and closed measurements) should be between 0.93″ and 0.125″.

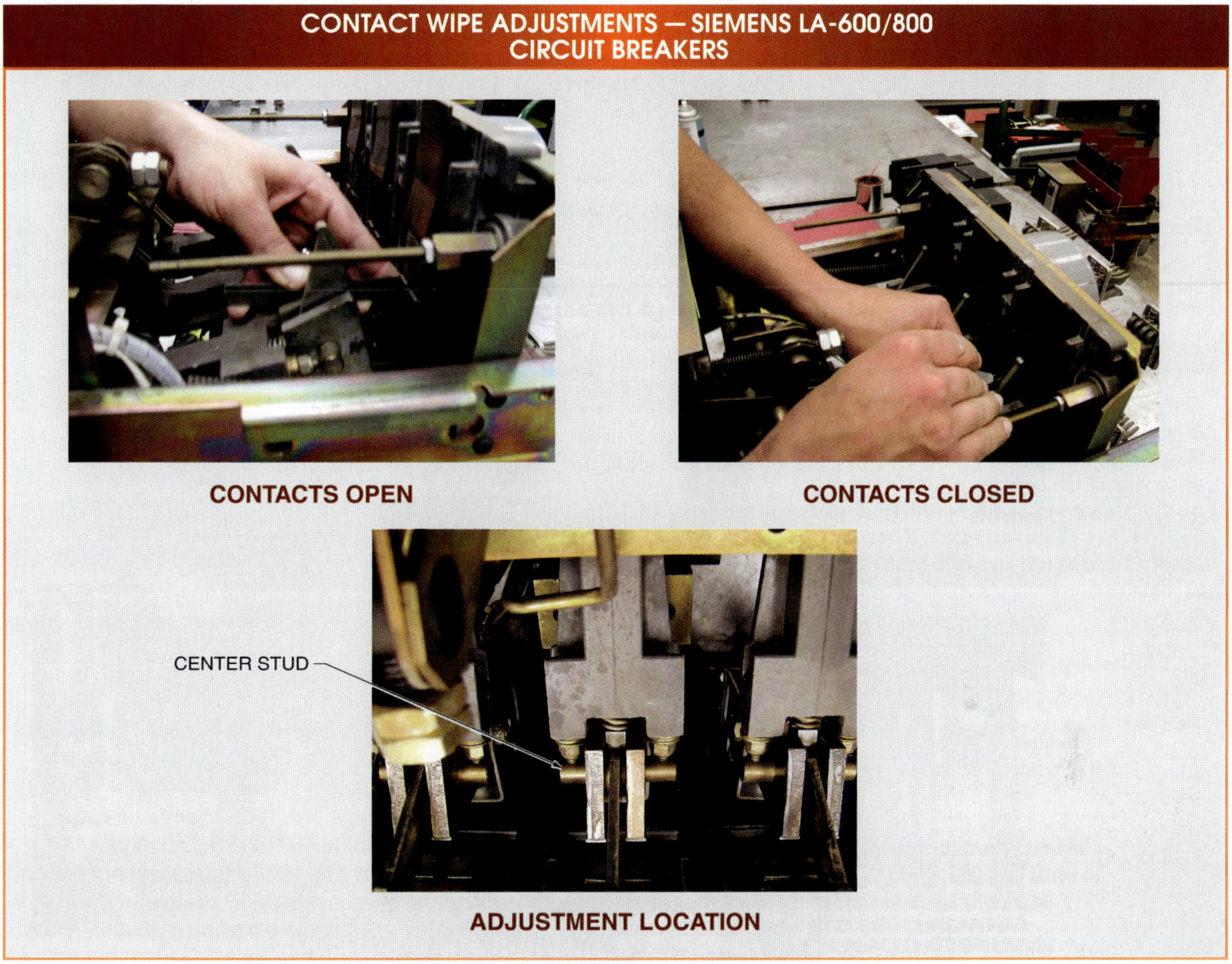

Shermco Industries

Figure 6-9. The contact wipe of a Siemens LA-600/800 circuit breaker is measured with a machinist's rule to obtain an accurate measurement and is measured from the leading edge of the moving contact to the backboard assembly.

Contact Gap Adjustments

Contact gap is the distance between the two closest points of the moving and stationary contacts in a circuit breaker. The OEM instruction manual should always be referenced prior to making any adjustments because these two points can vary with specific circuit breakers, depending on the make and model. **See Figure 6-10.**

Insufficient contact gap can cause a restrike condition during a fault, since minimum air gap clearance is required to extinguish an arc. Contact gap adjustments are performed on medium-voltage circuit breakers, vacuum circuit breakers, and some low-voltage circuit breakers.

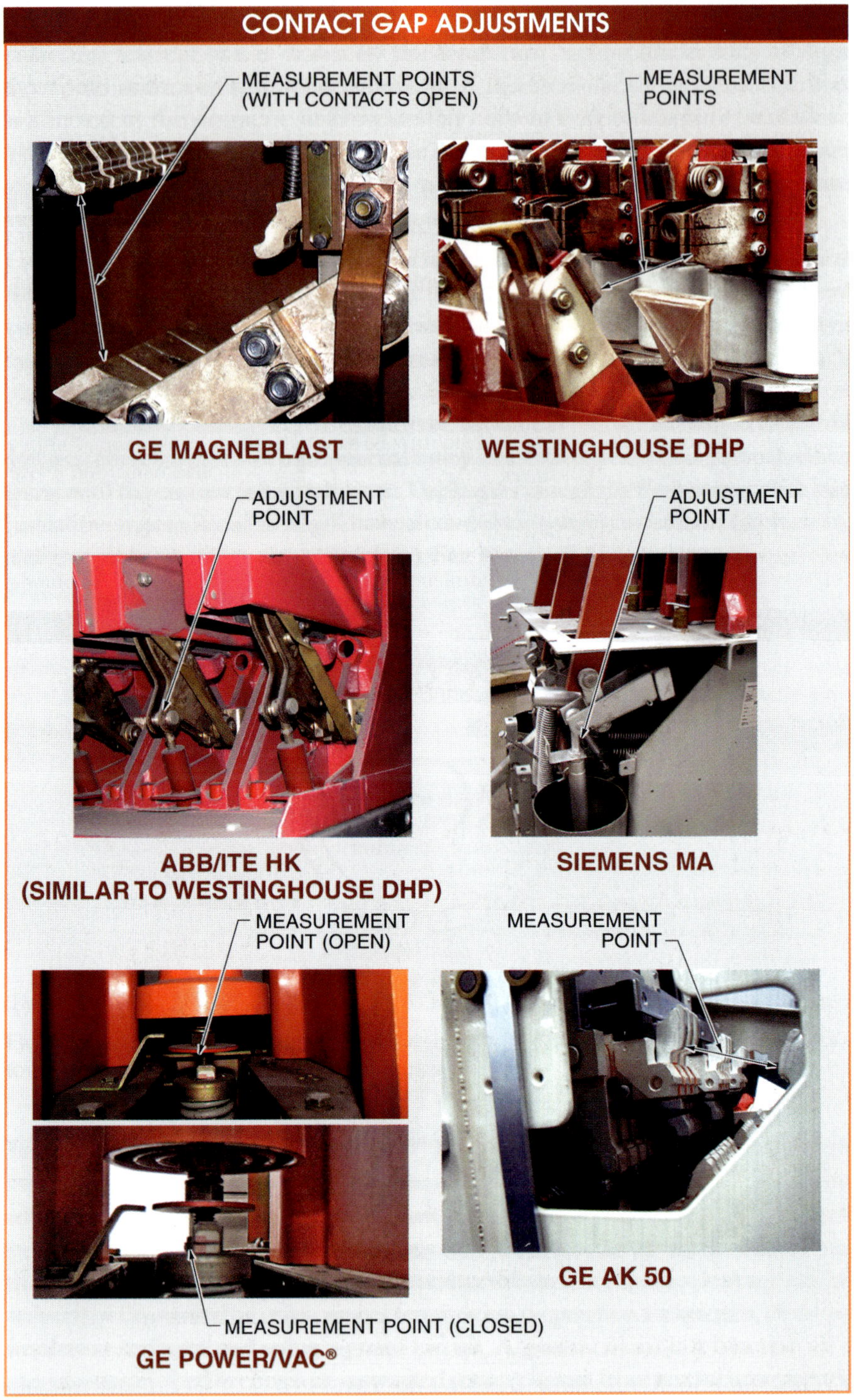

Shermco Industries

Figure 6-10. Each circuit breaker OEM has specific measurement points and adjustment methods. The OEM instruction book (IB) should be consulted for the specific circuit breaker being adjusted.

> **TECH FACT**
>
> *The gap adjustment on low-voltage power or medium-voltage air-magnetic circuit breakers is not usually a maintenance concern unless bent or deformed components are present. Nonetheless, it should be checked with the other adjustments.*

GE MagneBlast. Contact gap is measured on a GE MagneBlast circuit breaker between the rear edge of the moving primary contacts with the lowest part of the stationary primary contact. A GE MagneBlast circuit breaker has a separate adjustment for the contact gap and is adjusted with a spanner wrench. The spanner pins must fit a 5⁄16″ hole and be spaced a distance of approximately 2″ center to center. The locknut is loosened, and the adjusting plate is rotated to increase or decrease the gap. **See Figure 6-11.**

GE MAGNEBLAST CIRCUIT BREAKER CONTACT GAP ADJUSTMENT

Figure 6-11. The GE MagneBlast circuit breaker has a separate adjustment for the contact gap and is adjusted with a spanner wrench.

Westinghouse DHP. A Westinghouse DHP circuit breaker uses a lift rod on each phase to make all three adjustments. To adjust contact gap, the locknut at the top of all three contact adjustments is loosened and the contact is adjusted as needed.

ABB/ITE HK and Siemens MA. The ABB/ITE HK and Siemens MA circuit breakers are similar to the Westinghouse DHP in their construction and adjustments. For a Westinghouse and ABB/ITE circuit breaker, the adjustment is made at the top of the operating rod. For a Siemens circuit breaker, the adjustment is made at the bottom of the operating rod.

Contact gap is usually adjusted at the same point as the contact pressure and sequence. The adjustments are made via the operating rod on each phase. Increasing or decreasing the length of the operating rod changes the gap, pressure, and sequence all at the same time. Once the OEM tolerances for each adjustment cannot be met, the phase contact unit assemblies must be replaced. Contact assemblies should be replaced as a set, not just the stationary or moving contacts.

GE Power/Vac®. The factory procedure for measuring contact gap on a GE Power/Vac® circuit breaker requires a piece of masking tape to be affixed on the vacuum bottle insulator. A mark is made from a reference point, the circuit breaker is then closed, and a new mark is made from the same reference point. The masking tape is removed, and the distance between the two marks is measured with a caliper. A modified (non-factory) version of this procedure is to take a measurement from the top of the erosion indicator tab to the bottom of the insulator with the circuit breaker open, then closing the circuit breaker and retaking the measurement. The difference in the two measurements is the gap. **See Figure 6-12.**

TECH TIP

The contacts on vacuum circuit breakers are not visible, so some OEMs install indicators to ensure adequate gap is maintained.

CHECKING CONTACT GAP — GE POWER/VAC® CIRCUIT BREAKERS

CHECKING CONTACT GAP — GE POWER/VAC® CIRCUIT BREAKERS

GE POWER/VAC® CONTACT GAP WITH CIRCUIT BREAKER OPEN

GE POWER/VAC® CONTACT GAP WITH CIRCUIT BREAKER CLOSED

Shermco Industries

Figure 6-12. Checking contact gap on GE Power/Vac® circuit breakers requires taking a measurement from the deck where the operating rod passes through to the bottom of the insulator with the circuit breaker open then closing the circuit breaker and retaking the measurement.

Low-Voltage Power Circuit Breakers. Contact gap adjustment is usually not as much of a concern on low-voltage circuit breakers as it is on high-voltage, medium-voltage, or vacuum circuit breakers. The arc is not as persistent at low voltages. The GE AK50, AK75, and AK100 circuit breakers have a specific contact gap specification, but most other circuit breakers do not. Contact gap is measured between the two closest points of the moving and stationary arcing contacts. Contact gap is to be between 2⅝″ and 2¾″ and is adjusted by installing or removing shims on the crossbar. **See Figure 6-13.**

CHECKING CONTACT GAP — LOW VOLTAGE POWER CIRCUIT BREAKERS

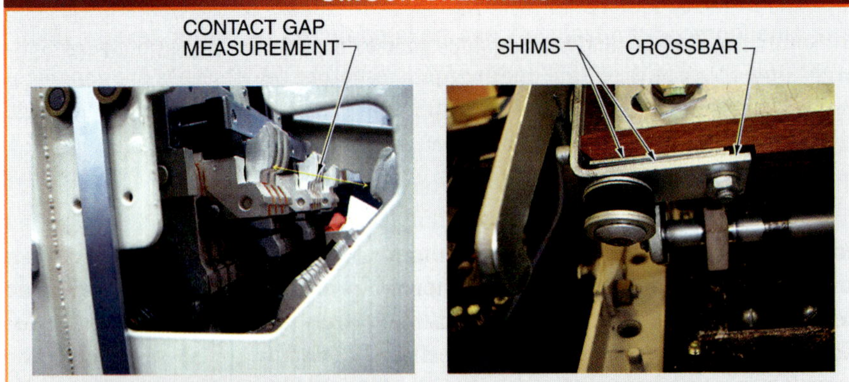

GE AK50 CONTACT STRUCTURE (AK75 AND AK100 SIMILAR)

GE AK50 CROSSBAR AND SHIMS (AK75 AND AK100 SIMILAR)

Shermco Industries

Figure 6-13. Contact gap on the GE AK50/75 and 100 is to be between 2⅝″ and 2¾″and is adjusted by installing or removing shims on the crossbar.

Contact Sequence or Make

Contact sequence is the difference in contact spacing as arcing (moving) contacts make contact with stationary contacts in a circuit breaker. Ideally, the contacts on all three phases should touch at the same time. If the arcing contacts do not make contact at the same time (± the manufacturer's tolerance), then the last one to make contact has an increased rate of erosion due to the extra loading placed on it. As the arcing contacts erode, the main contacts will begin to erode.

Contact sequence is measured by slow-closing the breaker until the first arcing contact just makes contact. The contacts on the other two phases must be within a specified distance, usually about 1⁄32″. The specifications on the other two phases for specific circuit breakers must be reviewed since the actual tolerance and contact sequence adjustment vary between OEMs. **See Figure 6-14.**

CONTACT SEQUENCE AND MAKE SPECIFICATIONS
GE MagneBlast: GE gives no specification for contact sequence for MagneBlast circuit breaker
Westinghouse DHP (Second Series): Westinghouse gives no specification for contact sequence for the type DHP circuit breaker
ABB/ITE HK: ABB states a maximum of 1⁄32″ between phases when the first arcing contact just touches; this is adjusted using the operating (lift) rod
GE AKR-30/50: GE gives no specification for contact sequence for the type AKR-30/50 circuit breaker
GE AK50: Contact sequence is adjusted to be a maximum of 1⁄32″ between contacts on same phase and no more than 1⁄16″ between phases by using crossbar adjusting plates for each phase contact assembly
Westinghouse DS/DSL: Westinghouse gives no specification for contact sequence for type DS/DSL circuit breaker
ABB/ITE K-Line: ABB states a maximum of 1⁄32″ between phases when first arcing contact just touches; this is adjusted using connecting link stud
Siemens LA: Siemens states maximum of 0.062″ between phases when first arcing contact just touches; this is adjusted using connecting link stud

Figure 6-14. The specifications for specific circuit breakers must be reviewed because the actual tolerances and contact sequence adjustments vary.

TECH FACT

At the Shermco Industries circuit breaker shop, the wipe is adjusted first, followed by the sequence, and then the gap is adjusted. Usually, one adjustment or another will be near the limits of the OEM-specified tolerances but still be within the tolerance range.

The final contact adjustment must be a balance between wipe, sequence, and gap. A properly adjusted circuit breaker must have all three adjustments fall within the OEM-recommended tolerances. All three adjustments are typically made at the operating link on low-voltage power circuit breakers or the operating rod on medium-voltage circuit breakers.

OPERATING MECHANISM ADJUSTMENTS

The operating mechanism usually does not require adjustment unless it is disassembled or has significant wear. It can, however, have lubrication issues over time due to the heat generated by the circuit breaker as current flows through it. When these situations occur, significantly more work in addition to the normal adjustments is required to ensure that the circuit breaker is functioning properly. Before beginning mechanism adjustments, the circuit breaker must be visually inspected for proper lubrication, corrosion, and wear on latch surfaces, bearings, and pivot points. Problems in these areas cause issues in circuit breaker adjustment. It is not recommended to perform these adjustments on a normally-operating circuit breaker, but clearances can be checked.

Each OEM requires a different procedure for adjusting its circuit breaker operating mechanisms. The primary mechanism adjustment is the trip latch (latch bite) adjustment. The trip latch adjustment aligns the prop and roller in the operating mechanism to allow the circuit breaker to latch when it is closed. When this adjustment is performed incorrectly, the circuit breaker can trip to the open position after a close is initiated (trip free).

GE MagneBlast Adjustments

The GE MagneBlast circuit breaker is different from other makes of circuit breakers in that it has a total of 13 separate adjustments, including driving pawl and contact adjustments. Other makes typically require about seven adjustments.

Driving Pawl Adjustment. A *driving pawl* is a circuit breaker component that advances the drive gear to charge closing springs. A *latching pawl* is a circuit breaker component that locks the drive gear in place when charging the closing springs. When adjusting a driving pawl, there are three different clearances to measure: between the top of the latching pawl and latching pawl stop block, between the driving pawl and drive gear tooth, and between the drive gear tooth and latching pawl. With the circuit breaker open and the closing springs discharged, the driving pawl is adjusted using the following procedure:

1. Manually charge the springs using a ½″ drive ratchet by rotating the eccentric counterclockwise until the outer latching pawl is on top of one of the drive gear teeth. Use a feeler gauge to check the clearance between the top of the latching pawl and latching pawl stop block. Clearance must be ≥ 0.047″.
2. Continue to rotate the manual charging handle counterclockwise until the closing springs are approximately 50% charged and the driving pawl has traveled through its return stroke. The driving eccentric must be at the 9 o'clock position and the clearance between the driving pawl and drive gear tooth must be ≥ 0.015″.
3. Rotate the manual charging handle until the driving pawl has advanced to its maximum travel. The driving eccentric must be at the 3 o'clock position and the clearance between the drive gear tooth and latching pawl must be ≥ 0.015″.

All three clearances are adjusted by moving the spring charging motor. The closing springs should be completely charged, and the spring blocking device is installed. The seven bolts securing the motor must be loosened. The spring closing motor must be moved straight forward or rearward. The motor is moved approximately twice the distance required for the adjustment. **See Figure 6-15.** When moving the motor, apply the following guidelines:

- If the latching pawl clearance is less than the minimum specified, the motor should be moved rearward.
- If the driving pawl clearance is less than the minimum specified, the motor should be moved forward.
- One bolt should be tightened on the right side of the mounting frame to ensure proper alignment.
- The remaining bolts should be tightened to the proper torque value.
- Since the circuit breaker has not been closed, the closing spring blocking tool can be removed.
- The clearances must be rechecked once the bolts are tightened.

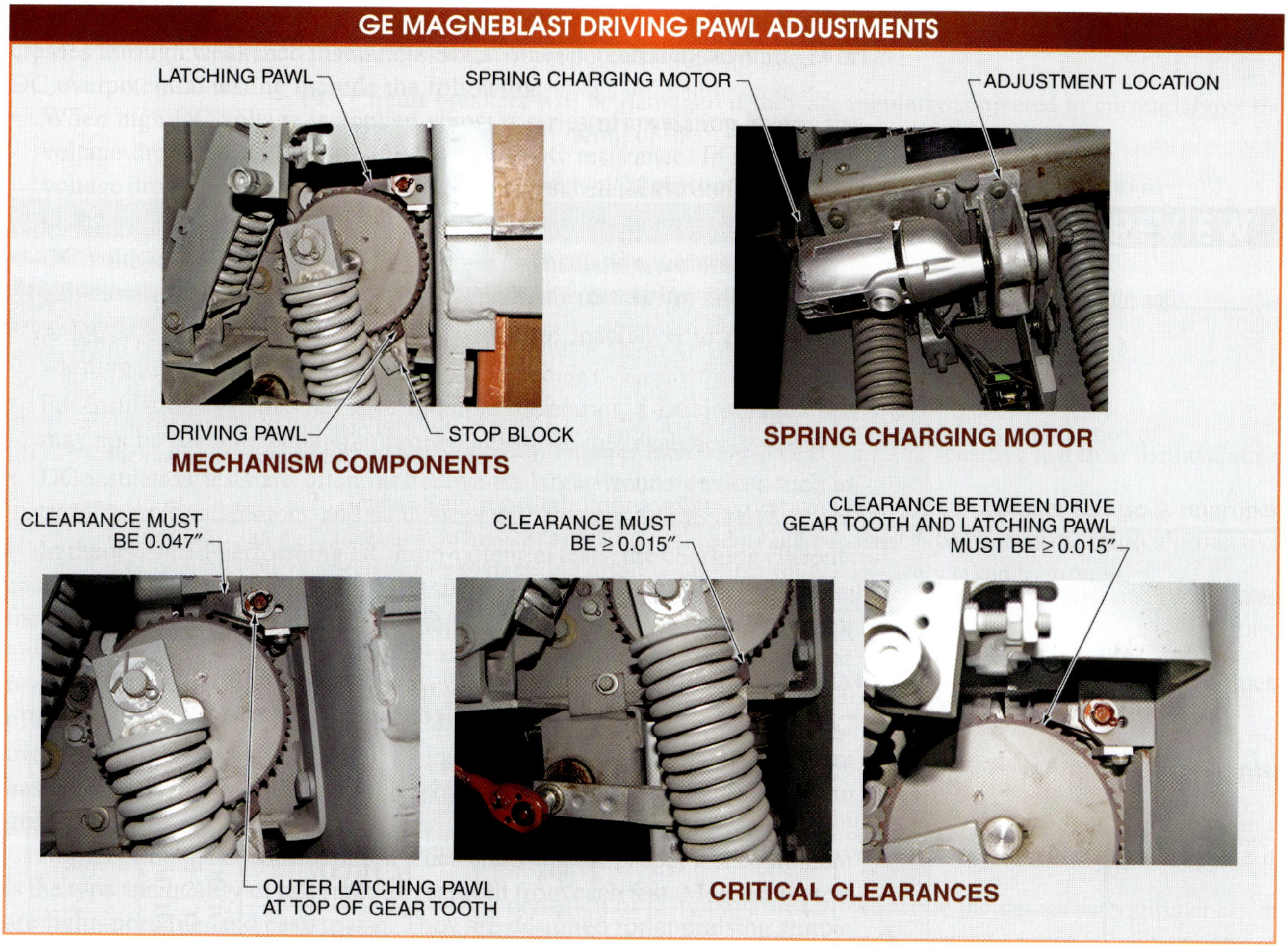

Figure 6-15. The driving and latching pawl adjustments ensure that the circuit breaker's closing springs will charge correctly.

Other Adjustments. There are several checks that can be made on various control circuit switches. These vary by the model number of the circuit breaker and should be performed during major maintenance. Most of these adjustments are made by bending the switch arms to obtain the necessary clearances. The most common types of switch adjustments include the following:

- trip armature travel: The trip armature should have a travel of 1⁄16″ to 5⁄16″ before the trip latch begins to move. It can be adjusted by moving the trip coil support and/or adjusting the trip arm adjusting screw. The locking screw can be loosened by moving the trip armature to the desired position, and retightening the set screw.
- closing latch wipe: There must be 3⁄16″to 1⁄4″ wipe between the closing latch and the roller. The roller is located on the back side of the switch cam and is not visible. It can be adjusted by loosening the set screw, moving the closing latch wipe to the desired position, and retightening the set screw.
- close latch monitoring switch: The closing latch must be fully reset and the close latch monitoring switch must be operated before the spring charging motor can operate. With the closing latch fully reset, there must be 1⁄32″ or less clearance between the switch striker arm and the switch mounting bracket. It is adjusted by bending the striker arm.

SAFETY FACT

Use long-handled tools when performing mechanism adjustments because fingers could get pinched between the contacts.

- latch check switch: An ohmmeter or light circuit should be connected to the latch check switch contacts. The closing springs are manually charged until the mechanism linkage resets. The trip latch must be rotated by pressing the manual trip lever. The trip latch will reset slowly. The contacts should make when the gap between the trip latch and the stop pin on the latch roller link is at 1⁄16″. There must be a minimum clearance of 1⁄64″ between the switch arm and the switch support. To adjust, the latch checking switch arm must be bent.
- positive interlock switch wipe: The positive interlock switch wipe prevents the circuit breaker from operating during raising or lowering in the cubicle. The switch rides on a side rail and opens the control circuit in the circuit breaker unless the circuit breaker is fully up or fully down. When the positive interlock switch is in its normal (reset or made) position, the clearance between the positive interlock switch arm and the switch mounting plate should be 1⁄32″ or less. Adjustment is made by bending the switch arm. **See Figure 6-16.**

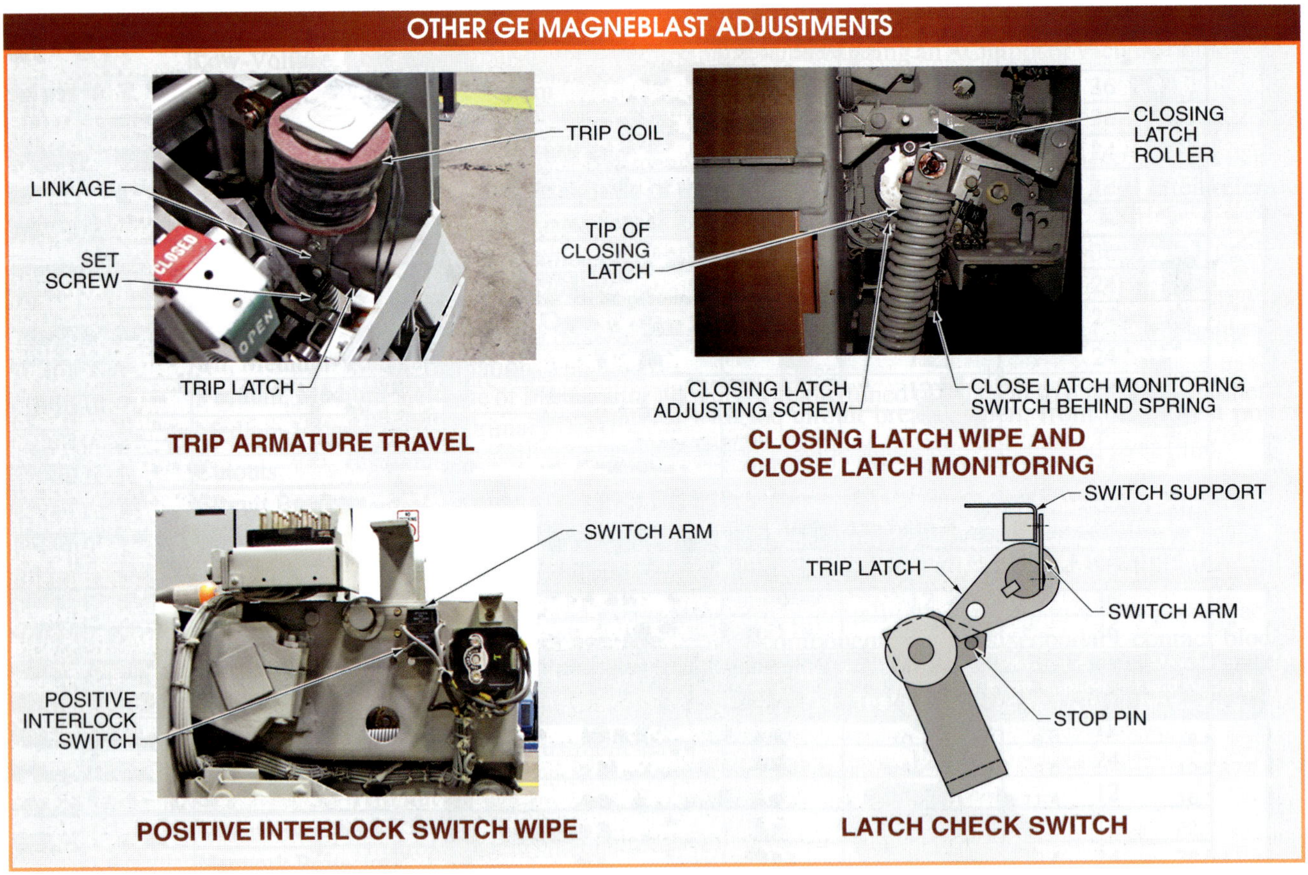

Figure 6-16. Other adjustments typically performed on GE MagneBlast circuit breakers include trip armature travel, closing latch wipe, and positive interlock switch wipe.

Westinghouse DHP Adjustments

The Westinghouse DHP circuit breaker has a similar appearance to that of the Siemens MA/MC and ABB/ITE HK but has its own unique adjustments. The most critical adjustments are made on the trip latch, anticlose interlock, and floor tripper.

Trip Latch Clearance Adjustment. A trip latch is used to hold the mechanism in a closed position. When the trip button is depressed, it releases the trip latch, which allows the contacts to open. **See Figure 6-17.** Trip latch clearance is adjusted using the following procedure:

1. Remove the cover to the trip coil and charge the closing springs.
2. Rotate the trip cam adjusting screw clockwise until the trip latch roller makes contact with the surface of the tripping trigger.
3. Manually raise the trip trigger about 1″. The tripping trigger must remain in the raised position.
4. Slowly rotate the trip cam adjusting screw counterclockwise until the trip latch roller and tripping latch just reset.

Note: There must be a small clearance between the trip latch roller and the tripping trigger when properly adjusted.

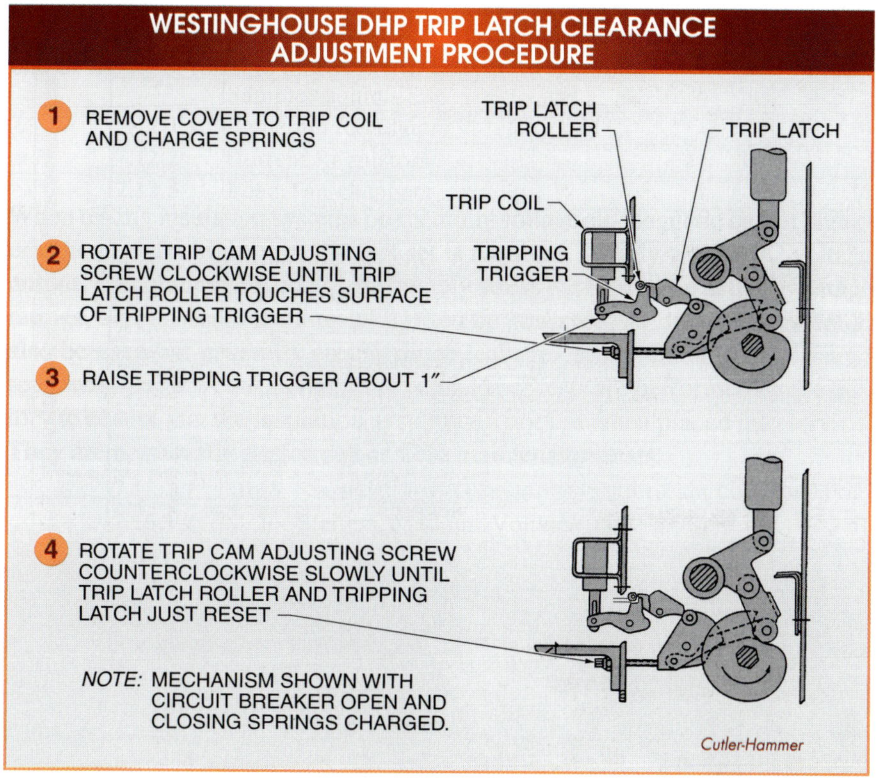

Figure 6-17. A trip latch is used to hold the mechanism in a closed position and is adjusted on Westinghouse DHP circuit breakers after removing the front cover and charging the closing springs.

Anticlose Interlock Adjustment. An anticlose interlock is used to prevent the circuit breaker from operating while it is being racked in or out. **See Figure 6-18.** Interlocks are safety features that prevent a worker from performing an action that could cause damage to the circuit breaker or personnel. The anticlose interlock is adjusted by using the following procedure:

> **CAUTION**
>
> *Closing springs are charged, and serious injury could occur if hands or fingers are in the mechanism.*

1. Remove the interphase barrier support pan (metal plate located on top front of circuit breaker) and the closing release coil cover.
2. Remove the anticlose interlock screw and clean it of all thread locking compound. *Note:* A propane torch is required to heat and soften the thread locking compound applied to the anticlose interlock screw threads.
3. Charge the closing springs and close the circuit breaker. Recharge the closing springs after closing the circuit breaker.
4. Apply new thread locking compound to the anticlose interlock screw. With the circuit breaker closed and the springs charged, the circuit breaker can be adjusted. (The anticlose interlock screw passes through the operating rod).
5. Slowly tighten the anticlose interlock screw until a clearance of 0.10″ to 0.30″ is obtained between the closing trigger and the closing latch roller.
6. Cycle the circuit breaker (open and closed) several times to ensure proper adjusment.

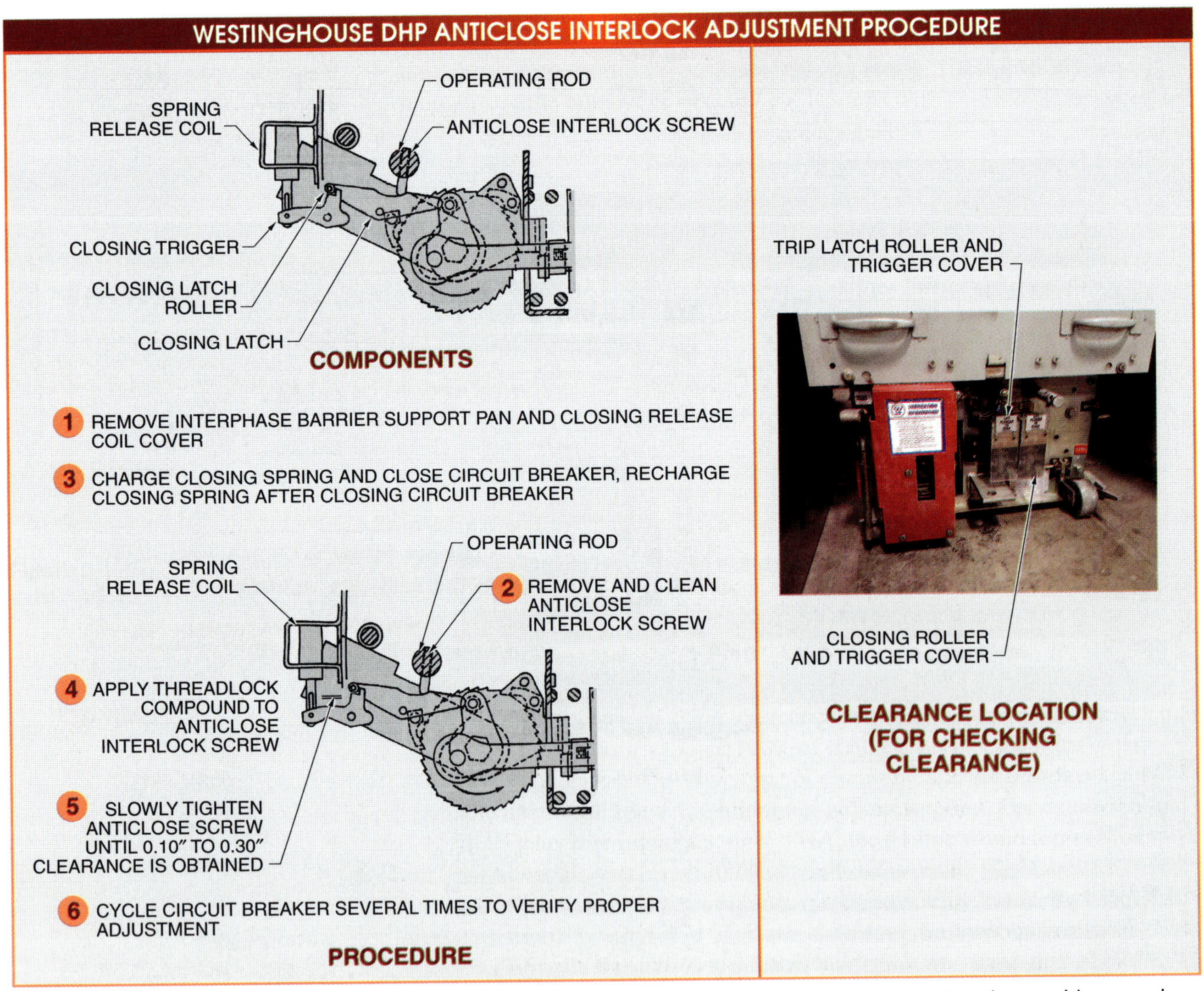

Figure 6-18. Anticlose interlocks are safety features that prevent workers from performing an action that could cause damage to the circuit breaker or personnel.

CASE STUDY — INTERLOCK FAILURE

An incident occurred when a technician racked a 4.16 kV circuit breaker that was in the closed position. As soon as the primary disconnects contacted the energized bus, the circuit breaker exploded. The technician spent one week in the hospital, but survived his injuries. He was wearing a 40 cal/cm^2 arc flash suit at the time. The actual incident energy is not known.

DESTROYED SWITCHGEAR

DESTROYED ARC FLASH SUIT AND HOOD

Conclusion: Arc flash protective clothing and PPE are not worn when something may happen. They are worn because there is no knowledge of when it may happen. Even though this technician suffered serious burns, he survived and was not disabled. Circuit breakers have interlocks to prevent this type of event from happening, but these interlocks are fragile. Lack of lubrication, inadequate maintenance, and improper adjustment can cause them to fail. For example, in the Shermco circuit breaker repair facility, about 40% of circuit breaker interlocks are not functioning when they are received for servicing

ABB Inc.

Floor Trippers. A *floor tripper* is an interlock that protrudes from the bottom of the circuit breaker and causes the circuit breaker to trip open and the closing springs to discharge when a circuit breaker is racked in or out of its cubicle. **See Figure 6-19.** *Note:* In order to perform this adjustment, a special hand tool (locally manufactured) must be used. A floor tripper is adjusted by using the following procedure:

1. With the circuit breaker open, charge the closing springs. Insert the hand tool under the release lever to ½″ level. The circuit breaker must not close at this point.

2. Continue to slide the hand tool under the release lever to ⅝″ level until the circuit breaker closes. *Note:* Both levers are located on right side of the circuit breaker, just left of the wheel when viewed from the front of the circuit breaker.
3. With the circuit breaker in the closed position, slide the hand tool under the tripping lever to ½″ level. The circuit breaker must not trip open.
4. Continue to slide the hand tool under the lever to ⅝″ level. The circuit breaker should now trip open. If the circuit breaker opens or closes at ½″ level, or does not open or close at ⅝″ level, it requires adjustment.

FLOOR TRIPPERS

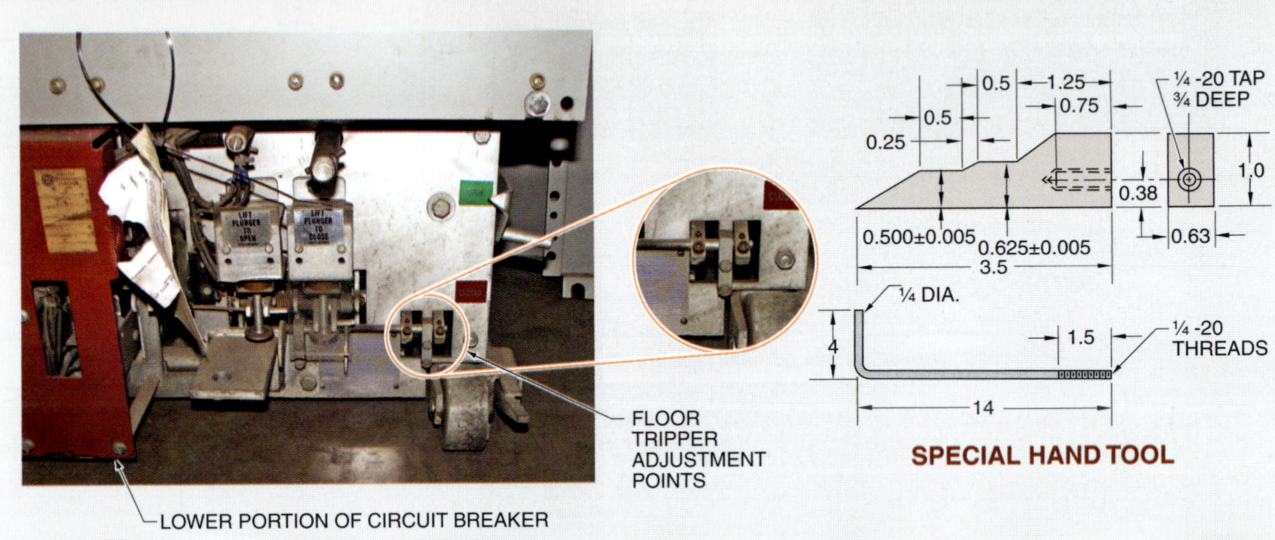

Figure 6-19. A floor tripper is located on the bottom of a circuit breaker and causes the circuit breaker to trip open and the closing springs to discharge when a circuit breaker is racked in or out of its cubicle.

ABB/ITE HK Adjustments

The most important adjustments for ABB/ITE HK circuit breakers are the latch check switch and control relay adjustments. **See Figure 6-20.**

Latch Check Switch Adjustment. If the circuit breaker does not close electrically, the latch check switch may be out of adjustment. To adjust the latch check switch, use the following procedure:

1. Connect an ohmmeter or light circuit to the contacts of the latch check switch.
2. Rotate the adjusting screw clockwise (in) until the contacts start to open. *Note:* There is often an audible "click" when the contacts change state.
3. Rotate the adjusting screw counterclockwise (out) until the contacts make.
4. Rotate the adjusting screw out one additional turn.

Control Relay Adjustment. The control relay adjustment is set by installing or removing shims between the control relay and control pan. When properly adjusted, there should be 0.06″ to 0.09″ clearance between the control device lever and the limit switch crank.

ABB/ITE HK CIRCUIT BREAKER ADJUSTMENTS

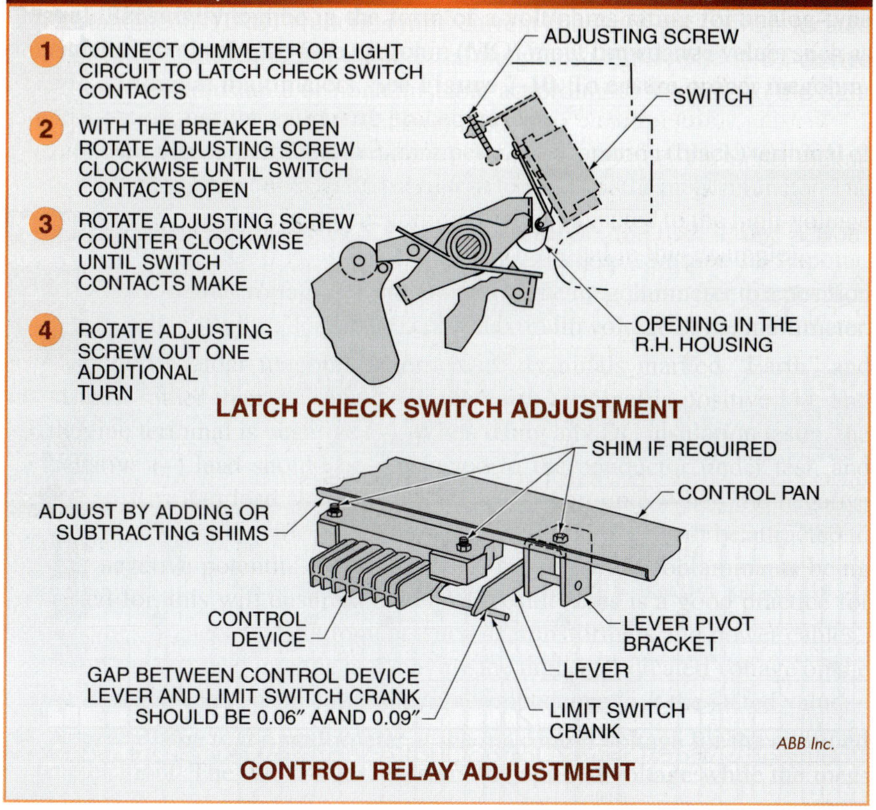

Figure 6-20. The most critical adjustments on ABB/ITE HK circuit breakers are made on the latch check switch and control relay.

GE Power/Vac® Circuit Breaker Adjustments

All mechanism adjustments on the GE Power/Vac® circuit breaker are made from the bottom. Care must be exercised to ensure that the circuit breaker is on a stable surface where the mechanism can be accessed. The best method is to use a lift truck to insert and remove the circuit breaker from its cubicle. The GE Power/Vac® circuit breaker has three important mechanism adjustments: the close coil plunger gap, trip coil plunger gap, and close latch stop bolt. **See Figure 6-21.**

Close Coil Plunger Gap. To perform the close coil plunger gap adjustment, the circuit breaker must have the closing springs charged and the gag plate installed. The close coil plunger gap is adjusted by lifting the lock tab from the adjusting nut and pressing in on the close button until resistance is felt. The gap between the close plunger button and the coil housing must be 0.25″ to 0.30″ for the ML-18-1 and ML-18-2 mechanisms and 0.35″ to 0.40″ for the ML-18-3 mechanism.

Trip Coil Plunger Gap. To perform this adjustment, the circuit breaker must have the closing springs discharged and the circuit breaker closed. The trip button should be pressed until resistance is felt. The gap between the trip button and the trip coil housing should be 0.20″ to 0.25″. Circuit breakers with the optional second trip coil installed are adjusted in the same manner.

Close Latch Stop Bolt. To perform this adjustment, the circuit breaker must have the closing springs charged and the circuit breaker open. The ¼″ × 20 adjusting bolt should be rotated clockwise until the closing spring discharges. The adjusting bolt should be rotated counterclockwise ½ turn (about three flats of the bolt head). The closing springs should be recharged. If the closing springs do not charge, the bolt head can be rotated counterclockwise an additional one to two flats.

GE POWER/VAC® CIRCUIT BREAKER ADJUSTMENTS

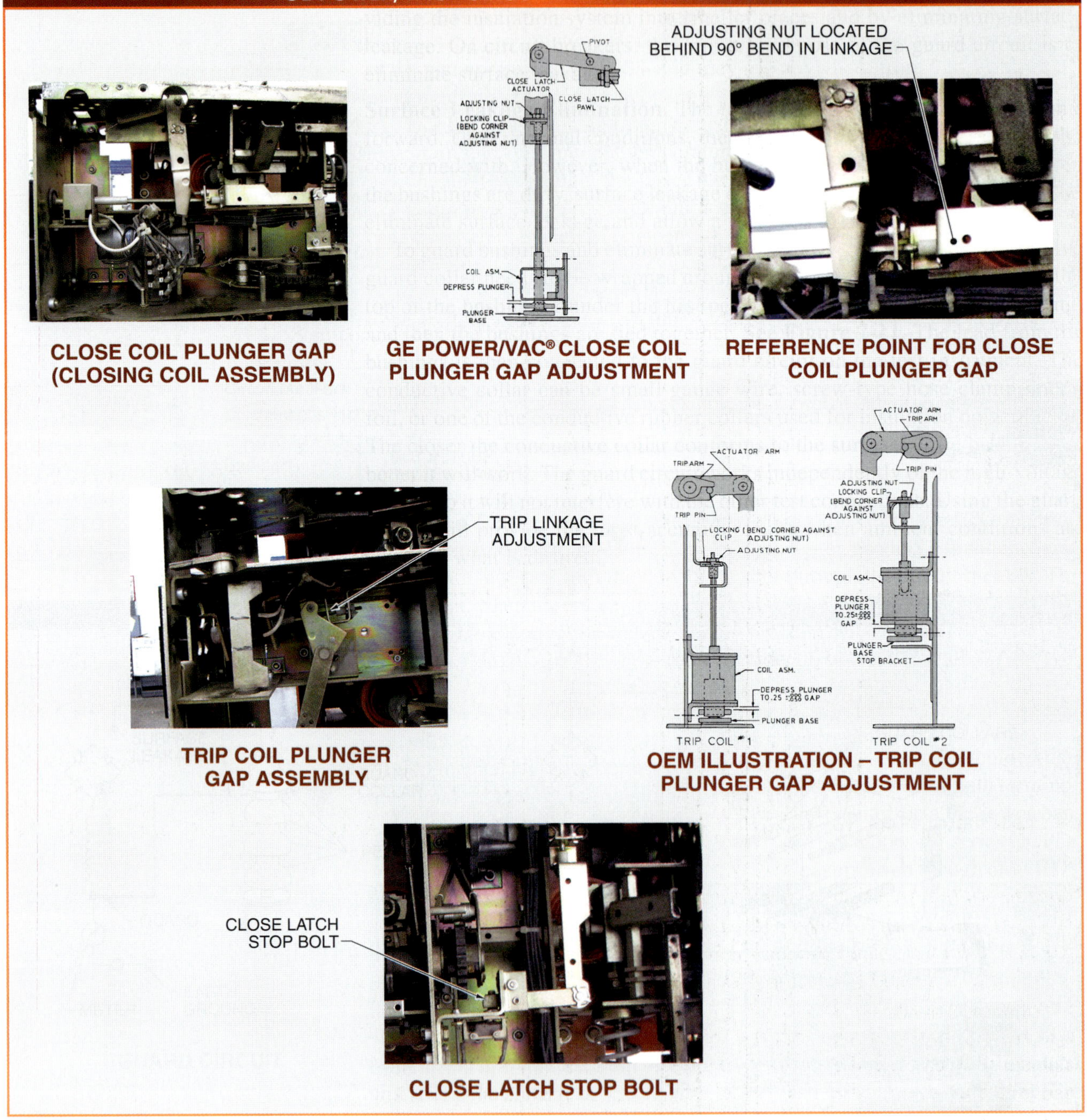

Shermco Industries

Figure 6-21. The most critical adjustments made on GE Power/Vac® circuit breakers are the close coil plunger gap, trip coil plunger gap, and close latch stop bolt.

GE AKR 30/50 Adjustments

There have been several designs of the GE-type AK circuit breaker, but the basics have remained fairly stable. Circuit breakers ending with the letter "U" in the model number have current-limiting fuses. The AKR/AKD/WavePro low-voltage power circuit breaker appeared in the 1970s but began to sell in significant numbers in the early 1980s. There was some overlap of the older AK and the AKR during this time. The main types of adjustments performed on these circuit breakers are with the buffer assembly and the trip latch. **See Figure 6-22.**

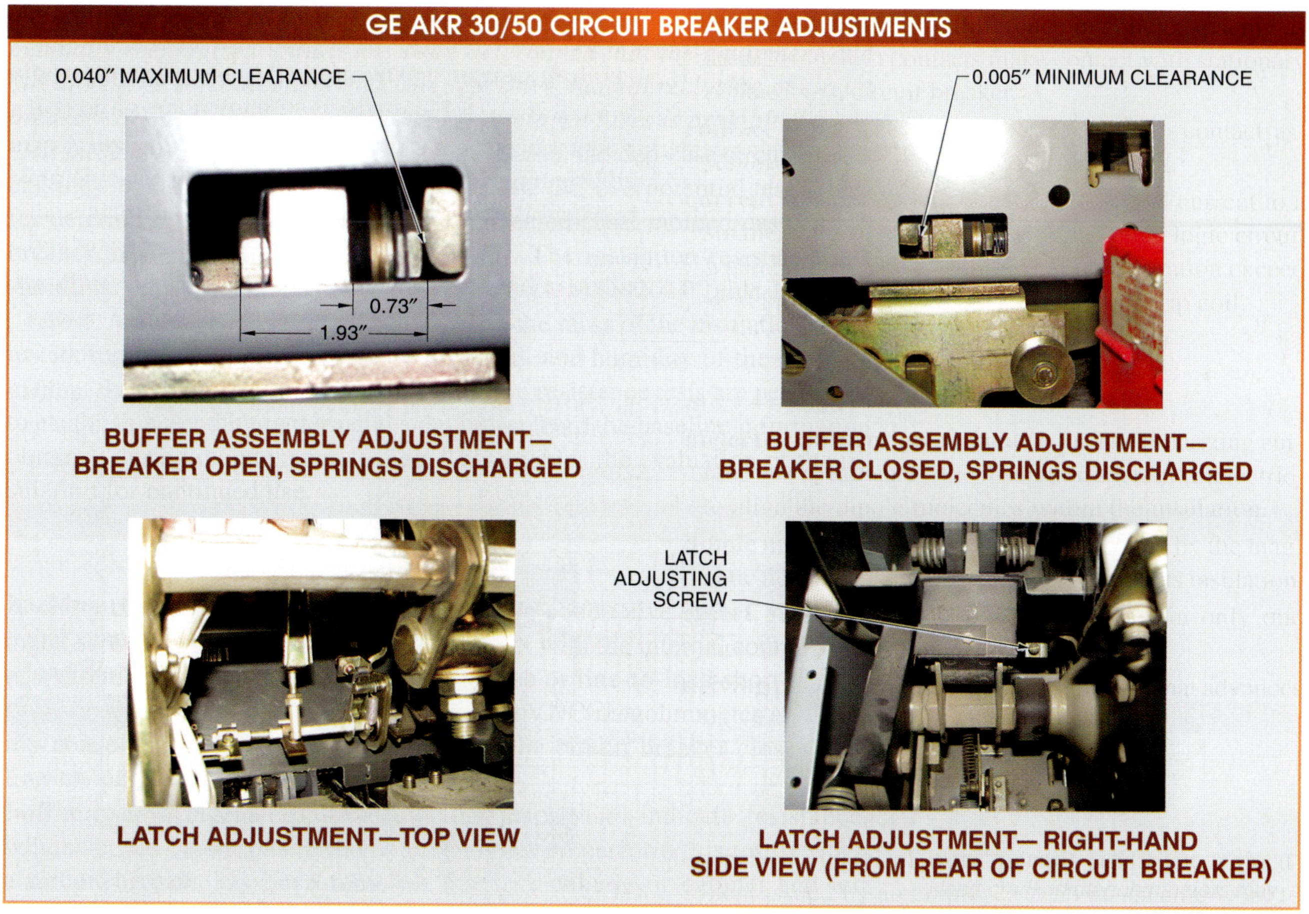

Shermco Industries

Figure 6-22. The primary adjustments performed on GE AKR 30/50 circuit breakers are with the buffer assembly and the trip latch.

Buffer Assembly Adjustment. A *buffer assembly* is a mechanical component that absorbs the shock of a mechanism when a circuit breaker opens and prevents the contact assemblies from overdriving when a circuit breaker closes. With the circuit breaker open and the closing springs discharged, there must be 0.040″ of maximum clearance between either of the end plates. When the circuit breaker is closed and the closing springs are discharged, there must be a 0.005″ minimum clearance between the end plate and buffer nut.

Both clearances are adjusted by tightening or loosening the buffer nut. Care must be taken not to overtighten the neoprene washers because they absorb the shock of the circuit breaker opening and can easily fail if overtightened. The overall dimensions of the clearances are adjusted by adding or removing spacers.

Trip Latch Adjustment. The trip latch is adjusted if the circuit breaker does not close or trip properly. The adjustment can only be checked by performing the adjustment procedure. To adjust the trip latch, use the following procedure:

1. Close the circuit breaker.
2. Use a long-shanked screwdriver and slowly rotate the adjusting screw clockwise, as viewed from the top of the circuit breaker.
3. Continue to rotate the screw until the circuit breaker trips (about 3½ turns).
4. Rotate the adjusting screw counterclockwise 3½ turns.

GE AK50 Adjustments

When performing adjustments on a GE AK50 circuit breaker, the circuit breaker must be in the open position and the closing springs charged to perform trip latch and trip latch centering adjustments. **CAUTION:** Since the closing springs are charged, use long-handled or long-shanked hand tools and keep hands clear of fast-acting components when working on the operating mechanism.

Trip Latches. The trip latch is the primary mechanism adjustment on a GE AK50 circuit breaker. The clearance between the trip latch and the roller is to be between 1/64″ to 1/32″ with the circuit breaker open and the closing springs charged. The clearance is adjusted by rotating the adjusting screw.

Trip Latch Centering. The trip latch must be centered on the roller. Centering is accomplished by using the latch buffer stop, which is located on the right side of the center frame section, adjacent to the trip latch and roller. **See Figure 6-23.**

GE AK50 CIRCUIT BREAKER ADJUSTMENTS

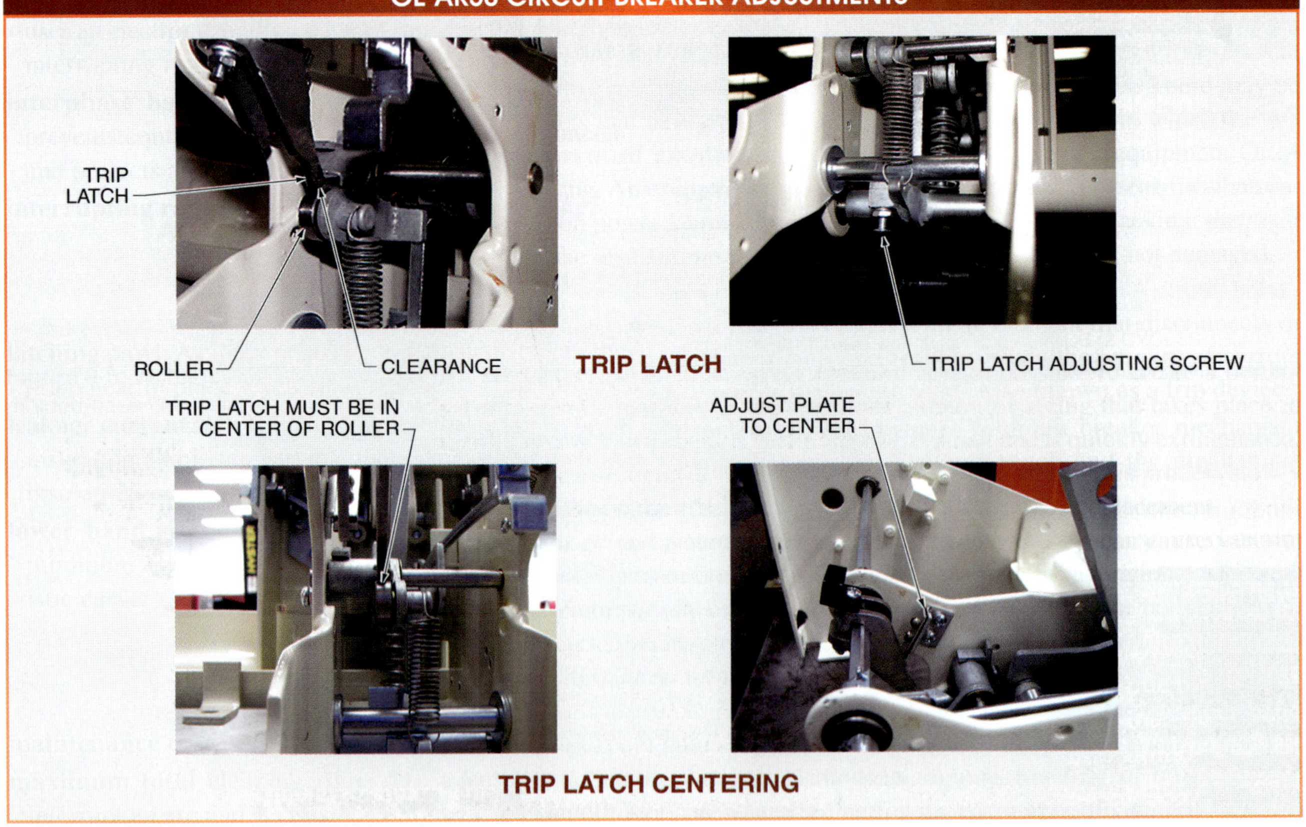

Figure 6-23. The main adjustments performed on GE AK 50 circuit breakers are the trip latch and trip latch centering.

Cutler-Hammer/Westinghouse DS Adjustments

The most important adjustments are the trip latch overlap, trip actuator, and circuit breaker open position stop. **See Figure 6-24.**

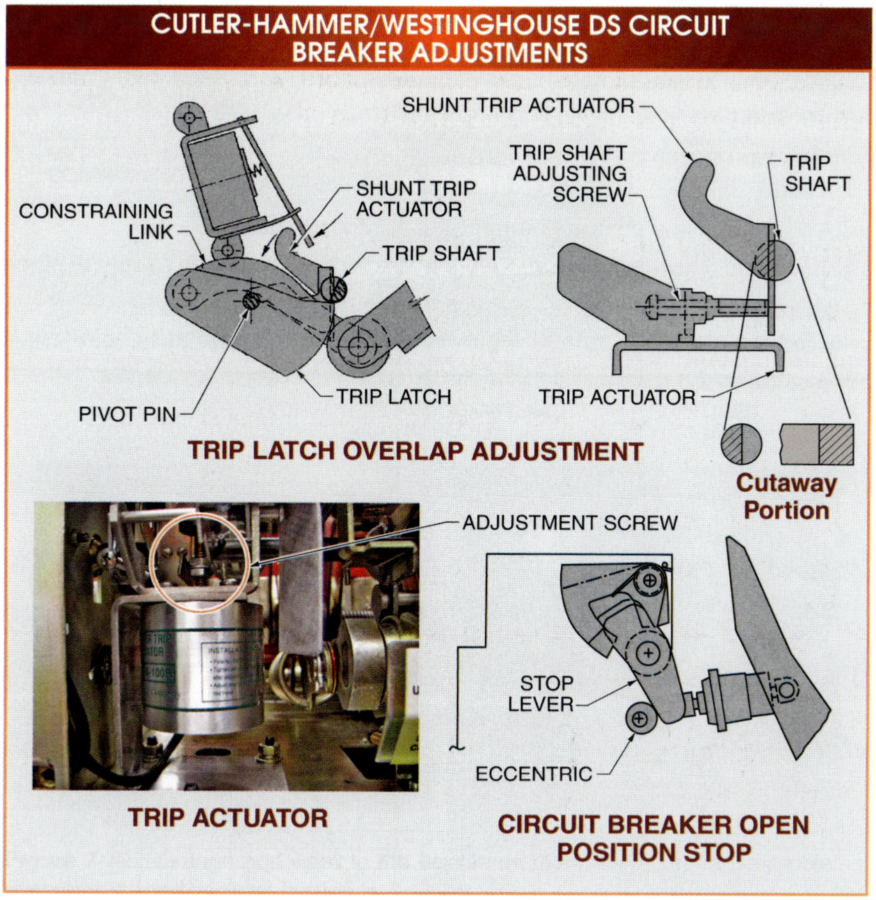

Figure 6-24. The most critical adjustments on Westinghouse/Cutler-Hammer DS circuit breakers are made on the trip latch overlap, trip actuator, and circuit breaker open position stop.

Trip Latch Overlap. The trip latch overlap adjustment sets the trip shaft in the proper relationship with the trip latch in order for the circuit breaker to trip properly. On a DS-style circuit breaker, the trip shaft rotates counterclockwise when the trip button is depressed, which allows the trip latch to release, tripping the circuit breaker open. To adjust a trip latch overlap, use the following procedure:

1. Close the circuit breaker.
2. Slowly rotate the adjusting screw clockwise until the circuit breaker trips.
3. Rotate the adjusting screw counterclockwise 3½ to 4 revolutions.

Trip Actuator. The trip actuator adjustment is similar for all types of circuit breakers, regardless of the OEM. To adjust the trip actuator, use the following procedure:

1. Rotate the adjustment screw (red circle) several revolutions counterclockwise.
2. Close the circuit breaker.

3. Rotate the adjustment screw clockwise until circuit breaker trips open.
4. Rotate the adjustment screw clockwise the specified number of turns. The AC-PRO™ circuit breaker retrofit kit requires four turns counterclockwise after the circuit breaker trips.

Breaker Open Position Stop. This adjustment is listed for the DS 206, DS 416, and DS 532 in earlier manuals, but it is limited to the DS 632 in later manuals. The circuit breaker open position stop adjustment ensures a positive reset for the mechanism when the circuit breaker opens. To adjust, use the following procedure:

1. Ensure that the circuit breaker is in the open position.
2. Loosen the stop position bolt nuts so eccentrics can be rotated.
3. Eccentrics must be loose enough to rotate but snug enough to stay in place once rotated.
4. Obtain a clearance of 0.005″ between the eccentrics and the stop levers.
5. Retighten the nuts.

Shermco Industries

Circuit breakers are used with medium-voltage metal-clad switchgear installations.

ABB/ITE K-Line Adjustments

The most important adjustments are on the trip latch and bar engagement, control relay, shunt trip device, and magnetic latch device. **See Figure 6-25.**

Trip Latch Engagement. The trip latch engagement adjustment sets the prop to the center of the roller. Improper adjustment results in a trip free condition (too little engagement) or a circuit breaker that does not trip (too much engagement). The trip latch engagement is adjusted by using the following procedure:

1. Rotate the adjustment screw counterclockwise several revolutions.
2. Close the circuit breaker.
3. Rotate the adjustment screw clockwise until circuit breaker trips open.
4. Rotate the adjustment screw counterclockwise two revolutions.

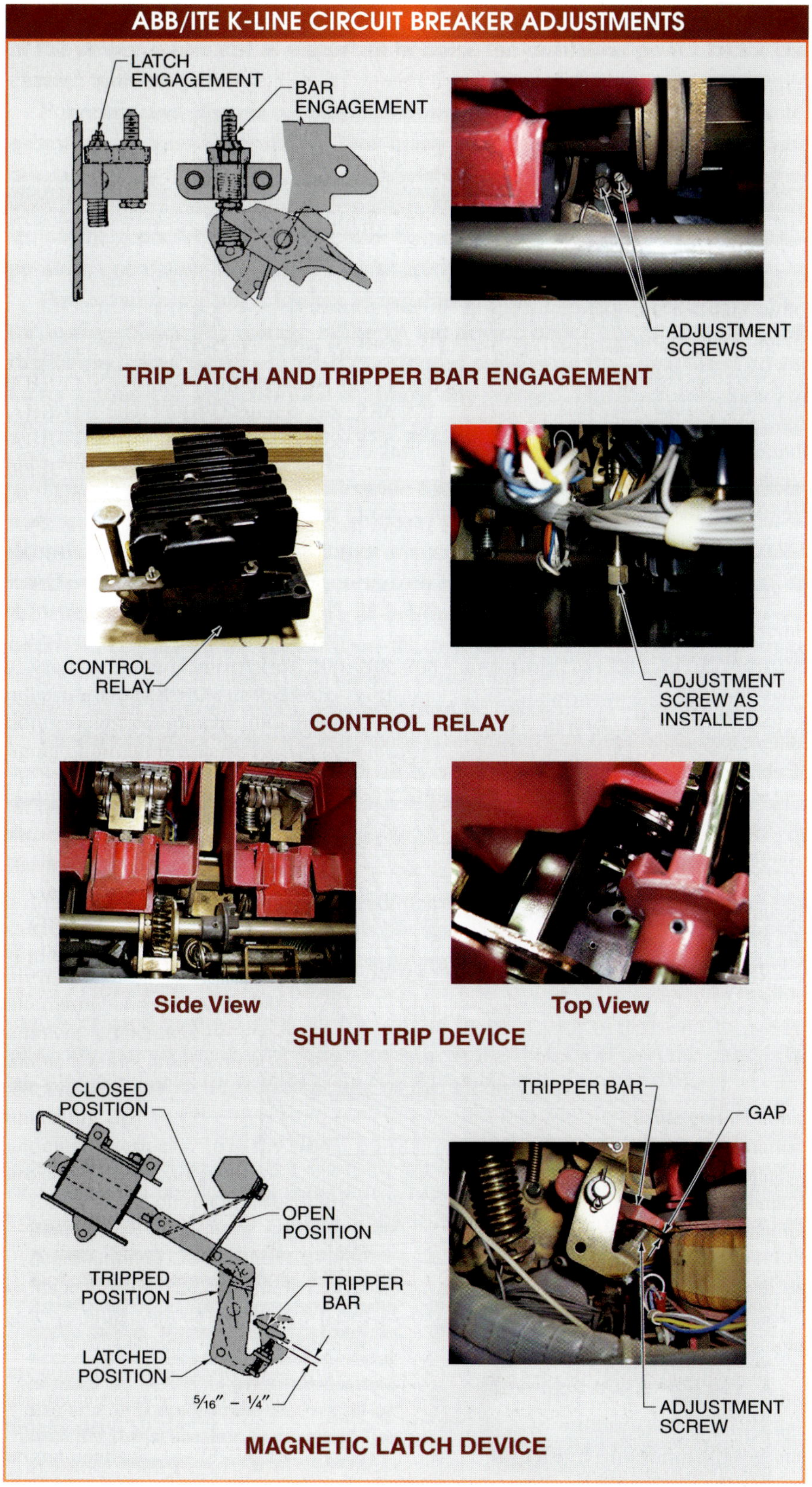

Figure 6-25. The most important adjustments for ABB/ITE K-line circuit breakers are the tripper latch and bar engagements, control relay, shunt trip device, and magnetic latch device.

Tripper Bar Engagement. The tripper bar adjustment ensures that a circuit breaker has a positive tripping action once the trip actuator is energized. The tripper bar engagement is adjusted by using the following procedure:

1. Rotate the adjustment screw counterclockwise several revolutions.
2. Close the circuit breaker.
3. Rotate the adjustment screw clockwise until the circuit breaker trips open.
4. Rotate the adjustment screw counterclockwise 3½ revolutions.

Control Relay Adjustment. The control relay houses the limit switch (LS), which cycles the spring charging motor, contains the Y relay, which actuates the antipump circuit, and contains the latch-release relay (X), which operates the closing release rod. The control relay actuator rod is adjusted by using the following procedure:

1. Rotate the adjustment screw counterclockwise several revolutions.
2. Charge the closing springs.
3. Manually push the adjustment screw up as far as it will go.
4. While holding the adjustment screw in the up position, rotate the adjustment screw clockwise until the circuit breaker closes.
5. Rotate the adjustment screw clockwise 1½ revolutions.

Shunt Trip Device Adjustment. The shunt trip device electrically opens the circuit breaker. This adjustment is usually only made if the shunt trip device has been replaced, or if it is malfunctioning. The shunt trip device is adjusted by using the following procedure:

1. Rotate the adjustment screw counterclockwise as far as it can be rotated (without forcing it).
2. Close the circuit breaker.
3. Push the lever connected to the armature of the shunt trip device (located on the bottom of shunt trip device) as far as it will go.
4. Rotate the adjustment screw clockwise until the circuit breaker trips open.
5. Rotate the adjustment screw an additional three revolutions clockwise on all circuit breakers without an electronic OCPD.
6. On small-frame circuit breakers (600 A and 800 A frames) using electronic OCPDs, rotate the adjustment screw clockwise one additional revolution (rather than three revolutions specified in Step 5).

Magnetic Latch Device Adjustment. Small-frame K-Line circuit breakers (with 600 A and 800 A frames) with an electronic OCPD have a magnetic latch device that is not adjustable. Large frame circuit breakers have a different style magnetic latch device that requires adjustment whenever the device is removed/replaced or if the trip latch or tripper bars have been adjusted. The magnetic latch device is adjusted by using the following procedure:

1. Close the circuit breaker.
2. Push upward on the tripper bar until slight resistance is felt, hold in that position, and do not trip the circuit breaker.
3. Measure the distance of gap between the tripper bar and the adjustment screw. The distance should be ¼″ to 5⁄16″.
4. Adjust the screw as necessary to obtain the specified clearance.

Siemens LA Adjustments

The most important adjustments on these types of circuit breakers are made on the trip latch engagement, trip actuator, and racking mechanisms. **See Figure 6-26.**

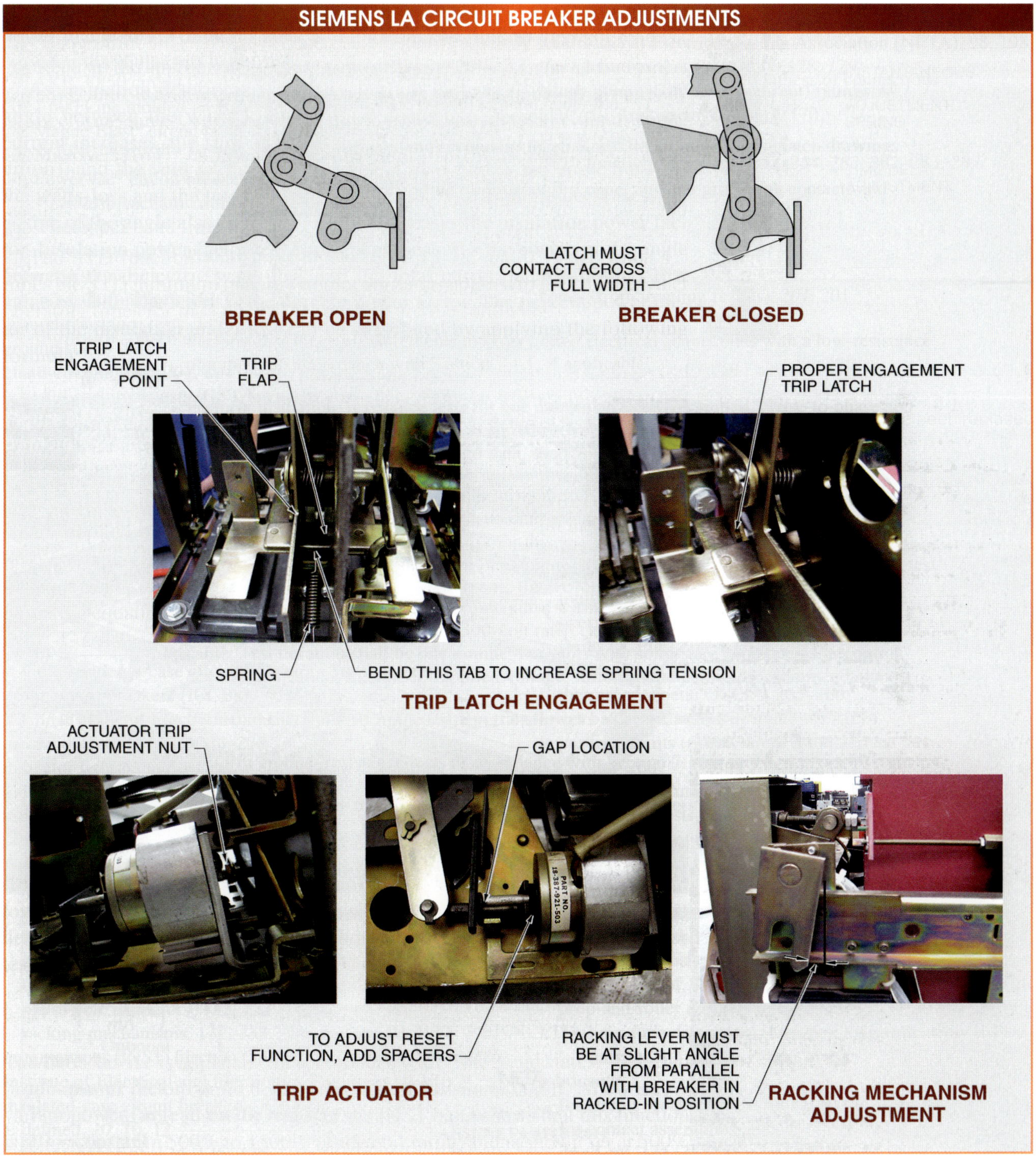

Figure 6-26. The most important adjustments on Siemens LA circuit breakers are the trip latch engagement, trip actuator, and racking mechanism.

Trip Latch Engagement Adjustment. With the trip latch engagement, there is no set procedure other than trial and error to achieve the proper spring tension. Inadequate spring tension causes the trip flap to pull away from the trip latch, while too much tension can cause a no-trip condition.

Trip Actuator Adjustment. The trip actuator causes the circuit breaker to trip once it receives a trip signal from the OCPD. Improper adjustment can lead to a no-trip condition. With the circuit breaker open and the closing springs charged, there must be an unspecified gap between the actuator rod and the trip paddle. This gap is adjusted using the nut on the end of the trip actuator shaft. If the trip actuator does not fully reset, spacers can be added at the appropriate location.

Racking Mechanism Adjustment. The racking mechanism must be properly adjusted in order for the circuit breaker to properly engage and disengage the bus. There must be a slight angle of the racking lever from a parallel position.

SUMMARY

The most commonly needed adjustments are on the contact assemblies. The operating mechanism usually only needs to be adjusted if there is wear or the circuit breaker is disassembled. The same type of adjustment often has different terminology in operation manuals, depending on the OEM.

Contact adjustments include overtravel or wipe, pivot point (on disassembled circuit breakers), gap, and sequence or make. On circuit breakers that have bayonet-style contacts, there is no provision for contact wipe, however, penetration and centering adjustments are made. Operating mechanism adjustments include latch and other breaker-specific adjustments.

DIGITAL RESOURCES
ATPeResources.com/QuickLinks
Access Code: 223436

chapter 6

CIRCUIT BREAKER ADJUSTMENT

CHAPTER REVIEW

Name ______________________________ Date ______________

True-False

T F **1.** The final contact adjustment must be a balance between wipe, sequence, and gap.

T F **2.** The racking mechanism of a Siemens LA circuit breaker causes the circuit breaker to trip once it receives a trip signal from the OCPD.

T F **3.** Bayonet contacts can have two adjustments performed on them: penetration and centering.

T F **4.** Before beginning operating mechanism adjustments, the circuit breaker must be visually inspected for proper lubrication.

T F **5.** To ensure consistent results, with butt contacts, all adjustments should be made by closing the breaker at normal speed.

T F **6.** If the arcing contacts do not make contact at the same time, then the first one to make contact has an increased rate of erosion.

T F **7.** For the GE MagneBlast, there must be 3⁄16″ to 1⁄4″ wipe between the closing latch and the roller.

T F **8.** If the driving pawl clearance is less than the minimum specified on a GE MagneBlast circuit breaker, the motor should be moved rearward.

T F **9.** Contact gap adjustment is usually not as much of a concern on low-voltage circuit breakers as it is on high-voltage, medium-voltage, or vacuum circuit breakers.

T F **10.** The contact gap of a GE MagneBlast circuit breaker is adjusted with a spanner wrench.

T F **11.** With a GE MagneBlast circuit breaker, there is no adjustment for arcing contact wipe.

T F **12.** The contact wipe of Cutler-Hammer VCP-W vacuum circuit breakers is more difficult to measure than that of other types of circuit breakers.

T F **13.** The arcing contacts of a GE AK 50 require a contact wipe between 1⁄16″ to 7⁄64″.

T F **14.** For an ABB/ITE HK circuit breaker, contact gap is usually adjusted at the same point as the contact pressure and sequence.

T F **15.** A lighting circuit can be used in lieu of a continuity tester to perform a contact wipe adjustment on a GE AKR/AKRD-50 circuit breaker.

T F **16.** To measure the contact wipe on the GE AKR/AKRD-30/50 circuit breaker, the circuit breaker is slow closed, and the gap between the back of the stationary contact and the contact pivot is measured.

T F **17.** To perform a trip coil plunger adjustment, the GE Power/Vac® circuit breaker must have the closing springs charged and the circuit breaker open.

T F **18.** GE considers the contact wipe adjustment to be a precision measurement.

T F **19.** To measure contact wipe on an ABB/ITE K-Line, the circuit breaker is slow closed until it latches, then a measurement is taken at the rear of the contact.

T F **20.** The most important adjustments for ABB/ITE HK circuit breakers are the latch check switch and control relay adjustments.

T F **21.** The trip actuator adjustment is similar for all types of circuit breakers, regardless of the OEM.

T F **22.** On an ABB/ITE K-line circuit breaker, the control relay houses the limit switch (LS).

T F **23.** The trip latch of a GE AKR 30/50 circuit breaker is adjusted if the circuit breaker does not close or trip properly.

T F **24.** The trip latch engagement adjustment for an ABB/ITE K-Line circuit breaker sets the prop to the far right of the roller.

T F **25.** A small-frame ABB/ITE K-Line circuit breaker with an electronic OCPD has a magnetic latch device that is fully adjustable.

Multiple Choice

_______________ **1.** Contact ___ provides a self-cleaning action of the contact surfaces.

A. wipe
B. gap
C. penetration
D. sequence

_______________ **2.** A(n) ___ is used to prevent a Westinghouse DHP circuit breaker from operating while it is being racked in or out.

A. buffer assembly
B. latching pawl
C. control relay
D. anticlose interlock

_______________ **3.** Insufficient ___ can cause a restrike condition during a fault.

A. overtravel
B. wipe
C. contact gap
D. contact penetration

_______________ **4.** All mechanism adjustments on the GE Power/Vac® circuit breaker are made from the ___.

A. bottom
B. top
C. right-hand side
D. left-hand side

_____ **5.** With a Westinghouse DS416 circuit breaker in the closed position, dimension A (the gap between the center piece of the stationary arcing contact assembly and the arcing contact) must be ___″ or greater.
A. 0.005
B. 0.010
C. 0.015
D. 0.020

_____ **6.** The primary mechanism adjustment is the ___ adjustment.
A. driving pawl
B. latching pawl
C. stop block
D. trip latch

_____ **7.** GE requires a minimum contact wipe distance of ___″.
A. 0.025
B. 0.050
C. 0.075
D. 0.100

_____ **8.** If the thickness of the main contacts of a Westinghouse DS416 or DS532 circuit breaker is less than ___″, the main contacts require replacement.
A. 1/32
B. 1/16
C. 1/8
D. 1/4

_____ **9.** With a GE MagneBlast circuit breaker, primary contact wipe is adjusted at the ___.
A. upper bushing
B. lower bushing
C. operating rod
D. spring cap

_____ **10.** Contact ___ is measured by slow-closing the breaker until the first arcing contact just makes contact.
A. centering
B. wipe
C. sequence
D. overtravel

_____ **11.** On an ABB/ITE K-line circuit breaker, the ___ adjustment ensures that a circuit breaker has a positive tripping action once the trip actuator is energized.
A. buffer assembly
B. tripper bar
C. racking mechanism
D. magnetic latch device

_____ **12.** On Westinghouse DHP circuit breakers, a(n) ___ is used to hold the mechanism in a closed position.
A. trip latch
B. buffer assembly
C. control relay
D. anticlose interlock

_______________ **13.** With Westinghouse DHP circuit breakers, the back of the moving arcing contact faces must be in line with the front of the stationary arcing contact faces ±___″.

A. 1⁄32
B. 1⁄16
C. 1⁄8
D. 1⁄4

_______________ **14.** With a GE MagneBlast circuit breaker, the distance from the main moving contact to the tip of the main stationary contact should have a minimum clearance of ___ ″.

A. 1⁄16
B. 3⁄16
C. 5⁄16
D. 7⁄16

_______________ **15.** Contact wipe (difference between open and closed measurements) on a Siemens LA 60/800 circuit breaker should be no more than ___″.

A. 0.093
B. 0.125
C. 0.250
D. 0.375

Completion

_______________ **1.** Contact ___ is the distance between the two closest points of the moving and stationary contacts in a circuit breaker.

_______________ **2.** ___ is the distance a circuit breaker mechanism travels once the contacts touch and the mechanism latches.

_______________ **3.** Contact ___ is measured by slow-closing the breaker and measuring how far the moving contact penetrates the stationary.

_______________ **4.** A(n) ___ is a mechanical component that absorbs the shock of a mechanism when a circuit breaker opens and prevents the contact assemblies from overdriving when a circuit breaker closes.

_______________ **5.** ___ is the additional compression of a preloaded spring that is used to apply force to the vacuum interrupter contacts and to provide opening kick-off force.

_______________ **6.** A(n) ___ pawl is a circuit breaker component that advances the drive gear to charge closing springs.

_______________ **7.** Contact ___ is the difference in contact spacing as arcing (moving) contacts make contact with stationary contacts in a circuit breaker.

_______________ **8.** A(n) ___ is an interlock that protrudes from the bottom of the circuit breaker and causes the circuit breaker to trip open and the closing springs to discharge when a circuit breaker is racked in or out of its cubicle.

_______________ **9.** A(n) ___ pawl is a circuit breaker component that locks the drive gear in place when charging the closing springs.

_______________ **10.** Contact ___ is the amount of overtravel a contact assembly has once it initially makes contact.

_______________ **11.** Trip latch centering on a GE AK50 circuit breaker is accomplished by using the ___, which is located on the right side of the center frame section, adjacent to the trip latch and roller.

_______________ **12.** The control relay adjustment of an ABB/ITE HK circuit breaker is set by installing or removing shims between the control relay and ___.

_______________ **13.** GE-type AK circuit breakers ending with the letter "___" in the model number have current-limiting fuses.

_______________ **14.** The contacts of a GE MagneBlast circuit breaker should make when the gap between the trip latch and the stop pin on the latch roller link is at ___″.

_______________ **15.** Nearly all medium-voltage air circuit breakers use a(n) ___ rod for adjusting contact wipe, sequence, and gap.

_______________ **16.** A GE MagneBlast circuit breaker is different from other circuit breakers in that it has a total of ___ separate adjustments.

_______________ **17.** For mechanical adjustments, a GE PowerVac circuit breaker requires adjustments to the close coil plunger gap, trip coil plunger gap, and ___.

_______________ **18.** A floor tripper is adjusted with a ___.

_______________ **19.** The latch check switch on an ABB/ITE HK circuit breaker is checked with a light circuit or a(n) ___.

_______________ **20.** With Siemens LA type circuit breakers, the most important mechanical adjustments are made on the trip latch engagement, trip actuator, and ___.

Short Answer

1. What are the purposes of contact penetration and centering adjustments?

2. What is the typical sequence of adjustments for the contacts of ABB/ITE HK circuit breakers?

3. When is a shunt trip device adjustment made?

4. Explain why contact wipe on vacuum circuit breakers is measured differently than on air circuit breakers.

5. What is the factory procedure for measuring contact gap on a GE Power/Vac® circuit breaker?

Matching

GE Power/Vac® Contact Assembly

_______________ 1. Spring cap

_______________ 2. Wipe indicator

_______________ 3. Operating rod

_______________ 4. Disk

_______________ 5. Lock nut

_______________ 6. Erosion indicator

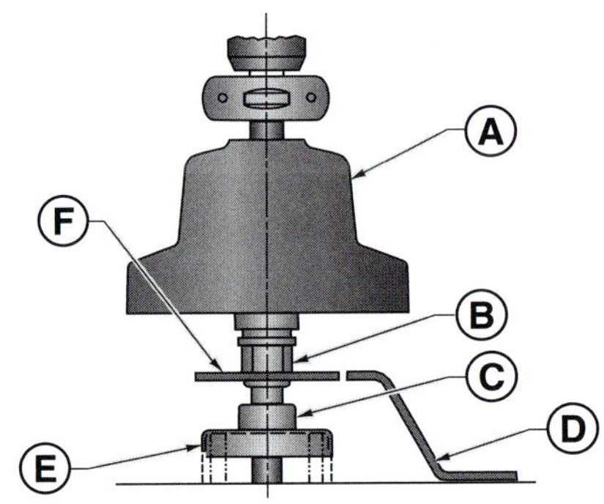

ABB/ITE K-600/800 Contact Assembly

_______________ 1. Torque hinge joint

_______________ 2. Adjustment stud

_______________ 3. Upper terminal current studs

_______________ 4. Measurement point (rear of contact)

_______________ 5. Moving main contact member

_______________ 6. Moving arc contact member

_______________ 7. Contact pressure adjustment gap

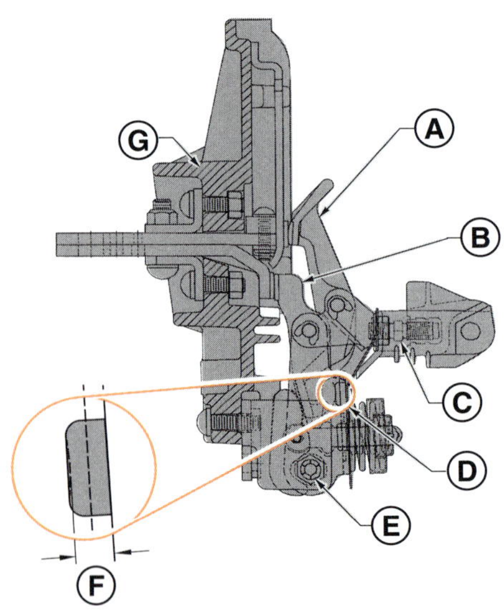

GE MagneBlast Mechanism Components

_______________ 1. Stop block

_______________ 2. Latching pawl

_______________ 3. Driving pawl

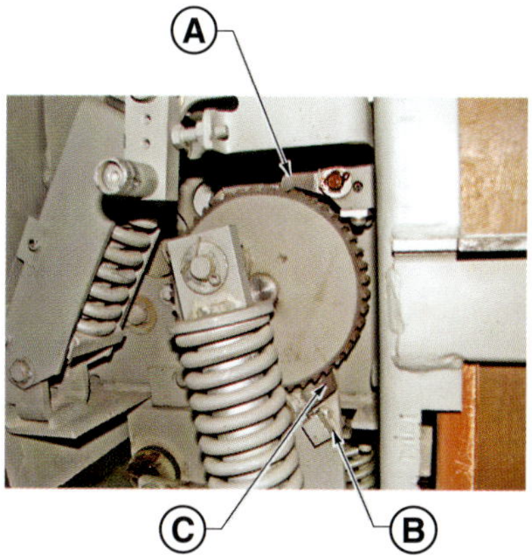

Westinghouse DHP Anticlose Interlock Components

_______________ 1. Closing latch roller

_______________ 2. Operating rod

_______________ 3. Closing trigger

_______________ 4. Anticlose interlock screw

_______________ 5. Closing latch

_______________ 6. Spring release coil

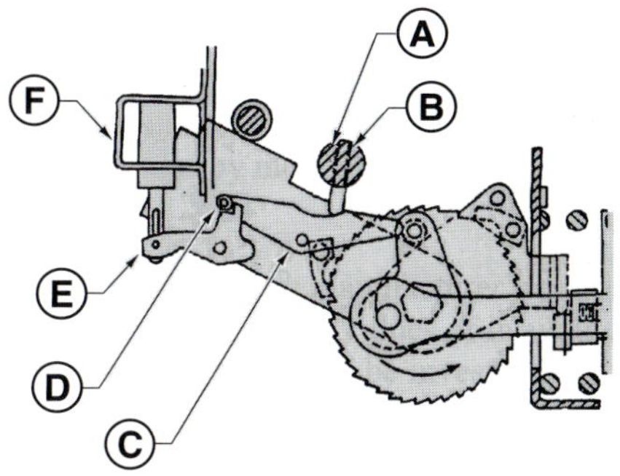

chapter 7

CIRCUIT BREAKER TESTING

There are several reasons to perform electrical tests on circuit breakers. One purpose of electrical testing is to determine whether the insulation is serviceable. Over a period of time, weather, contaminants, moisture, and electrical stresses cause insulation to slowly deteriorate. The process can take several years or, if there are chemical agents involved or the conductor carries more than its rated current, it can occur within a few months.

Current flow through electrical conductors deteriorates insulation, due to the heating that occurs from copper losses, which dries the insulation and causes embrittlement. Thermal degradation is cumulative, which means that all thermal damage adds to previous damage and eventually causes failure. Therefore, it may not cause any immediate noticeable damage.

Loose fasteners, damage that occurs while equipment is installed, manufacturing defects, and poor design are also reasons electrical testing is required. Electrical tests include the following:

- insulation resistance (megohmmeter)
- overpotential (hipot)
- insulation power (dissipation) factor
- contact resistance (micro-ohmmeter)
- time-travel analysis (timing)
- partial discharge (PD)
- vacuum bottle integrity test
- infrared thermography
- magnetron atmospheric condition test
- circuit breaker vibration signature analysis

Insulation resistance, overpotential, and insulation power factor tests determine the quality of insulation. The contact resistance test detects deterioration of the current path. The time-travel test detects changes in timing, which could indicate wear, binding, or other similar problems. PD testing is also an insulation test, but it tests emissions that are created when insulation begins to fail. PD tests can be performed while the equipment is energized and in service. PD can be monitored as part of predictive and preventive maintenance programs.

The vacuum bottle integrity test shows whether a vacuum bottle is safe to reenergize. Infrared thermography is a nonintrusive test that can identify loose connections, overheated components, and other types of hot spots. The magnetron atmospheric condition test is a field version of tests performed by the factory to determine the remaining life expectancy of a vacuum bottle. During circuit breaker vibration signature analysis, the vibrations created by circuit breaker operation are measured. A signature is developed that is used to detect problems and verify the timing of circuit breakers for first-time trip testing.

OBJECTIVES

- List five tests that can be performed on circuit breakers.
- State the advantages and disadvantages of direct current (DC) tests.
- List the three currents that are produced when testing insulation.
- Describe the difference between a DC hipot and a megohmmeter.
- State the recommended test voltages for a DC overpotential test on a circuit breaker.
- Describe one method of preventing failure while testing insulation using a DC hipot.
- Calculate the corrected insulation resistance using a temperature correction factors table.
- State the primary cause of failure in vacuum interrupters.
- Describe how an insulation power factor test set functions.
- State the tests that are performed on a circuit breaker using an insulation power factor test set.
- Determine whether a contact resistance test result is acceptable.
- State the five characteristics determined by the mechanical timing of a circuit breaker.
- Determine the DC overpotential test voltages for switchgear.
- State the advantages of partial discharge (PD) testing for switchgear.

KEY TERMS

- megohmmeter
- dielectric absorption current
- capacitance charging current
- leakage current
- guard circuit
- treeing
- predictive maintenance
- preventive maintenance

SAFETY

Before performing any electrical tests or maintenance work on circuit breakers, switchgear, and related equipment, the circuit breaker original equipment manufacturer (OEM) owner's/operator's manual, NFPA 70E®, the facility's electrical safety program, and OSHA regulations should be referenced to determine proper safety procedures and personal protective equipment (PPE) requirements. The major hazard from testing is electrical shock. High voltages are used to test insulation integrity, and contact with energized components could result in heart arrhythmia, electrical burns, or injuries from falling.

An arc flash from test equipment is generally not a risk since the available short-circuit current is low, and the output of the test equipment is typically well-protected by fuses. High-current test equipment may cause burns due to the heating of current-carrying parts but present almost no shock hazard.

DC VS. AC ELECTRICAL INSULATION TESTS

Electrical insulation used on circuit breakers, switchgear, and related equipment must be tested for adequate dielectric strength and deterioration. Tests can be performed using alternating current or direct current, depending on the characteristic of the circuit breaker test being performed.

Alternating current (AC) is current that reverses its direction of flow at regular intervals. *Direct current (DC)* is current that flows in only one direction. AC insulation tests are usually performed using insulation power factor or dissipation factor test sets. DC insulation tests can be performed using test instruments such as hipot testers and megohmmeters. Hipot testers measure insulation resistance by measuring leakage current. A *megohmmeter* is a high-resistance ohmmeter used to measure insulation deterioration on various conductors by measuring high-resistance values during high-voltage test conditions.

DC Voltage and Electrical Insulation

DC insulation testing is the earliest form of insulation testing, dating from the late 1800s. There are several advantages and disadvantages of performing DC insulation testing. Advantages of using DC test equipment include the following:

- lightweight
- portability
- ease of use
- direct reading of leakage current
- technicians are familiar with it
- low cost compared to AC test equipment

Disadvantages of using DC test equipment include the following:

- inaccuracies due to the manner in which measurements are taken
- application of high voltage in order to stress insulation for meaningful results
- test voltages must be over the circuit breaker's rated voltage to stress the insulation

DC insulation testing is performed by applying DC voltage of 500 V or higher across the insulation to be tested. Since all insulation has some leakage current through it, leakage current is directly measured with DC hipot testers in microamps or can be measured in megohms using a megohmmeter.

When DC voltage is applied across insulation, three currents are actually being measured: dielectric absorption, capacitance charging, and leakage current. **See Figure 7-1.**

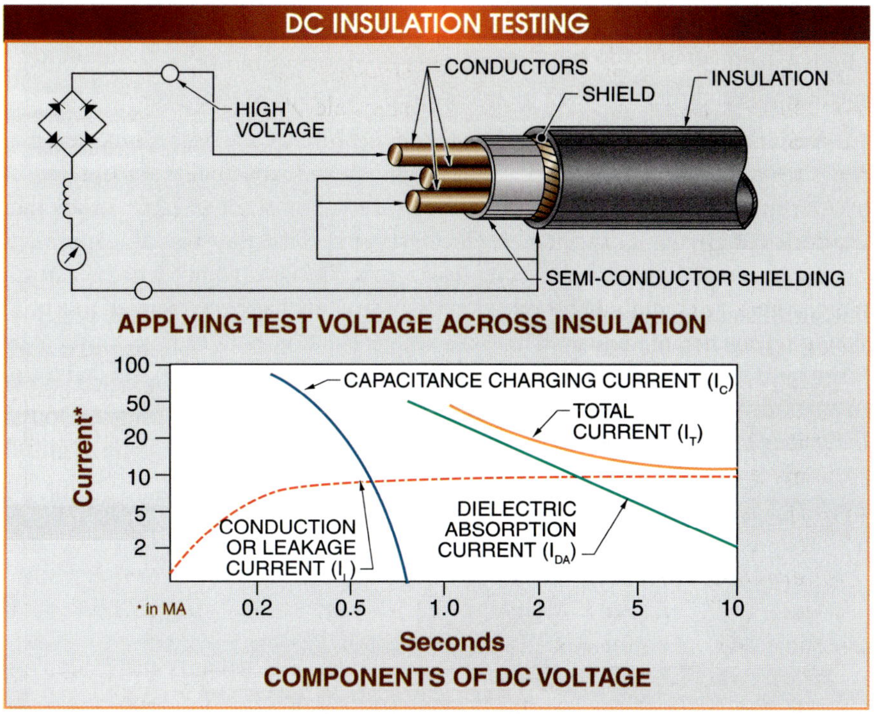

Figure 7-1. DC insulation testing is performed by applying DC voltage of 500 V or higher across the insulation to be tested.

APPLICATION OF DC INSULATION TESTS

DC insulation tests are generally only adequate for equipment rated below 15 kV. At higher voltages, depending on the mass of the insulation being tested, test results tend to be above the instrument's range. Modern test equipment is available that measures resistance in gigaohms and teraohms, but current is so low at these levels that even minute variations can affect the test results. IEEE Standard 43, *IEEE Recommended Practice for Testing Insulation Resistance of Rotating Machinery,* states the following for insulation resistance values above 5000 MΩ:

"12.2.2 Applicability of polarization index when IR_1 is greater than 5000 MΩ—When the insulation resistance reading obtained after the voltage has been applied for 1 min (IR_1) is higher than 5000 MΩ, based on the magnitude of applied direct voltage, the total measured current (I_T) can be in the sub-microampere range. At this level of required test instrument sensitivity, small changes in the supply voltage, ambient humidity, test connections, and other non-related components can greatly affect the total current measured during the 1-10 min interval required for a P.I. Because of these phenomena, when the IR_1 is higher than 5000 MΩ, the P.I. may or may not be an indication of the insulation condition and is therefore not recommended as an assessment tool."

Even though this standard is written for rotating machines, the underlying issues would be present regardless of what is being tested. The leakage current is so low at these values that almost anything can affect it, which renders the readings less meaningful.

Dielectric Absorption Current. *Dielectric absorption current (I_{DA})* is the polarizing current that is drawn by the insulation system (dielectric) to align the dipole molecules within the insulation. A *dipole molecule* is a molecule that is affected by the magnetic field created by current flow through its insulation. The current decreases to zero once the dipole molecules in the insulation are aligned with the current flow. Since this component is not used to evaluate insulation, readings must not be taken until the meter stabilizes.

Capacitance Charging Current. *Capacitance charging current (I_C)* is current that flows upon the application of DC voltage to charge the capacitance of the electrical system under test. Since insulation systems cover a conductor and are near another conductor or ground, a capacitance effect is created. Once the "virtual" capacitor's plates are charged, the current decreases to zero. The capacitance current starts at a high value and rapidly decreases after the insulation has been charged to its full voltage. This current is not used for evaluating insulation, and readings must not be taken until the current value stabilizes. The same capacitive effect occurs with any insulation system because it will have a conductor (plate), insulation (dielectric), and ground or another conductor (plate). **See Figure 7-2.**

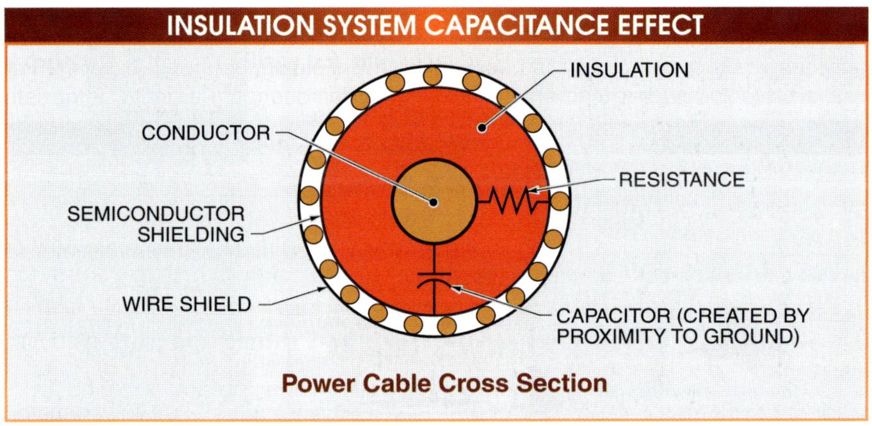

Figure 7-2. Since insulation systems cover conductors and are near other conductors or ground, a capacitance effect is created.

Leakage Current. *Leakage current (I_L)* is resistive current that flows through insulation and is the variable that is measured with an insulation resistance test instrument. The current path measured must be through the insulation, often referred to as the conduction or volumetric current. However, due to contaminants and other factors, there may also be leakage across the surface. Surface leakage can be reduced or eliminated by using proper termination preparation techniques, cleaning insulation surfaces, and using a guard circuit. A *guard circuit* is a function on a megohmmeter used to eliminate unwanted return current from test measurements.

Charging Current. The capacitance charging current and the dielectric absorption current are known collectively as charging current. At the beginning of the test, the meter on DC test equipment measures the total of the three currents (I_T). It cannot separate them. Since the dielectric absorption current and the capacitance charging current both decrease to zero during a DC test, only the leakage current is measured after a period of time. The amount of time required for the reading to stabilize to where only the leakage current is measured varies with the mass of the insulation and the type of device being tested.

Larger wound (inductive) devices or long cable runs tend to take more time to stabilize, while smaller nonwound (resistive) devices tend to stabilize quickly. Readings should stabilize in just a few seconds. Due to the high initial charging current, a megohmmeter initially reads low and then, as the charging current decreases, the megohmmeter reading increases. DC hipot testers, which measure microamps, provide the opposite meter indication starting at a high initial value and then decreasing as the charging current decreases.

CAUTION: Devices such as large transformers and long runs of power cable can maintain a possibly lethal charge after DC testing. Long cable runs can store a charge and even rebuild a charge after being grounded. Once testing is completed, larger wound devices or long cable runs must be grounded for a minimum of four times the test interval after testing. A test interval of 5 min requires the application of grounds for 20 min. Resistive devices, such as circuit breakers and switchgear, usually require little, if any, bleed-down time since they do not store a charge.

AC Voltage and Electrical Insulation

When AC voltage is applied across insulation, the same loss components are created as when DC voltage is applied across it. However, AC changes direction 60 times each second. The dipole molecules in the insulation never align, so the current required to align them does not decrease. The same is true for the capacitance charging current. Since the current is constantly changing direction, the plates never charge and the current demand stays constant. The microammeter on an AC hipot tester only reads the total current since the charging currents never decrease. **See Figure 7-3.**

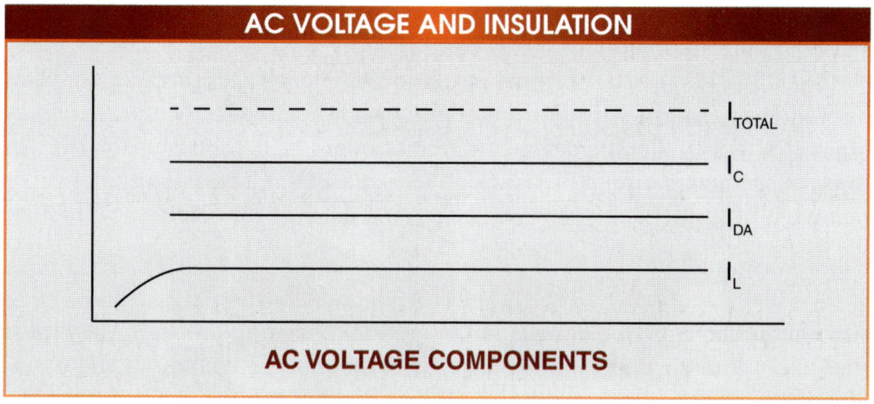

Figure 7-3. Since AC is constantly changing direction, the plates never charge and the current demand stays constant.

DC CIRCUIT BREAKER TESTS

DC circuit breaker tests can be performed for two purposes: determining insulation quality and determining the resistance of the current path. Circuit breakers used in oil and gas drilling or refining processes are subject to exposure to hydrogen sulfide (H_2S), and circuit breakers used near pulp and paper processes can be exposed to sulfur dioxide (SO_2). Circuit breakers used in water treatment or waste water facilities can be exposed to chlorine (Cl) and fluorine (F), both of which are extremely corrosive. Exposure to these chemicals can deteriorate electrical insulation and corrode copper current-carrying components. **See Figure 7-4.**

CIRCUIT BREAKER DEGRADATION

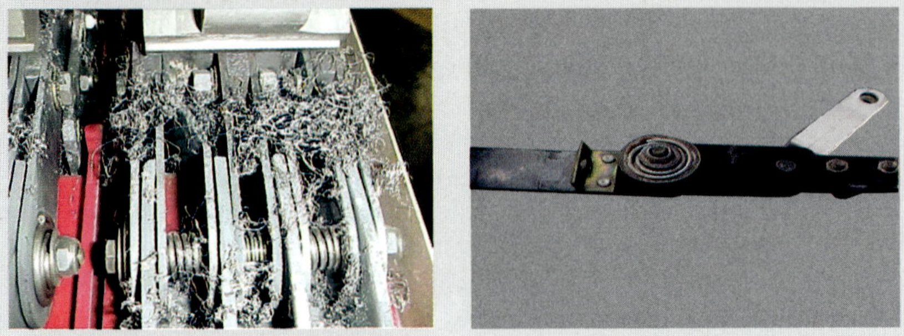

Silver Finger Growth Due to Hydrogen Sulfide (H_2S) Exposure

Corrosion Caused by Sulfer Dioxide (SO_2) to Silver Plating

Shermco Industries

Figure 7-4. DC circuit breaker tests are performed because circuit breakers used in certain applications are subject to degradation from exposure to hazardous chemicals.

Circuit breakers can also degrade due to exposure to environmental factors such as moisture, dirt, and temperature fluctuations. Moisture can cause rust and oxidation of metal components and devices. Dirt often contains metallic or other conductive materials and chemicals, especially in production areas. Sometimes these materials and chemicals can be combustible. Not only is the conductivity of these materials an issue but also they accumulate on excess lubrication around latches, pins, and pivots where they are drawn in and accelerate wear. Temperature fluctuations can cause condensation to form on metal components (causing degradation) as well as the loosening of fasteners caused by the alternation of 60 Hz current.

CAUTION: When performing insulation resistance tests on circuit breaker insulation or control wiring, do not apply high voltage to the spring charging motor. Verify that the device or circuit under test is properly isolated. On low-voltage power circuit breakers, either isolate the electronic trip device or remove it from service. Failure to do so could damage the electronic circuitry.

Arc Chute and Arc Extinguisher Degradation

Arc chutes are critical elements in the arc extinguishing process. They must stretch out the arc, cool it down in order to deionize it, contain it, and extinguish it, even during fault conditions. If the arc is not extinguished within a few cycles, it could cause a restrike. Restrikes often severely damage or destroy the circuit breaker and its associated switchgear. Even though the event lasts only a few cycles to 2 sec, the intense heat is more than enough to cause melting and destruction of any surrounding materials. **See Figure 7-5.**

When a circuit breaker interrupts fault-level current, carbon and metal spatter are forced upward into the arc chute. The intense heat of the arc deteriorates the arc chute, especially in the throat area. Carbon tracking and contamination can provide the platform for a restrike. Moisture ingress, especially in arc chutes containing organic materials, can actually create steam when an arc heats them.

This degradation of the arc chute is more easily found using a DC hipot tester rather than a megohmmeter. The larger transformer of the DC hipot tester can supply the leakage current on medium-voltage circuit breakers needed to stress the insulation. Megohmmeters are not capable of supplying sufficient current when testing medium-voltage circuit breakers. Megohmmeters should not be used for testing the primary insulation components, although they work well for low-voltage circuit breakers and for testing control wiring.

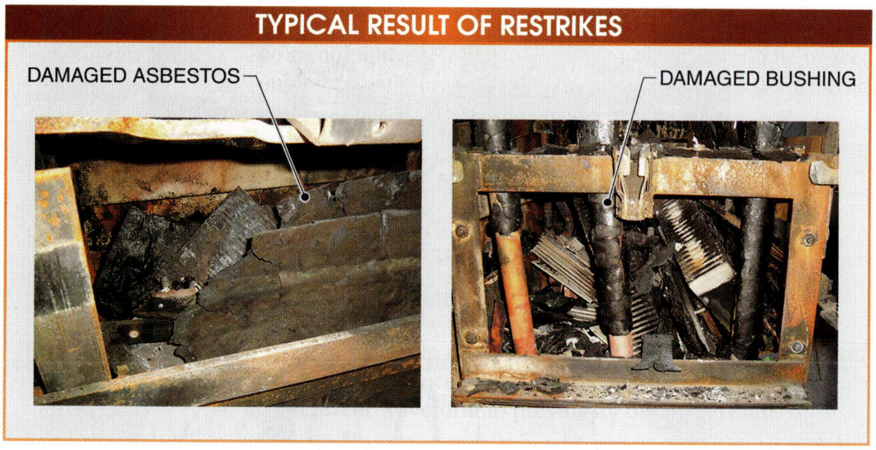

Shermco Industries

Figure 7-5. The integrity of arc chutes and extinguishers is tested to prevent damage to components from electrical arcs.

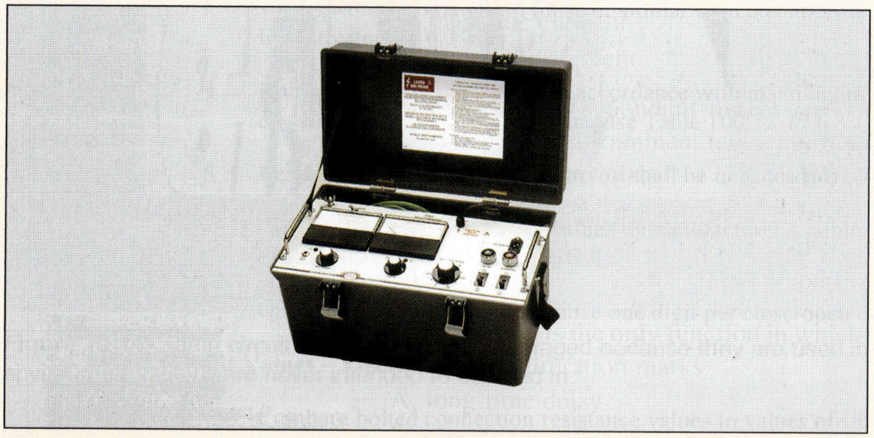

Megger Group Limited

Arc chute degradation is tested with a DC hipot tester because the larger transformer of the hipot tester can supply the leakage current needed to stress the insulation.

Pole Unit and Bushing Degradation

The pole units on medium-voltage air-magnetic circuit breakers can suffer the same type of degradation as arc chutes. Many bushings on these older circuit breakers were constructed of paper soaked in an epoxy, while other OEMs used older plastic resin or fiberglass insulation systems. Over a period of years, the insulation deteriorates due to heat, moisture, mechanical stress, dirt, and corrosion.

When insulation begins to deteriorate , it can form dielectric treeing or tracking. *Treeing* is irreversible internal degradation caused by the formation of conductive carbonized pathways. When this deterioration is on the surface of insulators, it is referred to as tracking. Modern circuit breakers use inorganic materials, such as porcelain, to reduce required maintenance and slow deterioration. **See Figure 7-6.**

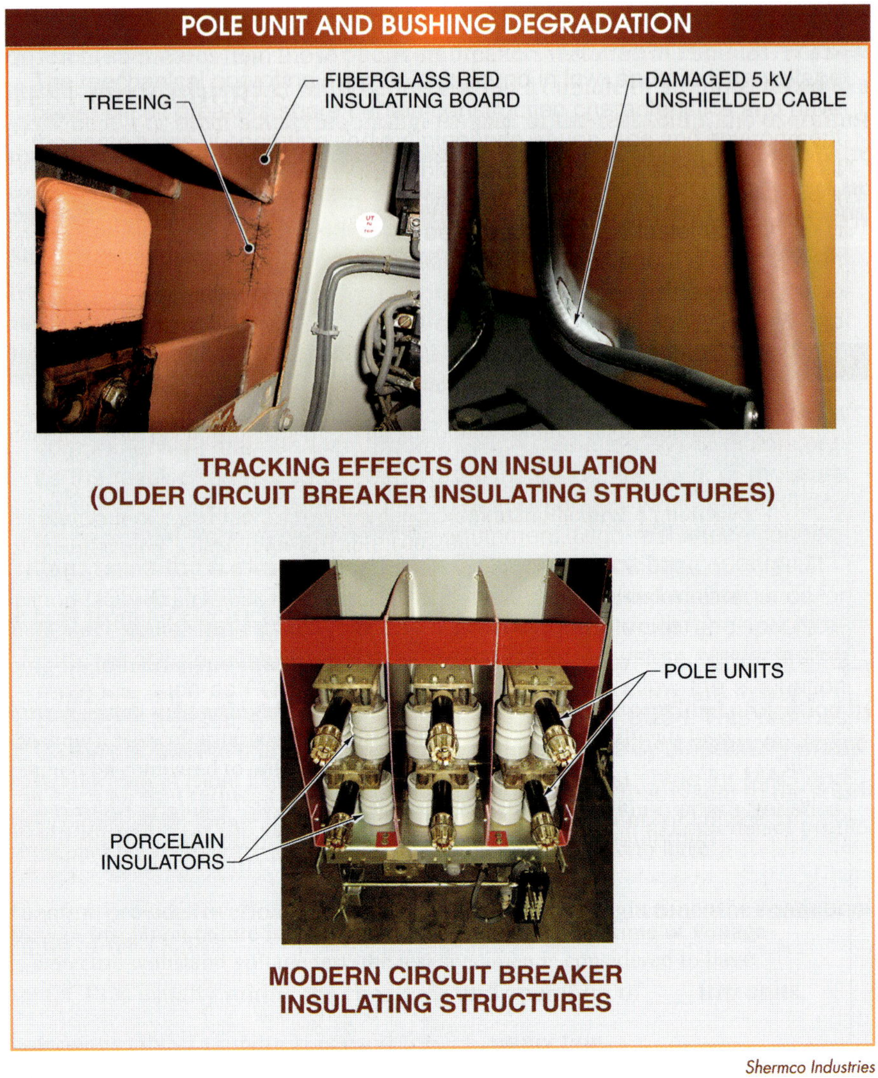

Figure 7-6. Modern circuit breakers use inorganic materials, such as porcelain, to reduce maintenance and to slow deterioration, such as that seen on older circuit breakers.

Test Equipment Selection

Selecting the test equipment to use (AC or DC, and megohmmeter or hipot) for performing specific test procedures is the first step in proper testing. An important consideration is the amount of damage to the insulation the test equipment can cause. When performing an AC overpotential test, the high charging currents are maintained throughout the test. Therefore, some deterioration of the insulation, although minor, occurs. When performing a DC overpotential test, the charging currents decrease rapidly and do not typically damage insulation. However, DC hipot testers can cause insulation to fail if used improperly.

Megohmmeters are nondestructive since they decrease their voltage output as the insulation resistance decreases due to the small output transformers used. This aspect makes it less likely that a megohmmeter will damage weak insulation. A DC hipot tester, however, has a larger output transformer, which gives it the capacity to maintain its voltage output even as current flow increases through weakened insulation. Some other potential disadvantages of DC overpotential testing include the following:

- When high DC voltage is applied across a series of insulation layers, the voltage drop of each layer is a function of DC resistance. In service, the voltage drop across each layer is a function of the dielectric constant. Some layers, especially in graded insulation, can be overstressed or understressed.
- DC voltage may not detect a void or gap in insulation, unless the void or gap has already caused some deterioration of the insulation around it.
- A DC overpotential test can cause weakened insulation to fail without warning.
- For insulation systems that have multiple dielectrics, a DC insulation test may not be accurate enough to properly evaluate the insulation system.
- DC insulation tests are often ineffective on larger wound devices such as transformers and motors, and on devices with voltage ratings above 15 kV.

In theory, when performing DC high-potential tests, the charging currents can be monitored during the test and, if they begin to cascade into failure, the test can be stopped before damaging the insulation. However, this is not always true because it assumes that everyone performing the testing has been properly trained. Suddenly applying high-voltage DC to insulation often causes the insulation to fail without warning. When performing a DC overpotential test, the voltage must be increased in several steps. Also, if there has been some treeing in cable, a DC hipot tester could cause that defect to grow during the test.

Another important consideration when choosing the proper test equipment is the type and quality of information received from each test. Megohmmeters are light, portable, and easy to use. They are designed for appraising simple insulation systems and low-voltage equipment. The main drawback of a megohmmeter is that it usually only shows gross defects in the insulation since it is not an overpotential test. The insulation is not stressed, and the possibility of damage to the device under test is extremely small. However, a megohmmeter typically will not indicate the presence of tracking, deterioration, or contaminants. In general, the higher the voltage the insulation can withstand during a test, the less likely it is to fail in service.

DC testing has been an accepted industry practice for many years. It is important that technicians understand the inherent limitations of all tests they perform and how to compensate for these limitations. Each test provides differing information about the electrical insulation being tested. Advantages of DC high-potential testing include the following:

- Because the charging current is low, DC test equipment is considerably smaller and lighter than equivalent AC test equipment.
- DC hipot testers display leakage current in amps (A). Megohmmeters directly display readings in megohms (MΩ). Either of these readings can be plotted and used to develop trends.
- If used properly, it is considered to be nondestructive testing.
- Tests are sensitive to small changes in insulation quality between test cycles.
- Technicians are familiar with it.

DC INSULATION TESTS

The common industry practice test voltage value for a DC overpotential test is twice the rated voltage plus 1000 V. The InterNational Electrical Testing Association (NETA) has developed specific recommendations for the test voltage to be applied that attempts to stress electrical insulation without causing unnecessary damage to the equipment under test. Although this test voltage value has the potential to damage insulation, it reduces the risk of damage to equipment and also indicates deterioration and tracking, which helps forecast service life. **See Figure 7-7.**

OVERPOTENTIAL TEST VOLTAGES

Recommended Overpotential Test Voltages for Switchgear Assemblies Table—100.2 ANSI/NETA MTS-2011

Type of Switchgear	Rated Maximum Voltage (kV)(rms)	AC	DC
Low-Voltage Power Circuit Breaker Switchgear	0.254/0.508/0.635	1.6	2.3
Metal-Clad Switchgear	4.76	14	20
	8.25	27	37
	15.00	27	37
	27.00	45	a
	38.00	60	a
Station-Type Cubicle Switchgear	15.50	37	a
	38.00	60	a
	72.50	120	a
Metal-Enclosed Interrupter Switchgear	4.76	14	20
	8.25	19	27
	15.00	27	37
	27.00	45	a
	38.00	60	a

Because of the variable voltage distribution encountered, the manufacturer should be contacted for recommendations before applying DC withstand tests to the switchgear. Voltage transformers above 34.5 kV should be disconnected when testing with DC. Refer to ANSI/IEEE C57.13-1993, IEEE Standard Requirements for Instrument Transformers, paragraph 8.8.2.

Recommended Overpotential Test Voltages for Circuit Breakers Table 100.19—ANSI/NETA MTS-2011

Nominal System (Line) Voltage (kV)	Insulation Class (kV)	AC Factory Test (kV)	Maximum Field Applied AC Test (kV)	Maximum Field Applied DC Test (kV)
1.2	1.2	10	6.0	8.5
2.4	2.5	15	9.0	12.7
4.8	5.0	19	11.4	16.1
8.3	8.7	26	15.6	22.1
14.4	15.0	34	20.4	28.8
18.0	18.0	40	24.0	33.9
25.0	25.0	50	30.0	42.4
34.5	35.0	70	42.0	59.4
46.0	46.0	95	57.0	80.6
69.0	69.0	140	84.0	118.8

Figure 7-7. NETA has developed specific recommendations for test voltage that attempt to stress electrical insulation without causing unnecessary damage to the equipment under test.

When a DC overpotential test is performed, the insulation under test is stressed. Sudden failure of the device or insulation is always possible. As insulation ages, it becomes weaker due to heating, overloading, and dielectric stresses, such as voltage transients, which can cause it to be more prone to failure. However, when insulation fails during testing, the damage is less than if it were to fail during normal operation.

CAUTION: Never apply full test voltage suddenly to insulation when using a DC hipot tester. Slowly increase the voltage in increments and keep the needle within the lower two-thirds of the microammeter scale. The ammeter needle should drop as soon as the voltage is adjusted to its value at each step. If the needle moves to the high end of the scale or if it continues to increase once the voltage is adjusted, insulation failure is imminent and testing must be discontinued immediately.

The effectiveness of DC overpotential tests vary according to whether the insulation system is composed of homogeneous insulation or a composite such as oil-filled and paper. DC voltage stresses insulation differently than AC voltage, and therefore DC tests can reveal defects that AC tests may miss. The reverse is also true, although the insulation power factor test has shown to be one of the most reliable indicators of insulation quality.

TECH TIP

OEMs of power equipment and cables often recommend testing with a megohmmeter prior to using a DC hipot tester. A megohmmeter can indicate whether insulation is damaged or weakened enough to fail immediately. If the insulation passes megohmmeter testing, then an overpotential test can be performed.

DC Overpotential Testing of Low- and Medium-Voltage Power Circuit Breakers

When testing insulation systems on medium-voltage air-magnetic circuit breakers, the use of a high-potential test set is required, usually with a DC voltage output. Megohmmeters do not adequately stress the insulation at these voltage ratings, and off-scale readings will often be observed. AC hipot testers could also be used but generally are not since leakage current cannot be measured separately and are considered proof tests. Proof tests are performed by the factory to ensure that the insulation system will not fail when placed into service. They are not usually performed as field maintenance tests.

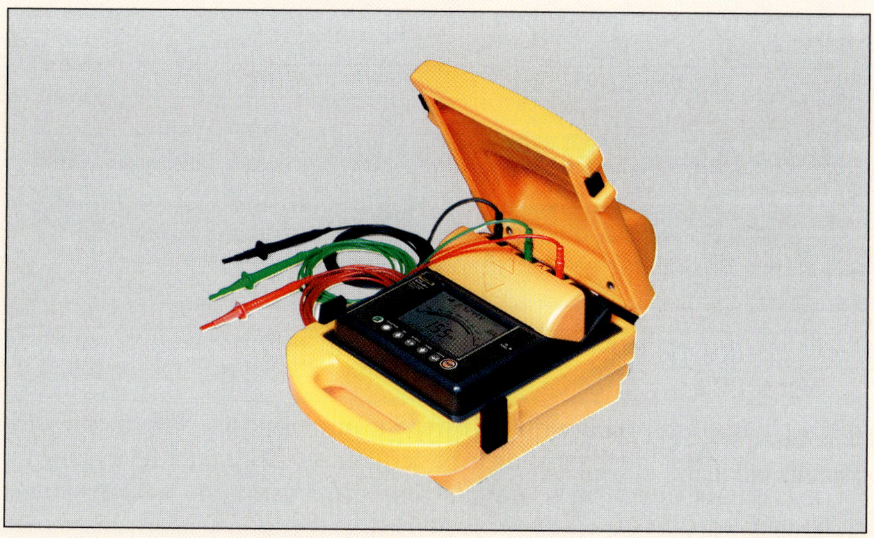

Fluke Corporation

OEMs of power equipment and cables often recommend testing equipment with a megohmmeter prior to using a DC hipot tester.

ANSI/NETA MTS-2011 TEST VOLTAGE RECOMMENDATIONS

As detailed in ANSI/NETA MTS-2011 Table 100.19, a 5000 V circuit breaker has 16,100 VDC or 11,400 VAC applied to it. There are no recommended minimum insulation resistance values or leakage current. In Table 100.19, it is recommended that if insulation does not indicate signs of failure during the 1-min test, it is considered satisfactory for continued service. A minimum insulation resistance value could be calculated, but it is not a good indicator of insulation quality. The value calculated does not indicate the serviceable value, which would be many times higher than the calculated minimum value. Recording the measured value, corrected for temperature, allows for trending of the test results.

The connections shown in Minimum Recommended Test Connections can be used, while the test voltages should be in accordance with ANSI/NETA MTS-2011 Table 100.19. The use of a megohmmeter may indicate moisture in medium-voltage circuit breakers, but it is only used to ensure that the circuit breaker is able to undergo additional tests. However, a megohmmeter works well for performing overpotential tests on low-voltage power circuit breakers. When 2500 V is applied to a 600 V circuit breaker for 1 min, there should be no evidence of failure.

Overpotential Test Voltages
Table 100.19 from ANSI/NETA MTS-2011

Nominal System (Line) Voltage (kV)	Insulation Class	AC Factory Test (kV)	Maximum Field Applied AC Test (kV)	Maximum Field Applied DC Test (kV)
1.2	1.2	1.2	6.0	8.5
2.4	2.5	2.5	9.0	12.7
4.8	5.0	5.0	11.4	16.1
8.3	8.7	8.7	15.6	22.1
14.4	15.0	15.0	20.4	28.8
18.0	18.0	18.0	24.0	33.9
25.0	25.0	25.0	30.0	42.4
34.5	35.0	35.0	42.0	59.4
46.0	46.0	46.0	57.0	80.6
69.0	69.0	69.0	84.0	118.8

MInimum Recommended Test Connections

Circuit Breaker Open	# Tests	Insulation Tested
Line Stab to Load Stab	3	Arc Chute
Circuit Breaker Closed	**# Tests**	**Insulation Tested**
Line Stab to Ground	3	Backboard—Stab to Ground
Phase-to-Phase	3	Backboard—Stab to Stab

Test Result Trending

Test result trending over a period of several maintenance intervals allows some measure of service life forecasting to be performed. This is done by looking for a sharp decrease in insulation resistance or a sharp increase in leakage current. As insulation ages, it slowly deteriorates. This causes the insulation resistance to slowly decrease or, if using a hipot tester to take readings, leakage current (μA) to slowly increase.

Test result data for determining trends is accumulated over several years as insulation resistance slowly decreases with time. **See Figure 7-8.** At the end of the trend period in this example, however, a knee, which is a significant downward change in the normal shape of the characteristic curve, appears on the trend characteristic curve. Actual insulation resistance can be above the minimum acceptable value, but when there is a sudden decrease in insulation resistance, major damage has occurred. A knee in a test trend almost always indicates that the piece of equipment or device being tested must be repaired or replaced immediately. If test results are not trended, the knee would not be detected and the equipment would probably fail in service.

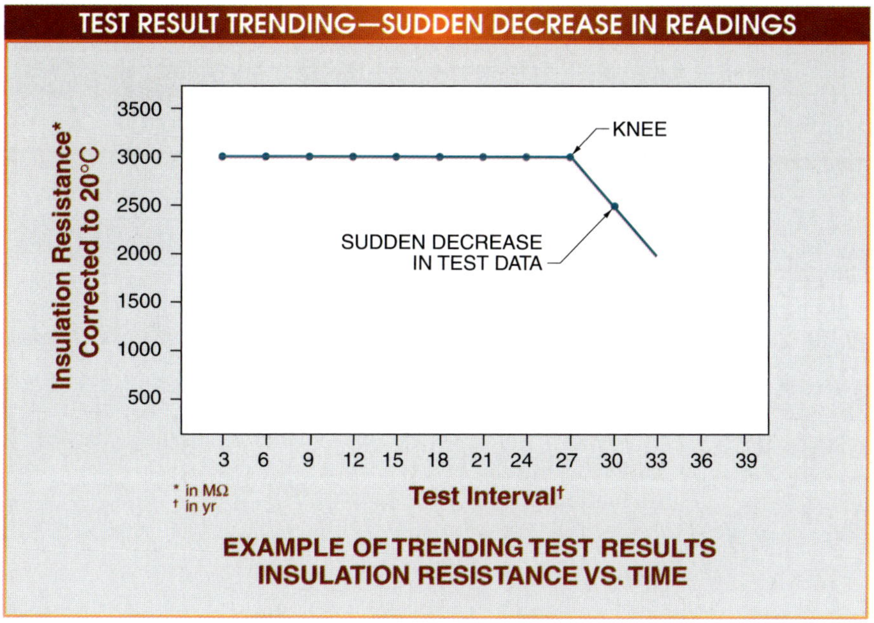

Figure 7-8. A sudden decrease in insulation resistance readings indicates that major damage has occurred.

Test results must be corrected to a base temperature of 20°C so they can be trended. Since temperature affects insulation resistance, the results would be erratic if they were not corrected. Uncorrected test results have little meaning since the insulation temperature varies each day and from year-to-year. By correcting to a 20°C base temperature, the effects of temperature changes are removed from the measured results. ANSI/NETA MTS-2011 Table 100.14.1 can be referenced for insulation temperature correction factors for DC insulation testing. **See Figure 7-9. See Appendix.**

The temperature of the circuit breaker under test is generally considered to be the ambient temperature. Temperature correction factors are less accurate the further the temperature is from 20°C. Electrical equipment and devices should not be tested when their temperature is below 0°C because ice can form in cracks in the insulation. Insulation would show as "good" on a test instrument, then when the insulation temperature increases above freezing, it would indicate "bad." The opposite is also true. High temperature correction factors (above 80°C) also tend to be inaccurate. Insulation at these temperature extremes should be tested when the insulation temperature is closer to 20°C.

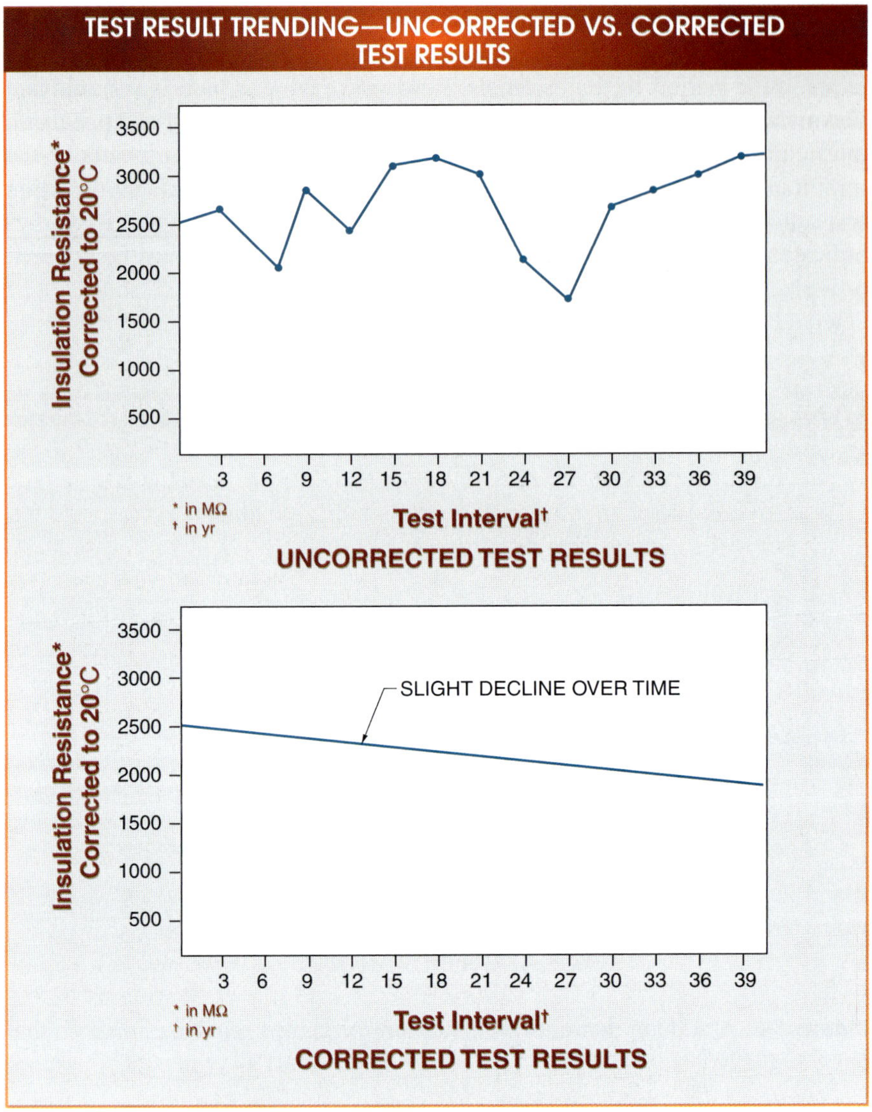

Figure 7-9. Since temperature affects insulation resistance, test results are erratic if they are not corrected.

For example, if insulation temperature is 40°C when tested and indicates an insulation resistance of 940 MΩ, the reading would have to be corrected for temperature. The temperature correction factor for 40°C is 2.5 for solid insulation. Therefore, 940 MΩ at 40°C actual temperature times a 2.5 temperature correction factor is 2350 MΩ, the corrected measurement.

Temperature can have a large effect on insulation resistance measurements. This is especially apparent when trying to trend test results from one year to the next. It is always a good practice to record the temperature at the time of the test, and the reading corrected to a 20°C base temperature.

Ensuring Proper Megohmmeter Function

If a megohmmeter is defective, due to either generating low voltage or an inaccurate meter, the time spent on testing is wasted. By using a multimeter, the function of a megohmmeter can be verified. Every multimeter has "sensitivity" or "input impedance" listed on the instrument faceplate or in its operator's manual. Sensitivity can be in the form of a volt/ohms rating for analog-type multimeters, or there may be a megohm (MΩ) input impedance value, such as 10 MΩ, for digital multimeters. **See Figure 7-10.** To ensure proper megohmmeter function, use the following procedure:

1. Connect a multimeter to a megohmmeter. The common (black) terminal of the multimeter is connected to the output terminal of the megohmmeter. The positive (+) terminal of the megohmmeter is connected to the high-voltage input terminal of the multimeter (1000 V terminal in this example).
2. Rotate the output voltage selector switch of the megohmmeter to a position where voltage output does not exceed maximum voltage of the multimeter.

   Note: With older megohmmeters (with terminals marked "Earth" and "Line," rather than "+" and "–"), the earth terminal is positive (+), and the line terminal is negative (–). When using any DC insulation tester, the negative (–) lead should be connected to the conductor under test, and the positive (+) lead should be connected to ground. When the negative lead is connected to the conductor under test, moisture will be attracted to that negative potential. Since moisture is one of the contaminants being tested for, this will ensure a better test result. This is a good practice for all types of electrical equipment, such as transformers and power cables.
3. Insert the test lead into the jack having the highest indicated voltage on the multimeter. This ensures that the input impedance is at the stated value.
4. Apply voltage to the multimeter at the maximum voltage for the specified meter input. The multimeter registers the applied voltage while the megohmmeter registers the input impedance of the multimeter.

TECH TIP

A megohmmeter is useful for evaluating low-voltage equipment and conductors. However, a megohmmeter should not be used as the only method to evaluate insulation quality on medium- and high-voltage devices. For devices above 15 kV, a megohmmeter is only adequate for an initial test or to verify that the primary insulation is good enough for further testing.

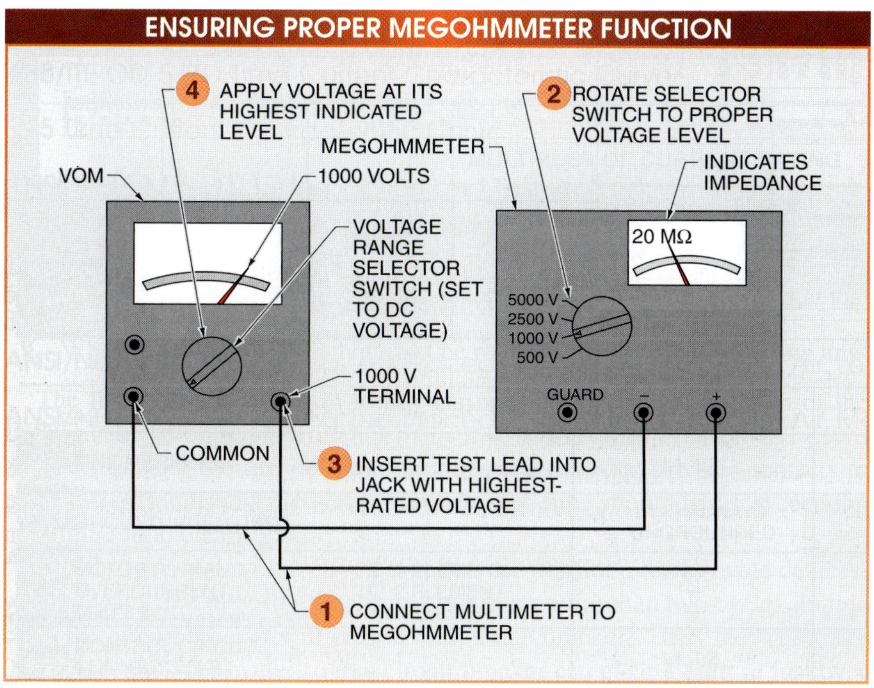

Figure 7-10. A multimeter or volt-ohm meter (VOM) can be connected to a megohmmeter to ensure proper megohmmeter function.

Guard Circuits

On many megohmmeters, there is a third terminal that is marked "G" or "Guard." Many field technicians do not understand the purpose of this very important circuit. The guard circuit is a means to eliminate unwanted return current from the measurements being taken. It is very similar to the unmetered return on DC high-potential test sets. Any return current that hits the guard bypasses the metering and is not measured. This allows an accurate measurement to be taken. There are two ways a guard can be used: by dividing the insulation system into smaller pieces and by eliminating surface leakage. On circuit breakers, the primary purpose of the guard circuit is to eliminate surface leakage.

Surface Leakage Elimination. The testing of circuit breakers is straightforward. Under normal conditions, there is not a lot of surface leakage to be concerned with. However, when the humidity is high (above 70%) or when the bushings are dirty, surface leakage can be an issue. The guard circuit will eliminate surface leakage and allow a more accurate test result.

To guard bushings and eliminate surface leakage, some type of conductive guard collar needs to be wrapped around the bushing usually either near the top of the bushing or under the first petticoat. This is done to each bushing, and then the bushings are tied together. **See Figure 7-11.** The lead from the bushings is then connected to the guard circuit on the test instrument. The conductive collar can be small-gauge wire, screw-type hose clamp, metal foil, or one of the conductive rubber collars used for insulation power factor. The closer the conductive collar conforms to the surface of the bushing, the better it will work. The guard circuit works independently of the high-voltage circuit, so it will not interfere with the other test connections. Using the guard circuit will provide the most accurate results when ambient conditions are less than what is desired.

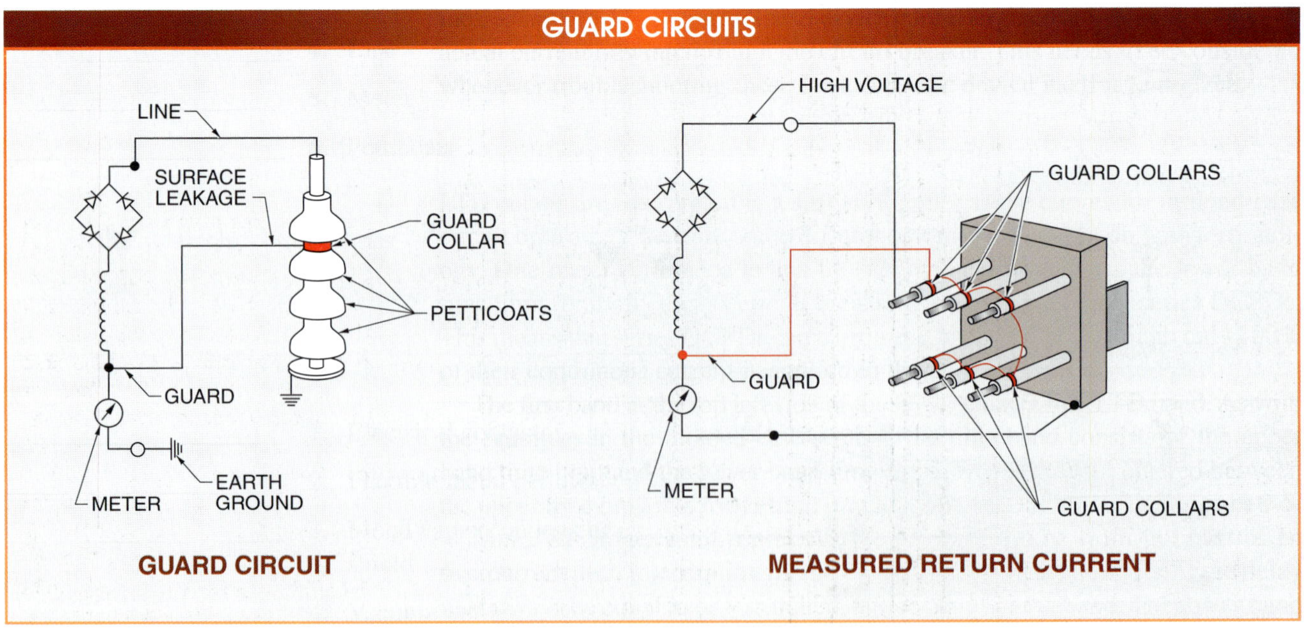

Figure 7-11. When a guard circuit is used, surface leakage travels to the guard collar and is directed above the meter. Any return current that goes above the meter is not measured. It is eliminated from the measurement. Any current that flows through the meter is measured.

DC Overpotential Testing of Molded-Case Circuit Breakers

The insulation resistance of molded-case circuit breakers must be tested, especially since the cases not only enclose and support the components but also are the main insulating component. They can deteriorate due to heat, moisture, and carbon tracking in the same manner as other insulating structures that contain intense heat during arc faults.

In-service circuit breakers must be inspected or tested after opening under fault conditions. Per Occupational Safety and Health Administration (OSHA) 29 Code of Federal Regulations (CFR) 1910.334—*Use of Equipment, "After a circuit is deenergized by a circuit protective device, the circuit may not be manually reenergized until it has been determined that the equipment and circuit can be safely reenergized. The repetitive manual reclosing of circuit breakers or reenergizing circuits through replaced fuses is prohibited."*

If circuit breakers are low-voltage, sealed-case, molded-case circuit breakers, they cannot be opened and inspected. The DC overpotential test is best for determining the overall condition of a sealed-case, molded-case circuit breaker that has interrupted a short circuit. The insulation resistance test should be performed for 1 min per ANSI/NETA MTS-2011 Table 100.19.

Note: Actual test results are dependent on the mass of the insulation being tested, temperature of the insulating material, and humidity of the environment at the time of test. In addition, insulation resistance tests are performed to establish a trending pattern and a deviation from the baseline information obtained during maintenance testing. This enables the evaluation of the insulation for continued use.

Safety Checks Using Megohmmeters

A safety check is performed to ensure that no conductive object, such as a metal screw, washer, or bolt, has made contact with the internal components of a circuit breaker to cause a phase-to-phase or line-to-load short circuit. A safety check is performed by using a 500 VDC megohmmeter and testing phase-to-phase, phase-to-ground with the circuit breaker closed, and line-to-load with the circuit breaker open.

The test voltage is applied only long enough to indicate resistance, which should be out of scale. The proper time to perform this test is when a circuit breaker has been removed from its cubicle or cabinet and left unattended or after maintenance work has been completed. Any resistance reading indicates that something is damaged or out of place inside the circuit breaker. Any problems must be corrected before placing the circuit breaker back into service.

Control Wiring

Control wiring inside the circuit breaker is critical to proper circuit breaker functioning. Therefore, the insulation of control wiring must be tested. Insulation rated for 300 V should have 500 V applied to it. Insulation rated for 600 V requires the application of 1000 VDC. The insulation resistance must be measured using a megohmmeter, and each wire must be tested to ground (frame of the circuit breaker on low-voltage and medium-voltage power circuit breakers). According to ANSI/NETA MTS-2011 a minimum insulation resistance of 2 MΩ for insulation resistance of control wiring is recommended.

TECH TIP

Hand-cranked megohmmeters may output less voltage than what they are rated for. Often, a 1000 V hand-cranked unit will only output 800 V to 900 V. The small reduction in output voltage from hand-cranked megohmmeters will not affect the test results appreciably. The reduction of voltage output in hand-cranked megohmmeters is due to clutch slippage on the hand-crank mechanism. This slippage prevents damage to the small generator inside the megohmmeter. The output voltage from battery-powered and line-powered units will be close to the specified voltage.

AC INSULATION TESTS

AC insulation tests for medium-voltage power circuit breakers, switchgear, and related equipment include the AC overpotential test, vacuum bottle integrity test, and insulation power factor test. AC insulation testing requires the proper evaluation of test results to determine what action, if any, needs to be taken.

AC Overpotential Tests

Charging currents do not decrease during AC overpotential tests and can mask the leakage current (I_L), which prevents taking a direct measurement of the leakage current component. AC overpotential tests are considered to be destructive since they create high charging currents that reduce insulation life. An AC overpotential test is known as a proof test. It is used by OEMs on new power equipment to verify that the device is safe to be placed into service or to "prove" that it is suitable for continued service. Any time a higher-than-rated AC voltage is applied to insulation, some insulation degradation occurs. The AC overpotential test is only performed on suspect electrical equipment and is not a normal maintenance test. The chance of failure when performing this test is fairly high.

One common exception to the use of AC hipot testers is when testing vacuum circuit breakers (VCBs). A vacuum bottle integrity test is a common field test that is usually performed with an AC hipot tester. DC voltage could be applied, but many DC hipot testers do not generate a true DC voltage output. Rather, output voltage spikes can be eight times the voltage indicated on a voltmeter.

Extremely high voltage spikes can damage a vacuum bottle by exceeding the safe voltage limits. Voltage spikes are caused by hipot testers that do not properly rectify and filter their outputs. There are several DC overpotential test sets that offer full-wave rectification and proper filtering. **See Appendix.** There may be other test sets that have a safe output voltage, so the manner in which the AC voltage is rectified into DC should be verified with the test equipment OEM prior to testing. An exception to the destructive aspect of AC testing is when using insulation power factor test equipment. With power factor testing, the rated voltage of the insulation is never exceeded, so the insulation is not damaged.

Vacuum Bottle Integrity Tests

An arc cannot occur in a perfect vacuum. It is impossible to create a perfect vacuum, so vacuum bottles have a minor amount of arcing that takes place in them. The arc drawn inside a vacuum bottle is small and is quickly extinguished, but it does cause some of the contact faces to be vaporized. This process slowly causes erosion on the contacts and eventually they require replacement.

It is damage and wear to the flexible metal bellows that can cause vacuum bottles to fail. The constant flexing over a large number of operations causes the bellows to embrittle and crack, which causes leakage. Since the bottle contains a vacuum, air leaks into the bottle and the vacuum is lost. Often a circuit breaker is able to safely interrupt load current since the load is removed from all three phases at the same time. Bottle failure usually occurs when the circuit breaker is closed, especially if it is closed into a circuit with a high inrush current or into a fault. Bottle failure can cause the bottle to rupture. **See Figure 7-12.**

A DC hipot tester can be used to perform a vacuum bottle integrity test, although most testing facilities prefer to use an AC hipot tester, which is usually what is recommended by the OEM. If a DC test instrument is used to perform a vacuum bottle integrity test, it must use full-wave rectification of its output. Other types of DC hipot testers must not be used unless it can be verified by the OEM of the test equipment that the equipment uses full-wave rectification.

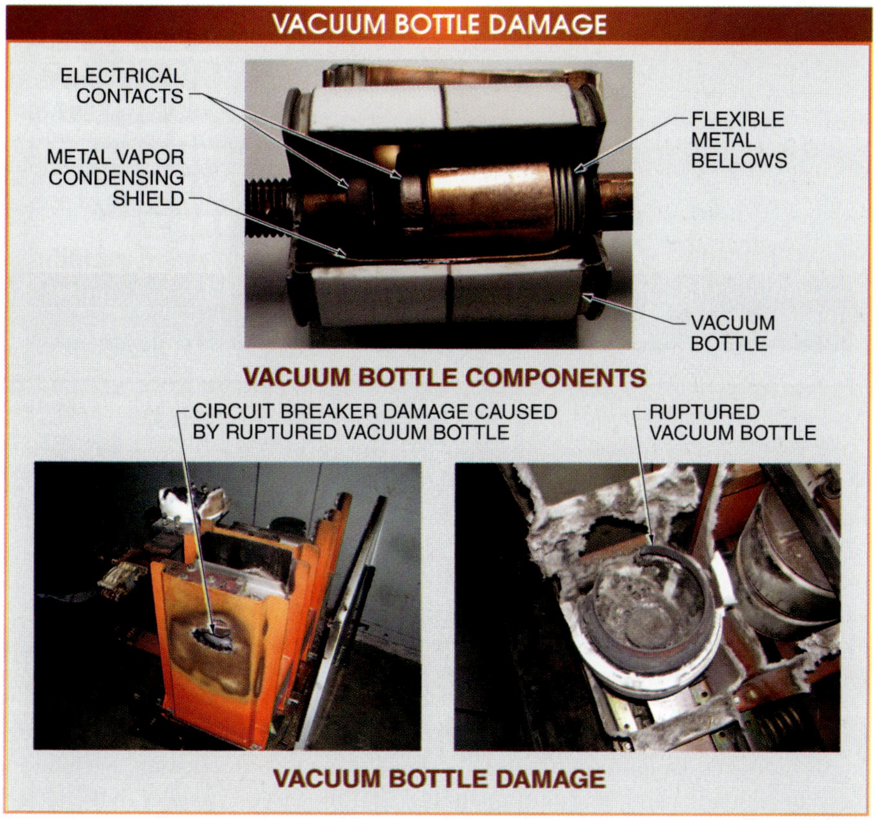

Shermco Industries

Figure 7-12. Damage and wear to the flexible metal bellows can cause problems, such as ruptured vacuum bottles.

Most vacuum bottle integrity testers are actually AC hipot testers with no ammeter and can have a fixed voltage output. The vacuum bottle integrity test is a go, no-go test, and normally lasts for 1 min. If the circuit breaker in the test set does not trip during the test, the vacuum bottle is suitable for continued service. The leakage current through the bottle is high enough to indicate failure of the vacuum within the bottle if the circuit breaker in the tester trips. The vacuum bottle integrity test does not provide an indication of remaining bottle life. It only ensures that the bottle has not failed at the time of test. The magnetron atmospheric condition (MAC) test provides information as to the remaining life of a vacuum bottle.

Test Equipment and Test Connections. Testing vacuum bottles is performed using the appropriate test connection. With the circuit breaker open, voltage is applied across each vacuum bottle. The ground lead of the tester is connected to the chassis of the circuit breaker, the high-voltage probe is connected to one side of the bottle, and the return lead is connected to the other side. Since this is a go/no-go test, failure is indicated by the tripping of the hipot tester. The voltage recommended by the OEM for the specific circuit breaker and bottle being tested should be applied. Common test voltages tend to vary by OEM and model depending on the specific vacuum bottle. **See Figure 7-13.**

VACUUM BOTTLE TEST CONNECTIONS

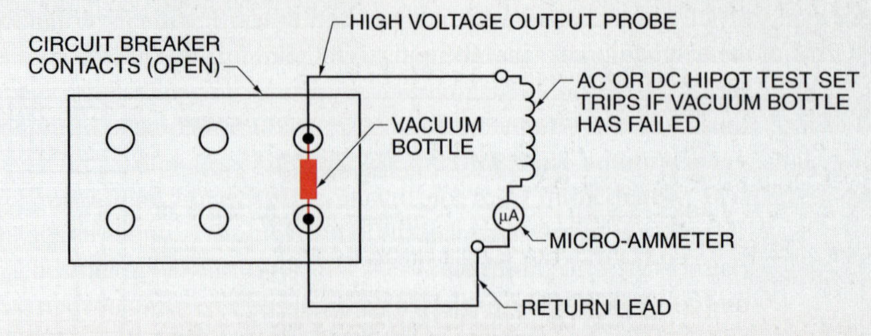

Test Voltages for Vacuum Bottle Integrity Tests

Manufacturer	Model or Type	Voltage Class*	AC Voltage*	Test Interval†
GE	PowerVac	≤ 15	36	5-10
Cutler-Hammer	VCP-W	≤ 15	27	60
ABB/BBC	VHK PV-	15	36	60
Powell	STD/CDR PV-	5	15	60
Powell	STD/CDR	8.25/15	27	60

* in kV
† in sec

Figure 7-13. Vacuum bottles are tested with the circuit breaker contacts open and with a hipot tester.

Evaluating Test Results. Because leakage current cannot be measured using an AC hipot or vacuum bottle tester, it cannot be evaluated. The only evaluation that can be made using this equipment is that if the circuit breaker in the tester trips, the vacuum bottle has failed. If the circuit breaker in the test set does not trip, the vacuum bottle is suitable for continued service. Although leakage current can be measured using a DC hipot tester, the same result criterion is used, and leakage current is typically not recorded.

> **TECH TIP**
>
> *GE technical bulletins GEK-86132 and GEK-39676 provide lists of DC hipot testers that use full-wave rectifiers and do not produce voltage spikes.* **See Appendix.**

Insulation Power Factor Tests

Insulation failure is usually unpredictable and hazardous, which can be dangerous to individuals and equipment nearby. The appreciable associated costs of repair, replacement, and service disruption must also be considered. An insulation power factor test is considered to be a more sensitive test than the DC overpotential test for determining insulation quality. It also helps to ensure that the electrical equipment and insulating materials in equipment, such as transformer stations, energized switchgear, and industrial facilities, are in proper condition.

General Power Factor Testing Considerations. As with many tasks, proper preparation is required for obtaining reliable test results when performing insulation power factor tests. Contamination caused by moisture, dust, or other conductive particles can give a false indication that the insulation has deteriorated or is failing. Surface leakage is one cause of false high insulation power factors. The error can be corrected by thorough cleaning. In applications where surface leakage is present, using a guard circuit may be necessary to eliminate the leakage current and obtain accurate results. The temperature of the device under test is important because the insulation power factor can change with temperature.

Porcelain and fiberglass components are not affected as much as ceramic, asbestos, or organic insulation, but many older circuit breakers use organic insulation. For best results, testing should be performed in warm, dry weather with the device temperature no less than 0°C (32°F). The ambient temperature should be recorded. Air and vacuum circuit breakers do not require any temperature correction since their results are recorded in watts-loss.

Power factor testing is known as insulation power factor or dissipation factor testing. Since the voltage rating of the device under test is not exceeded, insulation power factor testing is considered nondestructive. Insulation power factor testing can be performed at almost any voltage. However, it is standard procedure to apply the highest voltage possible, up to the rating of the insulation, on the device being tested.

Typically, 2.5 kV is usually adequate for devices tested in an outdoor substation up to about 161 kV. As the voltage rating of the substation increases, so does the test voltage. Higher voltages are necessary to overcome the electrostatic interference created by nearby energized equipment and lines. Depending on the clearances and configuration of the lines, it may also be necessary to use interference-cancelling circuitry on the test set or other means to eliminate interference. Electrostatic interference causes the meter to constantly change values, so an accurate measurement cannot be taken.

Insulation power factor measurement is an indirect method of measuring the losses (leakage current) inside a circuit breaker, transformer, or other electrical device. In an AC circuit, leakage current cannot be measured directly because the charging currents are constantly high and mask the much lower value of the leakage current.

When AC voltage is applied across good insulation (dielectric), leakage current through the insulation leads the applied voltage by an angle of nearly 90° since the device being tested appears to the test set as a capacitor. In a purely capacitive circuit, voltage leads current by 90°. The capacitor in this case is created by the conductor (the circuit breaker current path) acting as one plate, the circuit breaker insulation acting as the dielectric, and the chassis of the circuit breaker or ground acting as the second plate.

SAFETY TIP

Vacuum bottle failure can produce X-rays. Never apply voltages higher than those recommended by the circuit breaker OEM as this can cause vacuum bottle failure. Vacuum bottle failure can occur during the test. Use the circuit breaker as a shield when performing this test, and position the test set as far from the circuit breaker as practical. The potential amount of X-rays produced is small, but it is better to take precautions when testing. Always adhere to the warning labels affixed to vacuum bottles.

Shermco Industries

The test voltage and the leakage current are essentially in-phase with each other. In an electrical circuit, when the voltage and current are in-phase, power is created. Since the current is "leaking" through the insulation to ground, the watts produced are referred to as dielectric watts-loss. The "dielectric" is the insulation and "watts-loss" is the leakage current to ground through the dielectric times the applied voltage. Leakage current (I_L) is added to the capacitance current (I_C). The vector addition of these two currents results in the total current (I_T).

Two currents or voltages that are not in phase with each other cannot be added together directly. For example, in three-phase circuits, there can be three transformer windings interconnected to make either a wye- or a delta-configured winding. If each wye-connected winding's phase-to-ground voltage is 277 V and a phase-to-phase measurement is taken, the sum of the two windings will not be 554 V. It will be 480 V. The reason for this difference is that the voltages are not in phase with each other but are slightly out of phase, so they do not add directly. The same is true when performing an insulation power factor test. The capacitance current is not in phase with the leakage current, so they do not add directly.

The angle at which the total current leads the voltage is referred to as angle theta *(θ)*. The cosine of this angle is the power factor. When multiplied by 100, it is the percent power factor. **See Figure 7-14.**

INSULATION POWER FACTOR VECTORS

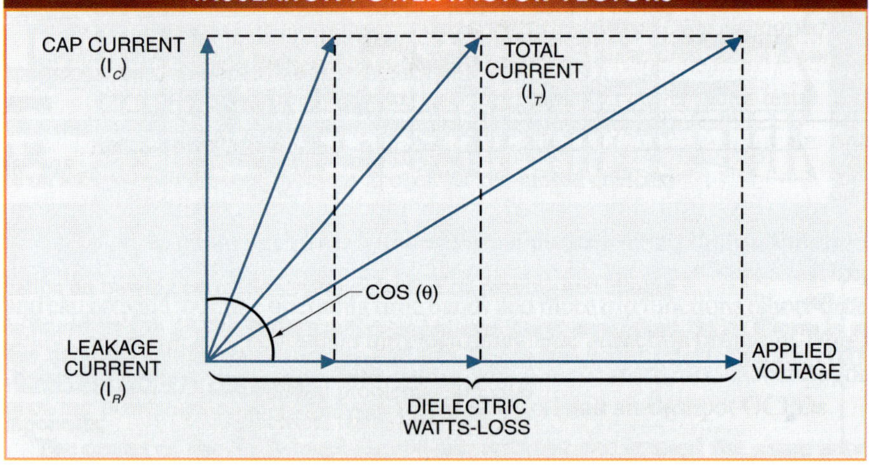

Figure 7-14. In an electrical circuit, the current "leaks" through the insulation to ground, and the watts produced are referred to as dielectric watts-loss. "Dielectric" is the insulation and "watts-loss" is the leakage current to ground through the dielectric times the applied voltage. As leakage current increases, the angle between the total current and the applied voltage decreases.

Due to impurities in insulation material and other variables, there is a certain amount of leakage current (loss) through all types of electrical insulation. Leakage is typically low, especially when the device is new. As a transformer, circuit breaker, or other electrical device ages, its insulation slowly deteriorates. This deterioration causes the insulation power factor to increase slowly until the device reaches the end of its service life. By measuring this loss of life through insulation power factor testing, any significant insulation problem can be detected. Problems can hopefully be detected early enough to remove it from service before it fails.

Perfect insulation theoretically has an angle of 90° between the total current and the dielectric watts-loss, which results in a power factor of 0% (cosine of 90° = 0). However, this condition does not occur in field applications. Depending on the type of insulation, it is more likely that insulation exhibits some loss through it, which results in a power factor of 0.5% or more. For most medium-voltage mineral-oil-insulated devices, an insulation power factor of 1% or less is typical. This translates into an angle of 89.427° between the dielectric watts-loss and the total current.

As insulation deteriorates, leakage current increases, which causes the angle between the watts-loss and total current to decrease. As leakage current increases, the sum of it and the total current add up (by vector addition), and the sum increases. However, it also causes the angle between the watts-loss and the total current to decrease. As this angle decreases, the cosine of the angle also increases, which increases the insulation power factor. Insulation power factor test equipment measures the cosine of the angle between the dielectric watts-loss and the total current and multiplies that value by 100. The result is the percent power factor. The percent power factor of the insulation under test can be calculated by applying the following formula:

$$pf = \frac{W_A \times 100}{E \times I_T}$$

where

pf = power factor (in %)

W_A = watts absorbed in insulation (in mW)

E = applied voltage (in V)

I_T = total current (in A)

Since watts is equal to voltage multiplied by current, the voltage cancels out in the equation leaving only:

$$pf = \frac{E \times I_L}{E \times I_T}$$

$$pf = \frac{I_L}{I_T}$$

where

E = applied voltage (in V)

I_L = in-phase loss current (in mA)

I_T = total current (in mA)

The in-phase loss current is very small compared to total current (I_T) in good low-loss electrical insulation. The power factor is the ratio of leakage current divided by total current or the cosine of angle theta *(θ)*. On Doble power factor test sets, power factor is calculated by applying the following formula:

$$pf = \left(\frac{W \times 10}{mA}\right) \times 100$$

where

pf = power factor (in %)

W = power (as read on the test set)

10 = constant

mA = total current as read on the test set

100 = constant

The capacitance of the insulation is also a useful diagnostic tool, especially when used in conjunction with the insulation power factor. Changes in the normal insulation capacitance may indicate such conditions as moisture, shorted layers of insulation, or open circuits in the capacitance network. An increase of insulation power factor accompanied by a marked increase in capacitance usually indicates excessive moisture in the insulation. An increase of insulation power factor with no increase in capacitance may indicate thermal deterioration or contamination by substances other than water.

Capacitance and dissipation factor (C and DF) equipment uses the tangent of the angle (dielectric loss angle) between the total current and the capacitance current. Dissipation factor and power factor are essentially the same up to a 10% power factor.

Preparing Circuit Breakers for Testing

Accurate test results and safe testing are often the result of good preparation and procedures. When preparing a circuit breaker for testing, the following minimum guidelines must be followed:

- The proper clearances must be obtained, and it must be ensured that the circuit breaker is in an electrically safe work condition before proceeding with work. The switchgear must be locked out, tagged out, and grounded if testing or maintenance is to be performed on it.
- Any "as-found" tests must be performed as required. These can include any of the DC tests as well as insulation power factor test.
- All insulation must be clean and dry.
- All bushings must be grounded and shorted. Bushings will have to be temporarily ungrounded during testing.

Testing Vacuum Circuit Breakers

Draw-out circuit breakers are usually tested separate from the switchgear. Substation-class oil, gas, and vacuum circuit breakers may be connected to bus or cable. Maintenance tests should be performed in the same way as the acceptance tests. If bus or cable were connected during the acceptance tests, maintenance tests should be performed in the same configuration. Anything connected to the circuit breaker is also tested. If the insulation power factor is not within expected values, the circuit breaker must be disconnected from all bus or cable and the tests must be run again.

There are usually two tests performed for insulation power factor values on draw-out air and vacuum circuit breakers: the guarded-specimen test (GST-L-guard) and the ungrounded-specimen test (UST). All tests are performed with the circuit breaker contacts in the open position. **See Figure 7-15.**

Evaluating Test Results—Vacuum Circuit Breakers. Doble® Engineering recommends performing a grounded-specimen test (GST-L-ground) on outdoor, sub-station-style vacuum circuit breakers. Doble Engineering recommends a percent power factor of 2% or less for the GST-L-guard on outdoor vacuum circuit breakers. The GST-L-guard and the UST should be performed on draw-out-style circuit breakers.

Circuit breakers are low-capacitance devices. If the measured current is less than 300 μA, the percent power factor should not be recorded, only the watts-loss.

If the measured current is greater than 300 μA, it should be tested at several different voltages, and the percent power factor and capacitance should be recorded. Otherwise, only the watts-loss should be recorded. The readings of the GST-L-guard should be compared to the readings of other bushings on the same circuit breaker tested and to previous test results. Results can also be compared to the Doble Engineering *Power-Factor Test Data Reference Book* and to other similar circuit breakers. The UST is used primarily to test the vacuum interrupter, although the interphase barriers are also tested. The vacuum interrupter is a very low-current specimen, so the watts-loss is used for evaluation. Typically, losses of 0.010 W (10 mW or less) should be measured.

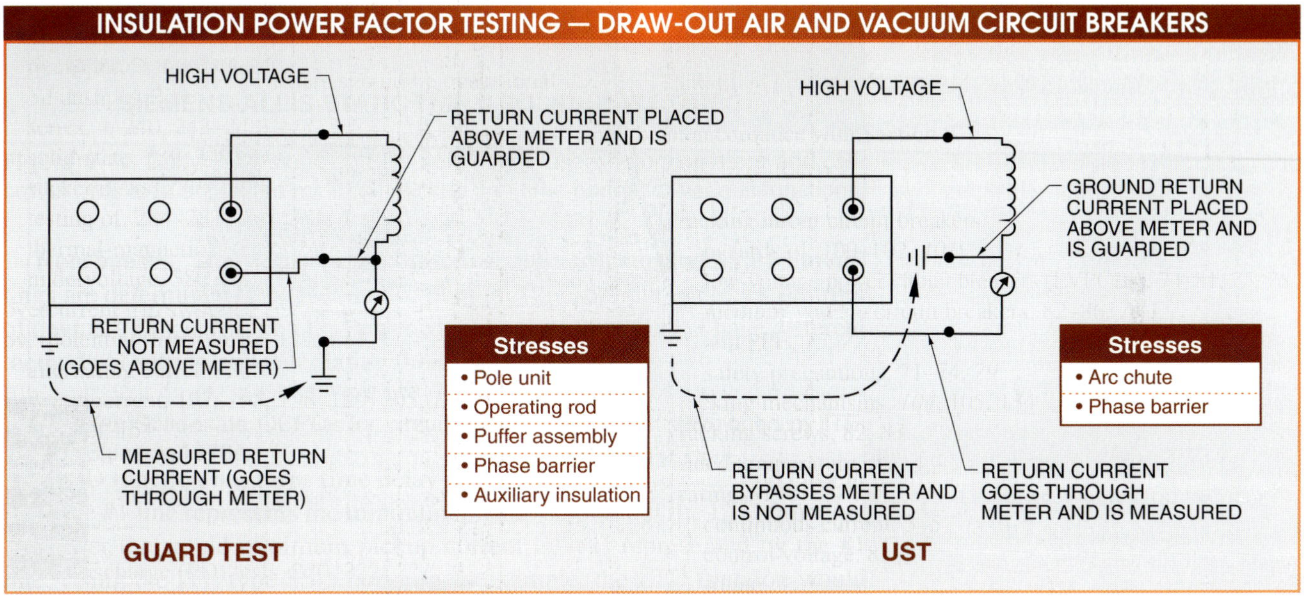

Figure 7-15. The two tests for insulation power factor on draw-out air and vacuum circuit breakers are the guarded-specimen test (GST-L-guard) and the ungrounded-specimen test (UST).

Insulation Power Factor Testing—Air-Magnetic Circuit Breakers. Testing air-magnetic circuit breakers is similar to testing vacuum circuit breakers. A GST-L-guard and a UST are performed. Preparation for testing and general evaluation of the test results are the same for both vacuum and air-magnetic circuit breakers. The UST is performed in the same manner that it was on the vacuum circuit breaker. The arc chute is the primary insulation structure being tested using the UST.

> **TECH TIP**
>
> *Doble® Engineering Company of Watertown, MA developed the first practical field test set for insulation power factor testing in 1929.*

Evaluating Test Results—Air-Magnetic Circuit Breakers. Insulation power factors tend to be higher with older air-magnetic circuit breakers due to the types of insulation systems used, even when they are dry and in good condition. Normally, watts-loss is evaluated on air circuit breakers. However, General Electric (GE) has a service advisory letter (SAL) that recommends a 5% maximum power factor, with 10% indicating dangerously deteriorated insulation. The Doble Engineering *Power-Factor Test Data Reference Book* can be referenced or results can be compared to previous tests and similar circuit breakers being tested. Watts-loss on air-magnetic and vacuum circuit breakers tends to be considerably low (usually less than 0.1 W), although this can vary depending on circuit breaker type and voltage. **See Figure 7-16.**

Note: The insulation power factor test results for bushings 1, 3, 5, and 6 are considerably higher than those for bushings 2 and 4. The watts-loss result on these bushings is also higher, which confirms the evaluation using watts-loss. The watts-loss result on the 5 to 6 test is also much higher than the other two results and is not consistent with expected results.

EXAMPLE INSULATION POWER FACTOR TEST RESULTS—MEDIUM VOLTAGE AIR CIRCUIT BREAKERS

Bushing #	Test kV	I (mA)	Watts	% PF
1	10	1.023	0.952	9.31
2	10	1.007	0.78	3.75
3	10	0.992	0.645	6.50
4	10	1.047	0.363	3.47
5	10	0.856	0.442	5.16
6	10	1.092	0.569	5.21
1 to 2	10	0.148	0.418	
3 to 4	10	0.154	0.453	
5 to 6	10	0.175	0.648	

Note: The inconsistent (shaded) test results indicate potential problems with the circuit breaker.

Figure 7-16. When performing insulation power factor tests on medium-voltage air circuit breakers, watts-loss and the power factor percent can be compared for consistency.

Test results that are not in the normal range can be investigated further by using the power factor tip-up test. With the power factor tip-up test, tests are conducted at several voltages and the results are plotted. **See Figure 7-17.** Some small increase in percent power factor can normally be seen. On 15 kV circuit breakers, the power factor at 2 kV is compared to the power factor at the maximum voltage that can be applied.

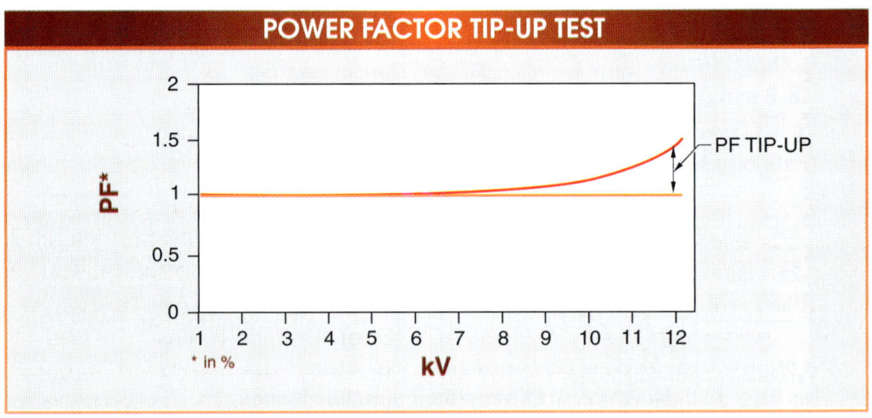

Figure 7-17. Insulation power factor test results that are not in the normal range can be investigated further by using the power factor tip-up test.

CONTACT RESISTANCE TESTS

A contact resistance test measures the conductivity of the entire current path, not just the resistance between the contact faces, which is what the name implies. There are numerous contact points within the current path, including the moving contact pivot, which is the cause of many maintenance problems. In order for the circuit breaker to carry the current it is rated for, all components within the current path must meet OEM specifications. Eroded contact faces, thick or hardened lubricants, loose or misadjusted parts, and inadequate contact pressure can cause the current path resistance to greatly increase and severely damage the circuit breaker.

Current flow through a conductor causes I^2R losses, which translates into heat. The higher the circuit resistance, the higher the I^2R losses and the more heat is produced. An increase in heating causes increased circuit resistance, and even more heat is created.

The moving contact pivot is one of the primary causes of increased current path resistance. The moving contact pivot is always lubricated at the factory and prevents binding and gouging of the metal components. As current flows through the circuit breaker, the heat (I^2R losses) causes the lubricant to dry out. Over a period of time, the lubricant can no longer perform its function. It dries and thickens, which slows the circuit breaker, then begins to flake off. As a result, metal-to-metal wear begins. **See Figure 7-18.**

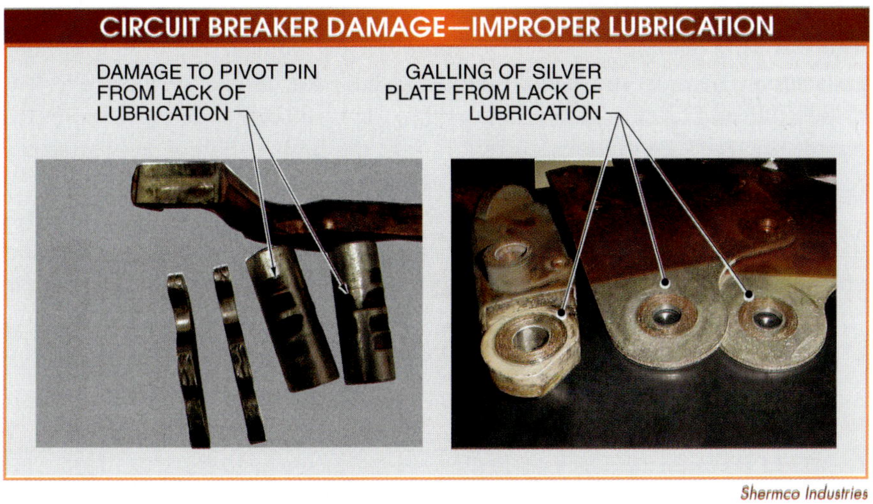

Figure 7-18. Thick or hardened lubricants can cause the current path resistance to greatly increase, which causes severe damage to circuit breakers.

The contact resistance test is performed using a micro-ohmmeter. DC current is passed through the current path, while the voltage drop is measured. Micro-ohmmeters convert this into a microhm or milliohm reading. Test set connections are made from line to load with the circuit breaker closed. Any resistance in the current path, such as the contacts, pivot point, or primary disconnects, is measured. As the lubricant dries, the resistance gradually increases. Erosion of the contacts also causes readings to increase, so an inspection will be required if readings are high or uneven.

The leads must not be connected to the primary disconnect fingers. Connections must be made directly to the bare metal surface just to the rear of the primary disconnect, unless it is connected to a piece of correctly sized bus. When making connections for a typical air-magnetic or vacuum circuit breaker, connecting C_1 and C_2 allows current to flow through the current path. P_1 and P_2 measure the voltage drop across the resistance (current path). P_1 and P_2 are connected to the inside of the C_1 and C_2 connections to eliminate measuring the resistance of the C_1 and C_2 connections. **See Figure 7-19.**

MICRO-OHMMETER TEST CONNECTIONS—CONTACT RESISTANCE TEST

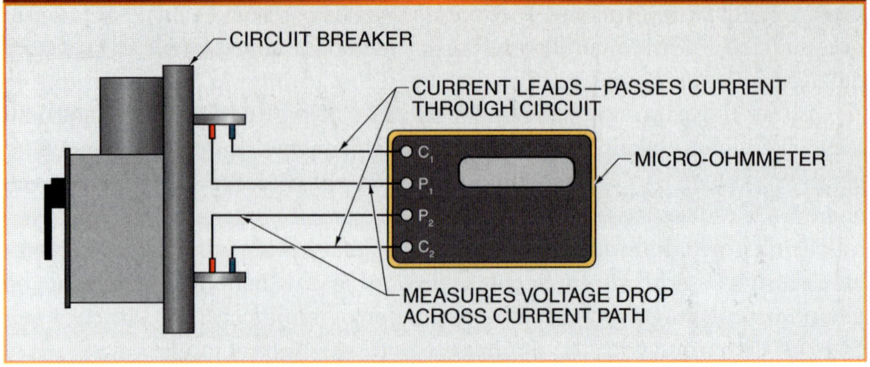

Figure 7-19. A contact resistance test is performed using a micro-ohmmeter. DC is passed through the current path while the voltage drop is measured.

Evaluating Test Results—Contact Resistance Testing

There is little documentation concerning what is considered to be too much resistance. The most prevalent theory is to have the contact resistance as low as possible and as even as possible within the three contacts. The smaller the circuit breaker's continuous current rating, the higher the contact resistance readings will be since the contact assemblies are smaller.

A 480 V, 4000-amp frame circuit breaker should give a reading of less than 30 µΩ if it is in good to excellent condition. Smaller-frame circuit breakers may give readings of 60 µΩ to 80 µΩ when they are in good to excellent condition. Molded-case circuit breakers can give contact resistance readings in the hundreds of microhms. Occasionally, an OEM will recommend values.

MAINTENANCE TIP

The circuit breaker shop at Shermco Industries uses the following "Rule of Thumb" for low-voltage power circuit breakers. Readings up to 100 µΩ are good, and readings that exceed 300 µΩ can indicate possible problems. For medium-voltage circuit breakers, the values need to be somewhat lower: 40 µΩ is good, and 200 µΩ indicates the need to investigate for problems. Any value in between may indicate that the lubricant is beginning to dry out, but the problem plateau has not been reached. The best time to perform this test is when the circuit breaker is new or recently relubricated, and future test results can be compared to them .

The circuit breaker should be retested annually, and the changes should be analyzed for trends. All circuit breakers of the same OEM, frame size, and type should have similar resistance values. Comparing tests between similar circuit breakers often reveals a circuit breaker that is starting to have issues. Performing this test also exercises the circuit breaker, which needs to be done as well.

The 50% rule is employed in ANSI/NETA MTS-2011. According to the rule, the highest reading should not be more than 50% of the lowest value. For example, a circuit breaker has readings of 125 µΩ, 199 µΩ, and 167 µΩ. The lowest reading is subtracted from the highest (199 µΩ – 125 µΩ = 74 µΩ). The result, 74 µΩ, is more than 50% of the lowest reading, so the circuit breaker shows signs of lubrication failure and requires service. If the circuit breaker is not serviced at this point, metal-to-metal wear could begin. This would cause the circuit breaker to slow down, go into a trip-free condition, or even seize.

MECHANICAL TIMING

Whenever an operating mechanism is disassembled or repaired, it may be necessary to perform a timing test. The timing test is not required during normal maintenance, although some companies perform it as part of their normal major maintenance tasks. During the test, the following circuit breaker characteristics can be checked:

- operating time (velocity or speed)
- wipe
- travel (displacement or stroke)
- contact bounce
- overtravel

Depending on the make, model, and options ordered, the specific time-travel analyzer can show other measured and calculated values. There is a wide variety of test sets available. It is important to be familiar with the test equipment and software prior to using it in the field. Modern time-travel analyzers have a transducer connected to the operating mechanism that is used to measure travel while test leads are connected to line-side and load-side bushings. The leads connected to the bushings capture when the contacts open and close and also provide some measurement of contact bounce. **See Figure 7-20.** Typical tests performed with a time-travel analyzer are the following:

- open (trip)
- close
- open-close
- close-open
- open-close-open

> **SAFETY TIP**
>
> *Always test for the absence of voltage before connecting a microohmmeter. If the test instrument is connected to an energized circuit, injury and/or equipment damage may occur.*

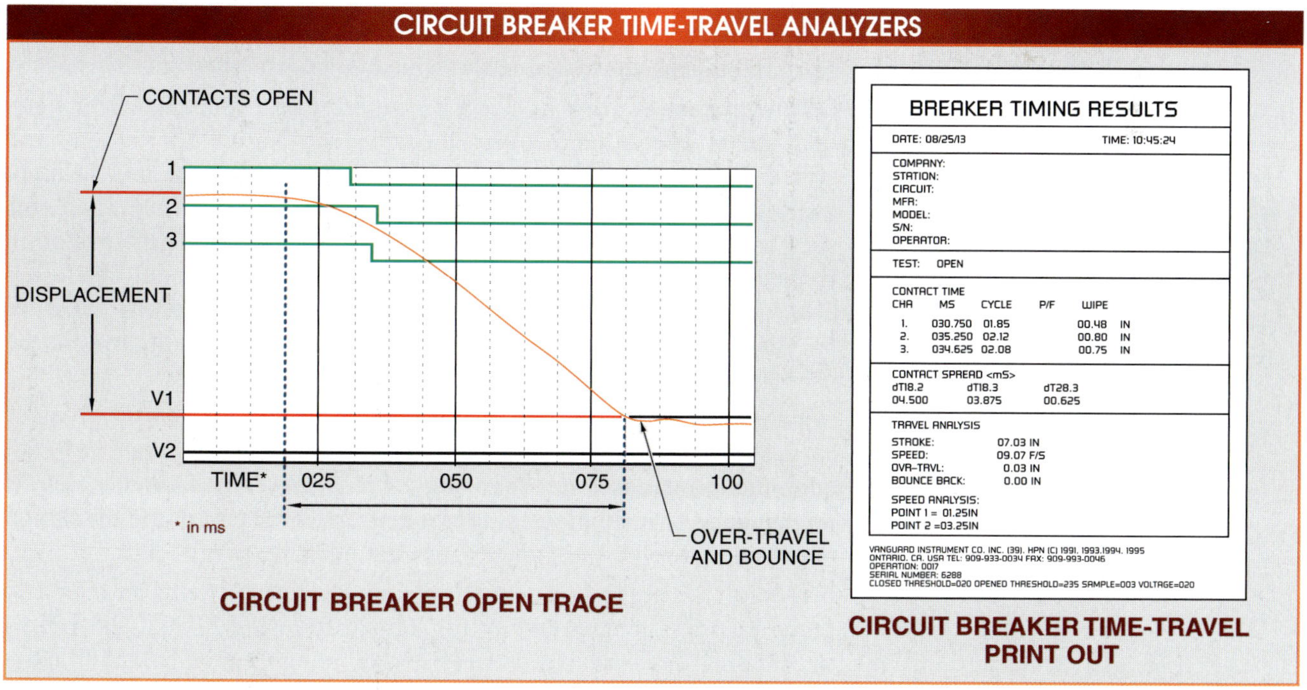

Figure 7-20. Circuit breaker time-travel analyzers have a transducer connected to the operating mechanism that is used to measure travel. Contact leads are connected to line-side and load-side bushings. The leads connected to the bushings capture when the contacts open and close and also provide some measurement of contact bounce.

SWITCHGEAR INSULATION TESTS

Switchgear and circuit breakers are usually maintained at the same time, especially in actual field applications. Most end-user companies have difficulty scheduling the downtime and human resources required to perform maintenance in the manner that OEMs recommend. Therefore, it is usually performed every two to three years. When maintenance is performed, the power system is shut down, the switchgear is isolated and grounded, all available equipment is serviced, and any device showing problems is replaced or taken off site and repaired.

Switchgear is often overlooked in terms of its importance for the proper functioning of the circuit breaker. If the switchgear has corrosion or mounting problems or has tracking from dust and corona, failure of the power system could result. Per ANSI/NETA MTS-2011, the following tests and maintenance practices are recommended to be performed on switchgear:

- resistance tests—All bolted connections are tested using a micro-ohmmeter. Connections that are clean and tight should have a maximum of 20 μΩ of resistance across them. Per ANSI/NETA MTS-2011, the 50% rule should be used for this test. Bolted connections on any phase should not be more than 50% higher than the other two phases. As alternatives, the torque on each bolt can be checked using a calibrated torque wrench, or a thermographic survey can be performed. **See Figure 7-21.**
- insulation resistance tests—The bus insulation is tested phase-to-phase and phase-to-ground for 1 min using a megohmmeter. This test ensures that the switchgear is not close to imminent failure. If the insulation resistance test result is not in accordance with ANSI/NETA MTS-2011 Table 100.1, the overpotential test should not be performed. Technicians must be aware that control power transformers connected to the system will also be measured. If they are grounded, this will appear as a ground during the test, and they must be ungrounded.
- DC overpotential tests—Once the insulation resistance has been verified, the overpotential test can be performed. The higher voltages and greater capacity of the hipot tester will stress the insulation, and the hipot tester will find issues that the megohmmeter will not find. Problems such as tracking, moisture intrusion, and weak or damaged insulation will be detected and can be repaired before it fails. It should be tested phase-to-phase for 1 min with the other phases grounded. If there are no OEM-supplied specifications, recommended test voltages can be used as detailed in ANSI/NETA MTS-2011 Table 100.2.
- evaluation of the test results—This consists of watching for sparking or arcing during the test or observing the ammeter to ensure that it will not continue to increase after the voltage has been removed at any step. If no evidence of insulation failure or distress is observed, the insulation is considered to be satisfactory for continued service.
- insulation resistance of control wiring—Control wiring is tested to ground using a megohmmeter for 1 min. To test, 500 VDC is applied to 300 VAC class wiring and 1000 VDC is applied to 600 VAC class wiring. It must be ensured that solid-state devices and controls are properly isolated or manufacturer recommendations are followed prior to performing this test. Test results should be compared to previous test

results or have a minimum insulation resistance of 2 MΩ, per ANSI/NETA MTS-2011.

- ground resistance tests—A low resistance to ground ensures that protective devices will operate correctly and that faults are safely taken to ground. Problems in this area create safety issues, and damage to power system components can result.
- calibration of meters and protective relays—Watt-hour meters and protective relays should be calibrated, and any adjustments or repairs should be made as required. Often the protective relays are calibrated on a regular schedule, regardless of whether the switchgear is tested. This creates issues where the protective relays function properly, but the circuit breaker cannot function properly due to problems with the switchgear.
- control power transformers—Insulation resistance should be tested phase-to-phase and phase-to-ground for 1 min using a megohmmeter. Table 100.5 of ANSI/NETA MTS-2011 should be consulted for recommended test voltages. Test results should agree with the OEM's recommendations or Table 100.5 of ANSI/NETA MTS-2011. **See Appendix.**

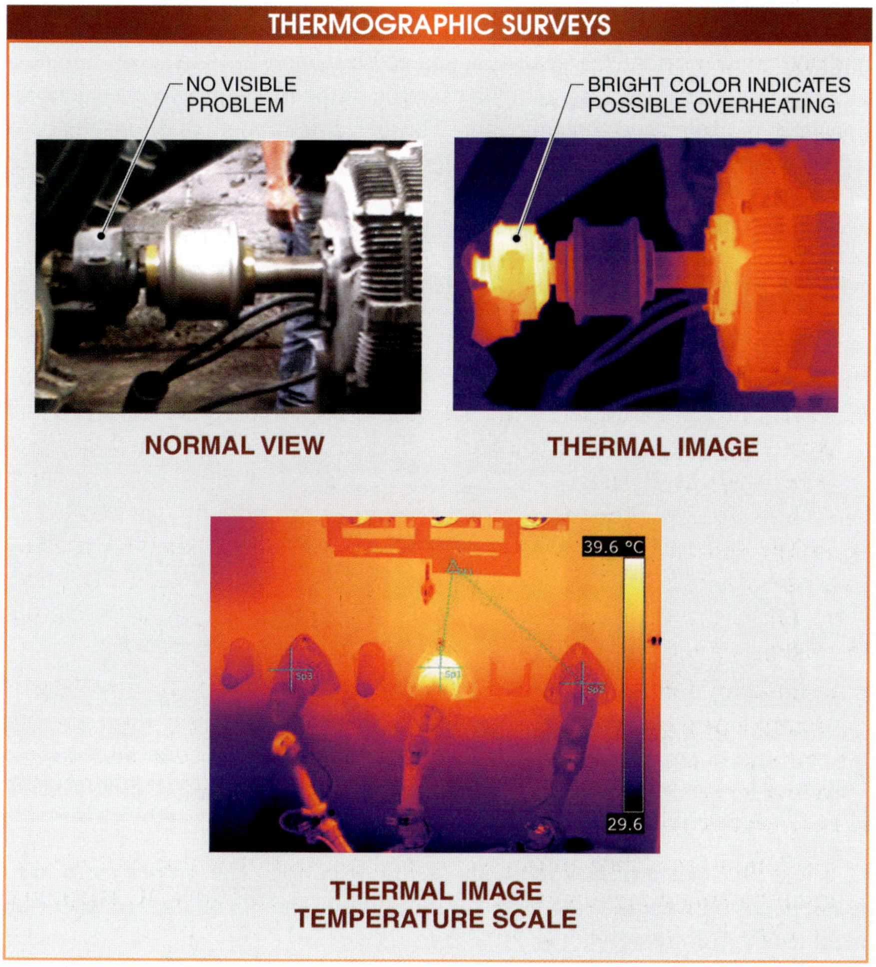

Fluke Corporation

Figure 7-21. A thermographic survey can be performed with a thermal imager to locate problems with electrical equipment that may not be visible with the naked eye.

PARTIAL DISCHARGE TESTS

Partial discharge (PD) is defined in IEC 60270 Section 3.1 as "*a consequence of local electrical stress concentrations in the insulation or on the surface of the insulation…*" PD is also referred to as corona or corona detection. PD testing is one of the newer methods of predictive maintenance for medium- and high-voltage equipment. PD allows the detection of insulation breakdown while the device or equipment is in service. For PD testing, either acoustic and/or other sensors are placed inside the equipment permanently or portable detectors are used. When insulation deteriorates, it can create sound, light, heat, and chemical reactions. PD testing can detect one or multiple emissions in order to detect insulation breakdown.

PD testing is used primarily to detect problems developing in power cables and accessories such as connection boxes and terminations. It can also be used to detect problems developing in switchgear. PD occurs when the insulation on power cable develops voids or electrical treeing. Voids can be created and water penetration into the cable can occur during the manufacturing process or by improper installation of the cable. Heat stress and overvoltage on a power system can accelerate the deterioration of power cable due to heat, chemical reactions, and other types of energy and can cause the voids to grow.

Continued erosion of the insulation causes channels that develop into electrical trees, sometimes referred to as water trees. The electrical trees can continue to grow until they finally bridge the insulation and cause failure. In switchgear, failure is often in the form of tracking along the surface of the insulation due to deterioration, heat, moisture, and contamination. The dielectric strength of the insulation is reduced, and the tracking continues until it reaches ground. **See Figure 7-22.** During PD, the arcing does not actually bridge the gap between the conductor and ground. Insulation failure is the primary reason for equipment failure. **See Appendix.**

INSULATION EROSION – ELECTRICAL TREES

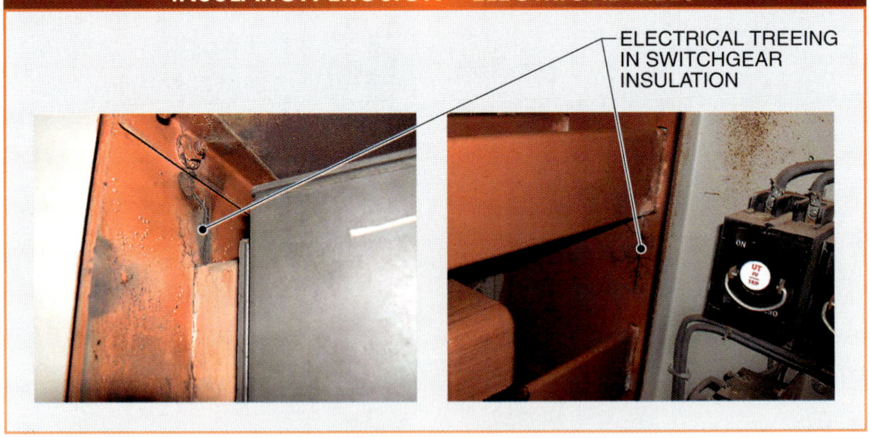

Shermco Industries

Figure 7-22. Continued erosion of insulation causes channels that develop into electrical trees, which can continue to grow until they finally bridge the insulation and cause failure.

There are three primary methods of PD detection. The proper method to use depends on what is being tested or monitored. The PD method of detection includes the following:

- capacitive discharge detection—It is used for unshielded components such as dry-type transformers, instrument transformers, and switchgear. Sensors are placed on the outside of switchgear and enclosures to detect problems.

- inductive discharge detection—It is used for metal-enclosed equipment such as shielded cables and terminations, oil-filled transformers, motors, and generators. Sensors are placed next to cables to detect problems.
- acoustic emission discharge detection—It is used for dry-type and oil-filled transformers, switchgear, and unshielded cables. Sensors are placed on the outside of switchgear and enclosures to detect problems.

Partial failures in the insulation (voids and treeing) create microsparking or arcing across a small section of the surface or within the insulation. As a result, certain emissions are created that can be detected by various types of sensors. One of the most common types of sensors used is the acoustic (noise) sensor.

PDs create radio frequency signals. One early type of detector was made by using an AM radio tuned off-station so that only static was heard. The static would increase in intensity when brought near a PD source. This method was crude and could not pinpoint the source, but it did provide some indication of a problem developing. Today's PD detectors can be portable or installed permanently to provide constant monitoring of systems. They can also determine PD location accurately.

By trending test results over a period of time, small increases in PD emissions can be detected, and the equipment can be removed from service and repaired before it fails. By permanently installing PD sensors and performing continuous monitoring, critical equipment can be trended and defects can be detected early. **See Figure 7-23.** *Note:* Equipment does not have to be opened up or shut down in order to perform PD testing.

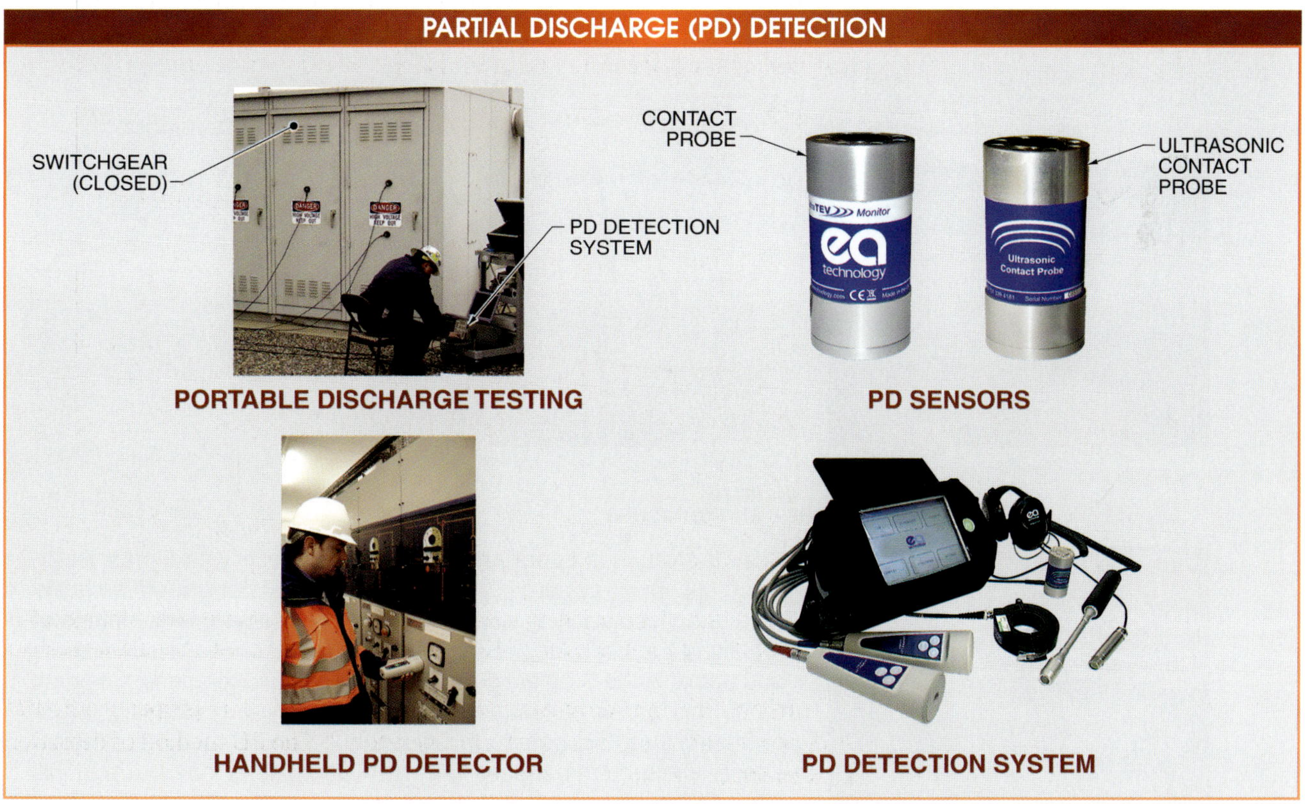

EA Technology Ltd.

Figure 7-23. Modern PD detectors can be portable or can be installed permanently to provide constant monitoring of the system. They can also determine PD location accurately.

Monitoring PD allows equipment to be removed from service only when the monitoring reveals a problem. This is referred to as predictive maintenance. *Predictive maintenance (PdM)* is the monitoring of wear conditions and equipment operation characteristics for comparison against a predetermined tolerance to predict potential malfunctions or failures. Electrical equipment can also be tested on a regularly scheduled basis for PD emissions. This type of maintenance is referred to as preventive maintenance. *Preventive maintenance (PM)* is scheduled work required to keep equipment in peak operating condition.

PM for PD involves increasing the voltage on the tested electrical device to determine the inception voltage. The inception voltage is the voltage in which the PD is first detected. Defects, such as voids, in the insulation have small electrical discharges across them. This release of electrical energy (sparking and discharges) is detectable. The problem in field usage is being able to separate background noise from the noise of electrical discharges. Software has been developed that separates background noise from the electrical discharges, which makes field application of PD testing practical. As insulation deteriorates, PD activity increases and can be trended over time.

CASE STUDY—INSULATION BREAKDOWN

An industrial company's maintenance electrician noticed a strong smell of ozone in the switchgear room located in the basement of one of their buildings. Because the maintenance electrician knew that this was not a good sign, he called a NETA-certified test company to investigate the cause. It was discovered that there was severe tracking at several locations within the switchgear. The tracking was caused by poor repairs performed a year earlier by another company. This 13.8 kV switchgear was close to failure due to the breakdown of the insulation. The ozone smell was just one indicator.

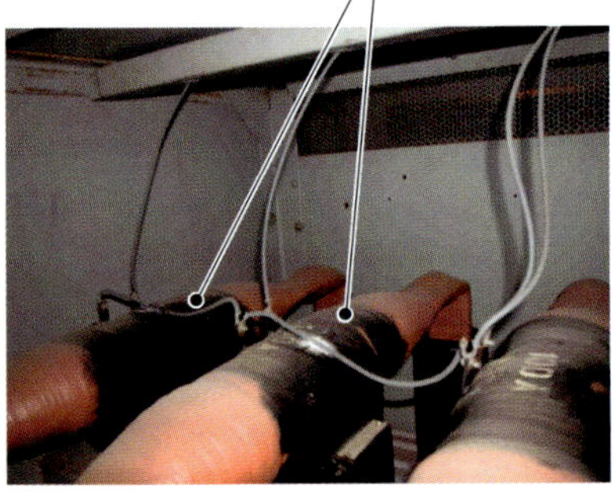

Conclusion: It is important to carefully choose who works on an electrical power system. The first company that performed the faulty repairs noticed tracking when maintaining the equipment. Instead of replacing the fiberglass redboard, technicians chose to cut corners and address the problem by painting it with insulating enamel paint. Since this repair technique did not address the underlying cause of the tracking, the deterioration continued until it reached a critical phase. Once cut, fiberglass redboard will absorb moisture and will need to be replaced within five to seven years. Topical treatment does not fix the problem.

MAGNETRON ATMOSPHERIC CONDITION TESTS

The magnetron atmospheric condition (MAC) test is a new test used in the industry for determining the remaining life of a vacuum bottle. The vacuum bottle integrity test, the predecessor to the MAC test, only indicates whether the vacuum bottle has failed. It does not provide any indication of its remaining life. OEMs assumed that vacuum bottles would be replaced midlife of the circuit breaker, which was estimated to be about 40 years. That would indicate a projected vacuum bottle life of about 20 years.

Tens of thousands of vacuum bottles were manufactured during the 1980s and 1990s. Many of these bottles are past or nearing the end of their projected life. Failure of these bottles is becoming more frequent as they age. Older bottles are especially prone to failure.

Early designs of vacuum bottles were evacuated through a brass fitting and then crimped. Newer bottles are sealed while the vacuum is still being drawn, which considerably improves the projected life of the bottle. The bottle material is not a significant factor in life expectancy, but the presence of the crimped fitting is significant. **See Figure 7-24.**

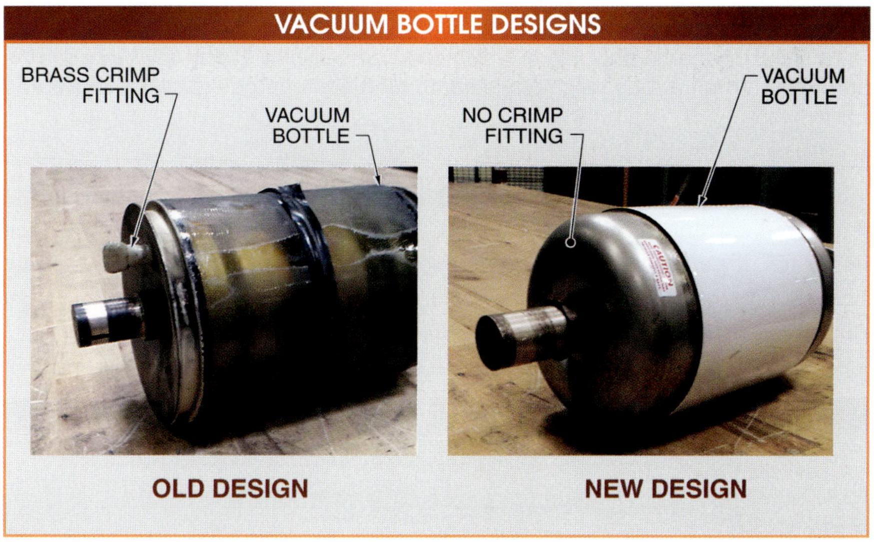

Shermco Industries

Figure 7-24. Early designs of vacuum bottles were evacuated through a brass fitting and then crimped. Newer bottles are sealed while the vacuum is still being drawn, which improves the projected bottle life considerably.

The MAC test is performed by vacuum bottle OEMs to verify that their new vacuum bottles will have an adequate life span and to reduce the replacement costs associated with defective bottles. The factory test equipment is large and does not lend itself well to field applications. One test equipment OEM produces a field-portable test set that replicates the factory test by using flexible magnetic field coils.

The MAC tester operates on the Penning discharge principle. According to the Penning discharge principle, charged particles (ions) can be created by applying high voltage across an open vacuum interrupter. When a strong magnetic field is introduced, the charged particles will move and produce current across the open contacts. The current is directly proportional to the pressure inside the vacuum bottle. The test is repeated 40 times to develop a characteristic curve. **See Figure 7-25.**

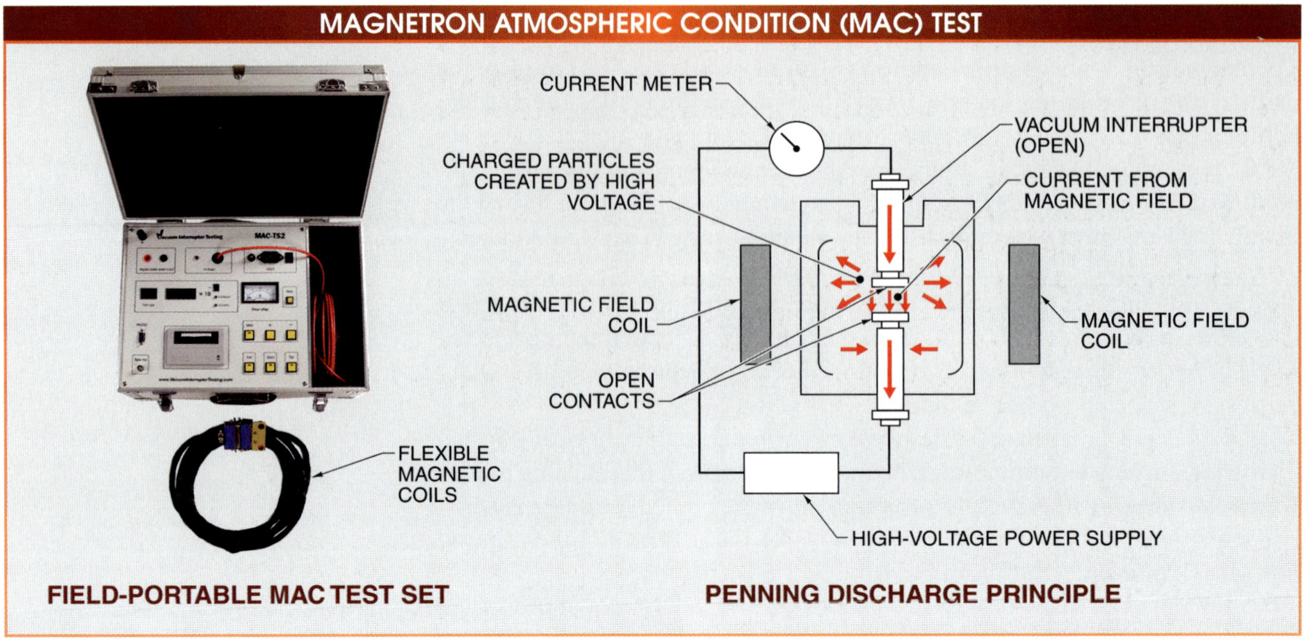

Vacuum Interrupters, Inc./Group CBS

Figure 7-25. The MAC test is based on the Penning discharge principle in which charged particles are created by applying a high voltage across an open vacuum interrupter. When a magnetic field is introduced, the charged particles move and produce current across the open contacts.

The pressure inside a vacuum bottle can be determined by comparing the generated current to a known pressure-ionization curve such as the Paschen curve for dry air. Vacuum interrupter remaining life is determined by taking the leakage rate for the years it has been in service and projecting that leakage rate over time. **See Figure 7-26.**

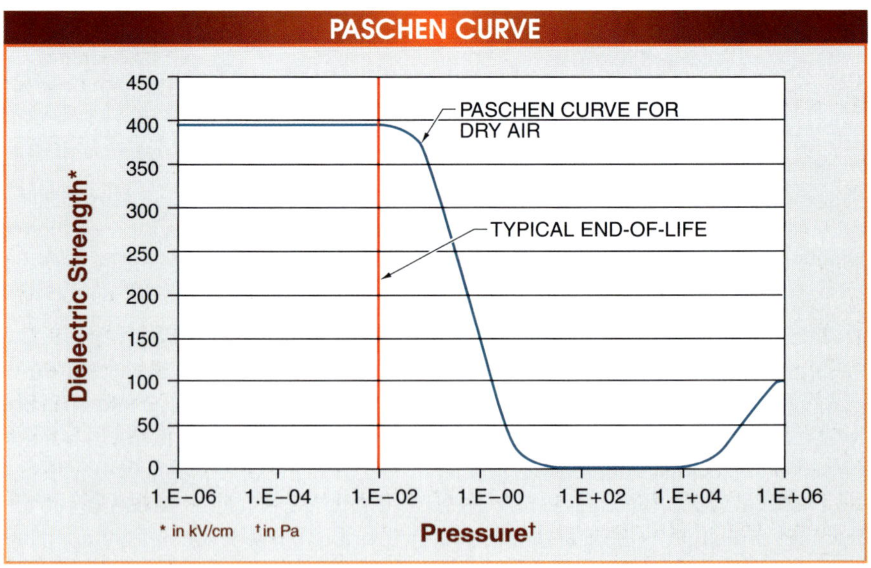

Vacuum Interrupters, Inc./Group CBS

Figure 7-26. The pressure inside a vacuum bottle can be determined by comparing the generated current to a known pressure-ionization curve, such as the Paschen curve for dry air.

Once an interrupter reaches an internal pressure of 10^{-2} Pa, it is at the end of its life and should be removed from service. This indicates that enough air has been absorbed by the vacuum bottle that its dielectric strength is compromised. Even on vacuum bottles that test above the minimum pressure requirement, the rate at which the pressure is changing could also indicate the need to replace the bottle. Even though ceramic and glass are extremely hard, they are still permeable to air and, over time, allow air to enter the vacuum bottle. This is true regardless of the OEM or voltage class of the equipment.

The advantage of using flexible field coils in the test set is that they can be applied to vacuum bottles while still being installed in the circuit breaker carriage. With vacuum circuit breakers such as GE PowerVac, the coils can be wrapped around the outside of the barrier and still achieve satisfactory test results.

Vacuum Interrupters, Inc./Group CBS

Vacuum bottles can be tested in place without disassembly with a magnetron atmospheric condition (MAC) test set.

CIRCUIT BREAKER VIBRATION SIGNATURE ANALYSIS

One issue that has received attention in the last two cycles of NFPA 70E is first trip testing. The issue is that once a circuit breaker has been opened and removed from its cubicle, it is no longer virgin. The mechanism has been operated and would no longer respond as it would during a fault. There are test sets specifically designed for first trip testing, but these test sets tend to be complicated and expensive. An alternative is circuit breaker vibration signature analysis (a vibration analyzer). In actual field tests, it has shown to be fairly accurate in predicting which circuit breakers may respond within OEM specifications if required to trip after being in service for several years.

The vibration analyzer uses the accelerometer installed on the Apple® iPhone® and an app that can be downloaded from the App Store. The Apple iPhone has a built-in three-axis accelerometer that is used for games and other apps. At this time, the Android-capable app has not been released.

Using magnets on a spare phone case, the Apple iPhone is attached to the front cover of the circuit breaker's operating mechanism. The app is opened, the circuit breaker is tripped, and the springs are recharged and closed while still in the cubicle.

The vibration from each operation is recorded, and the signature is uploaded to a database for comparison to known signatures. If there are variations outside what is considered acceptable, the test result is flagged and a possible cause is provided for an additional fee. **See Figure 7-27.**

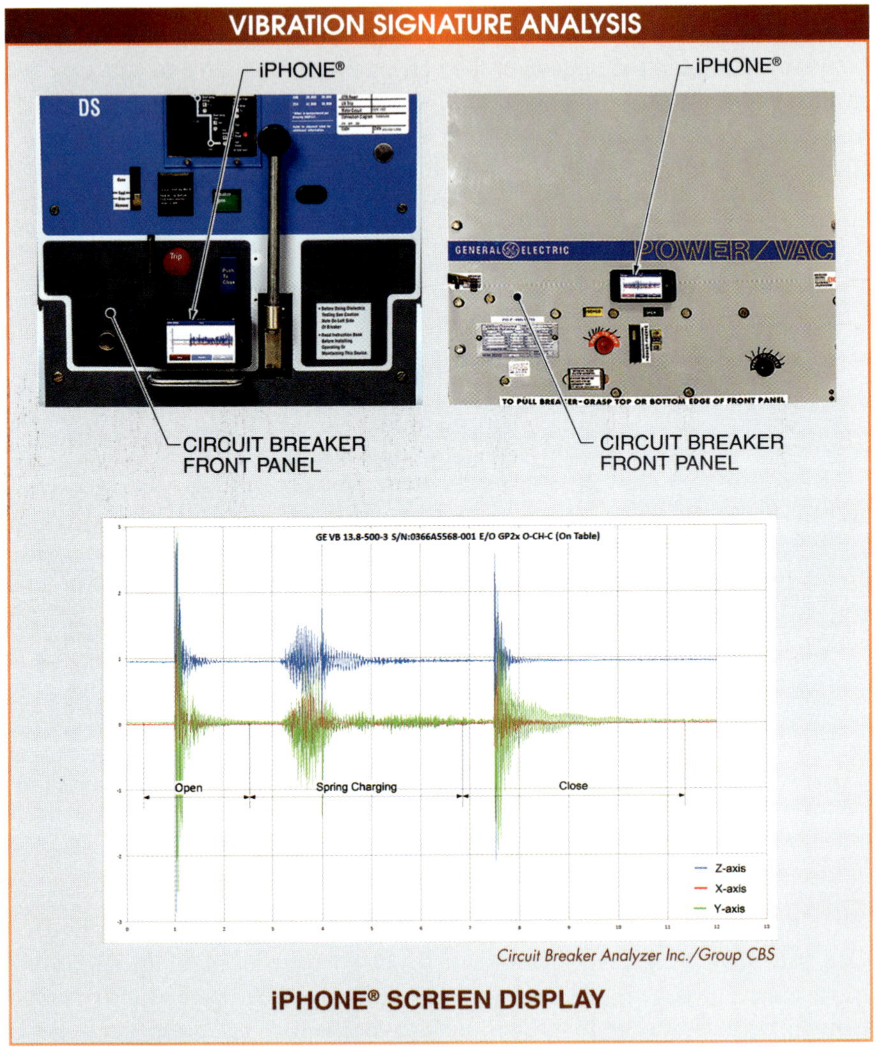

Figure 7-27. When using a device such as an iPhone® to perform a vibration signature analysis, the mounting location of the device is important for obtaining accurate results.

The vibration analyzer test is based on simple physics. During operation, a circuit breaker is subject to stress, and it deforms. This deformation is minor and is in the form of vibrations. After it springs back to its original shape, the circuit breaker tends to travel past its initial position and then back again. The oscillation/vibration is damped by the mass and shape of the circuit breaker, so it passes very quickly. Since a circuit breaker has a large number of moving parts, the vibration signature will be complex and is unique to the OEM, style, and frame size.

The vibration signature is evaluated by comparing the evaluation and baseline signatures. **See Figure 7-28.** A circuit breaker signature will deviate from the known good signature in predictable ways. The way in which the signature varies from the known sample indicates the issue. A large variation between the baseline signature and the evaluation signature indicates a possible problem or malfunction with the circuit breaker.

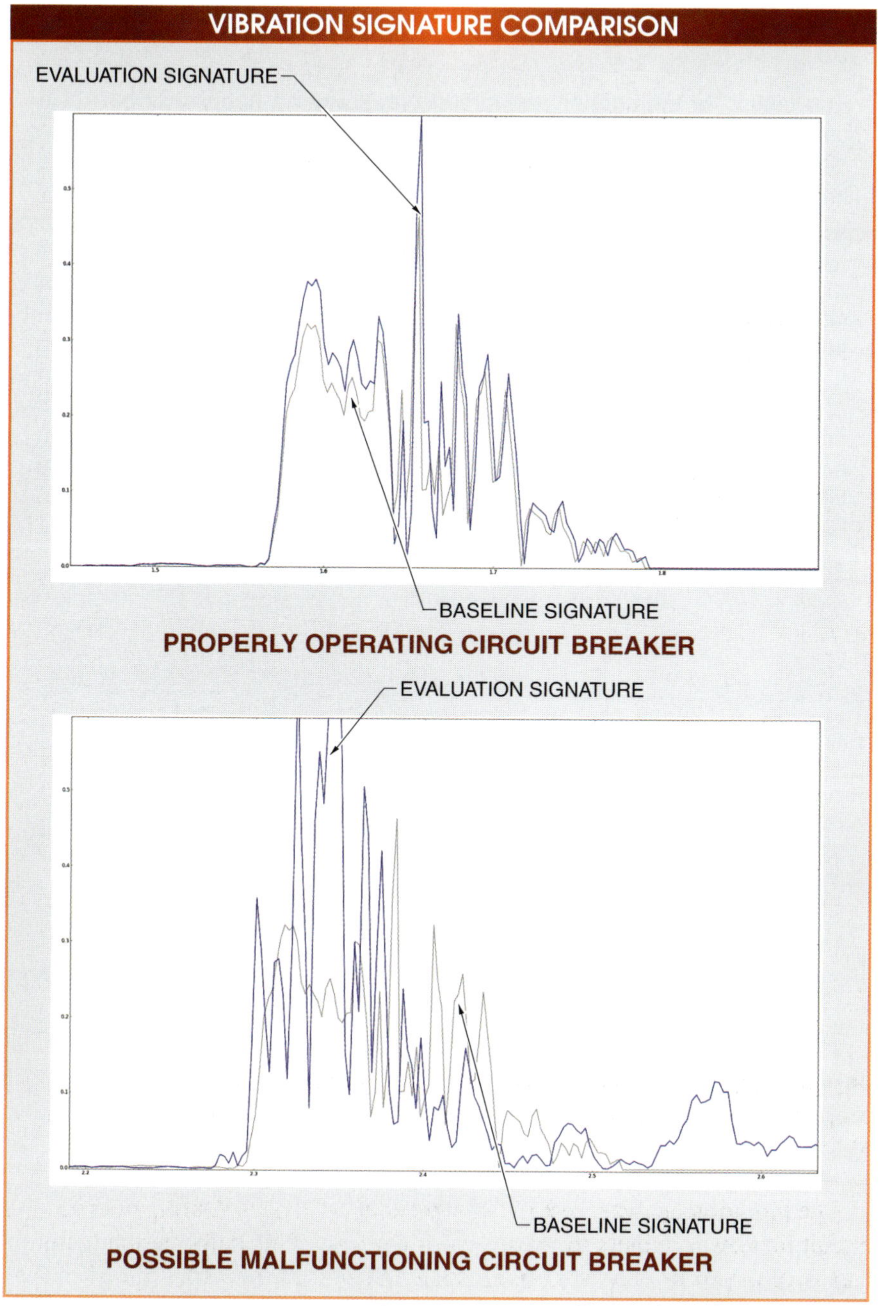

Circuit Breaker Analyzer Inc./ Group CBS

Figure 7-28. Large variations between the baseline signature and the evaluation signature indicate a possible circuit breaker problem or malfunction.

Trial tests performed on circuit breakers sitting on a bench revealed markedly different signatures than those of circuit breakers installed in cubicles. For best results, the circuit breaker must be installed in a cubicle or strapped down to a work bench or other stable mounting. As part of the test setup, the user must specify which position the circuit breaker is in so that the appropriate signature from the database can be used. The test is valid on most circuit breakers up to the 15 kV class. While data does not exist for station circuit breakers, it is possible to gather a database of their signatures.

SUMMARY

The testing of switchgear and circuit breakers is a necessary part of a comprehensive maintenance program. High insulation power factors can be the result of damaged or deteriorated insulation, tracking, or moisture intrusion. Any of these problems could result in unsafe working conditions for personnel working on or near the equipment. High-resistance connections can also be a safety issue since arcing or explosions could result.

ANSI/NETA MTS-2011 contains the minimum required maintenance for circuit breakers and switchgear. The recommended intervals and specifications will ensure reliability and safety for the electrical system. Newer testing methods, such as PD testing, and thermal imaging allow the evaluation of electrical systems while the equipment is still energized. This reduces downtime and the costs associated with maintenance by pinpointing what equipment is most likely to fail. Other technologies, such as for MAC and vibration analysis, provide methods for accurately testing characteristics, but they are not practical or require expensive test equipment.

DIGITAL RESOURCES
ATPeResources.com/QuickLinks
Access Code: 223436

chapter 7

CIRCUIT BREAKER TESTING

CHAPTER REVIEW

Name ______________________________ Date ______________

True-False

T F **1.** The DC overpotential test is considered to be a more sensitive test than the insulation power factor test for determining insulation quality.

T F **2.** During a switchgear insulation test, a high resistance to ground ensures that protective devices will operate correctly and that faults are safely taken to ground.

T F **3.** In-service circuit breakers must be inspected or tested after opening under fault conditions.

T F **4.** Damage and wear to the flexible metal bellows can cause vacuum bottles to fail.

T F **5.** During the timing test, circuit breaker characteristics that can be checked include wipe and travel.

T F **6.** Degradation of the arc chute is more easily found using a megohmmeter rather than a DC hipot tester.

T F **7.** PD allows the detection of insulation breakdown while the device or equipment is in service.

T F **8.** AC overpotential tests are considered to be nondestructive.

T F **9.** The capacitance charging current and the dielectric absorption current are known collectively as leakage current.

T F **10.** Hipot testers measure insulation resistance by measuring alternating current.

T F **11.** Many bushings on older circuit breakers were constructed of paper soaked in an epoxy.

T F **12.** Dielectric absorption current and the capacitance charging current both decrease to zero during a DC test.

T F **13.** DC test equipment is the same size as AC test equipment.

T F **14.** Megohmmeters should be used for testing the primary insulation components.

T F **15.** An arc flash from test equipment is generally not a risk.

T F **16.** When insulation fails during testing, the damage is more severe than if it were to fail during normal operation.

T F **17.** Electrical equipment and devices should not be tested when their temperature is below 0°C because ice can form in cracks in the insulation.

T F **18.** When performing a DC overpotential test, the voltage must be increased in several steps.

T F **19.** Proof tests are typically performed as part of field maintenance testing.

T F **20.** DC hipot testers can only provide measurement readings in megohms.

T F **21.** On circuit breakers, the primary purpose of the guard circuit is to eliminate surface leakage.

T F **22.** An arc cannot occur in a perfect vacuum.

T F **23.** Only high-end, bench-top multimeter models are available with "input impedance" or "sensitivity" functions.

T F **24.** A knee in a test trend almost always indicates that the piece of equipment or device being tested must be repaired or replaced immediately.

T F **25.** The effectiveness of DC overpotential tests vary according to whether the insulation system is composed of homogeneous insulation or a composite such as oil-filled and paper.

T F **26.** Leakage current can be measured and evaluated using an AC hipot or vacuum bottle tester.

T F **27.** The arc chute is the primary insulation structure being tested using the UST test.

T F **28.** Doble Engineering recommends a percent power factor of 5% or less for the GST-L-guard on outdoor vacuum circuit breakers.

T F **29.** Two currents or voltages that are not in phase with each other cannot be added together directly.

T F **30.** The vacuum bottle integrity test provides an indication of remaining vacuum bottle life.

T F **31.** An early type of PD detector was an AM radio tuned off-station so only static was heard.

T F **32.** As insulation deteriorates, PD activity decreases and can be trended over time.

T F **33.** Insulation failure is the primary reason for equipment failure.

T F **34.** Equipment must be opened up or shut down in order to perform PD testing.

T F **35.** Dissipation factor and power factor are essentially the same up to a 10% power factor.

Multiple Choice

__________ **1.** The contact resistance test is performed using a(n) ____.

A. megohmmeter
B. micro-ohmmeter
C. AC hipot tester
D. DC hipot tester

__________ **2.** The ____ test is best for determining the overall condition of a sealed-case, molded-case circuit breaker that has interrupted a short circuit.

A. DC overpotential
B. stress
C. insulation power factor
D. proof

_______________ **3.** DC insulation testing is performed by applying DC voltage of ___ V or higher across the insulation to be tested.
A. 200
B. 300
C. 400
D. 500

_______________ **4.** Most vacuum bottle integrity testers are actually ___ with no ammeter and can have a fixed voltage output.
A. DC hipot testers
B. AC hipot testers
C. multimeters
D. megohmmeters

_______________ **5.** A(n) ___ is performed to ensure that no conductive object, such as a metal screw, washer, or bolt, has made contact with the internal components of a circuit breaker to cause a phase-to-phase or line-to-load short circuit.
A. overpotential test
B. insulation test
C. safety check
D. vibration signature analysis

_______________ **6.** When testing devices such as large transformers and long runs of power cable, a test interval of 5 minutes requires the application of grounds for ___ minutes.
A. 10
B. 15
C. 20
D. 30

_______________ **7.** An AC overpotential test is also known as a(n) ___ test.
A. dielectric
B. hipot
C. integrity
D. proof

_______________ **8.** When performing insulation tests on control wiring, insulation rated for 300 V should have ___ V applied to it.
A. 500
B. 750
C. 1000
D. 1200

_______________ **9.** When performing a UST, typically, losses of ___ W should be measured.
A. 0.001
B. 0.010
C. 0.10
D. 1.0

_______________ **10.** Insulation is also known as a(n) ___.
A. capacitor
B. dielectric
C. ground
D. inductive device

__________ **11.** The vacuum bottle integrity test is a go/no-go test and normally lasts for ___ min.
A. 1
B. 2
C. 5
D. 10

__________ **12.** In a purely capacitive circuit, voltage leads current by ___°.
A. 45
B. 90
C. 120
D. 180

__________ **13.** AC changes direction ___ times per second.
A. 6
B. 10
C. 30
D. 60

__________ **14.** According to ANSI/NETA MTS-2011, a minimum insulation resistance of ___ MΩ for insulation resistance of control wiring is recommended.
A. 1
B. 2
C. 3
D. 4

__________ **15.** Once testing is completed, larger wound devices such as transformers or long power cable runs must be grounded for a minimum of ___ times the test interval after testing.
A. two
B. three
C. four
D. five

__________ **16.** With switchgear, bolted connections on any phase should not be more than ___% higher than the other two phases.
A. 25
B. 50
C. 75
D. 100

__________ **17.** PD detection methods include capacitive, inductive, and ___ discharge detection.
A. acoustic emission
B. conductive emission
C. impedance
D. resistive

__________ **18.** For contact resistance tests, a 480 V, 4000-amp frame circuit breaker should give a reading of less than ___ μΩ if it is in good to excellent condition.
A. 0.3
B. 3
C. 30
D. 300

_______________ **19.** For best results, insulation power factor testing should be performed in warm, dry weather with the device temperature no less than ___.

A. –17.78°C (0°F)
B. 0°C (32°F)
C. 4.4°C (40°F)
D. 10°C (50°F)

_______________ **20.** General Electric (GE) has a service advisory letter (SAL) that recommends a ___% maximum power factor, with ___% indicating dangerously deteriorated insulation.

A. 1; 5
B. 3; 5
C. 5; 10
D. 10; 20

Completion

_______________ **1.** ___ is scheduled work required to keep equipment in peak operating condition.

_______________ **2.** ___ is irreversible internal degradation caused by the formation of conductive carbonized pathways.

_______________ **3.** The common industry practice test voltage value for a DC overpotential test is twice the rated voltage plus ___ V.

_______________ **4.** ___ is defined in IEC 60270 Section 3.1 as "*a consequence of local electrical stress concentrations in the insulation or on the surface of the insulation...*"

_______________ **5.** Insulation test results must be corrected to a base temperature of ___°C so they can be trended.

_______________ **6.** According to the rule employed in ANSI/NETA MTS-2011, the highest contact resistance reading should not be more than ___% of the lowest value.

_______________ **7.** ___ current is the polarizing current that is drawn by the insulation system (dielectric) to align the dipole molecules within the insulation.

_______________ **8.** The ___ of a megohmmeter is a means to eliminate unwanted return current from the measurements being taken.

_______________ **9.** The magnetron atmospheric condition (MAC) test is a new test used in the industry for determining the remaining life of a(n) ___.

_______________ **10.** A(n) ___ is a high-resistance ohmmeter used to measure insulation deterioration on various conductors by measuring high-resistance values during high-voltage test conditions.

_______________ **11.** ___ current is current that reverses its direction of flow at regular intervals.

_______________ **12.** ___ current is resistive current that flows through insulation and is the variable that is measured with an insulation resistance test instrument.

_______________ **13.** NETA stands for the ___.

_______________ **14.** Partial discharge is also referred to as ___.

_______________ **15.** ___ is an indirect method of measuring the losses (leakage current) inside a circuit breaker, transformer, or other electrical device.

______ **16.** DC insulation tests are generally only adequate for equipment rated below ___ kV.

______ **17.** A(n) ___ is a function on a megohmmeter used to eliminate unwanted return current from test measurements.

______ **18.** As insulation deteriorates, leakage current increases, which causes the angle between the watts-loss and total current to ___.

______ **19.** The ___ test provides information as to the remaining life of a vacuum bottle.

______ **20.** A(n) ___ is a molecule that is affected by the magnetic field created by current flow through its insulation.

______ **21.** As the lubricant on a circuit breaker dries over time, the resistance gradually ___.

______ **22.** ___ is the monitoring of wear conditions and equipment operation characteristics for comparison against a predetermined tolerance to predict potential malfunctions or failures.

______ **23.** According to the ___ principle, charged particles (ions) can be created by applying high voltage across an open vacuum interrupter.

______ **24.** PD occurs when the insulation on power cable develops ___.

______ **25.** Perfect insulation theoretically has an angle of ___° between the total current and the dielectric watts-loss.

Short Answer

1. List five advantages and two disadvantages of using DC test equipment for insulation testing.

2. Explain why higher voltages are required when performing power factor tests in outdoor substations.

3. When performing power factor tests in outdoor substations, what conditions would require the use of interference-cancelling circuitry?

4. Explain why megohmmeters should not be used for testing medium-voltage circuit breakers.

5. What are the main causes of insulation deterioration over a time period of years?

6. List four typical circuit breaker timing tests performed with a time-travel analyzer.

7. Explain the criteria used to determine the remaining life of a vacuum interrupter and when it should be removed from service.

8. Explain how a vibration analyzer operates.

9. What are the four items that must be referenced prior to performing any electrical tests or maintenance work on circuit breakers, switchgear, and related components to determine proper safety procedures and personal protective equipment (PPE) requirements?

10. What are the two main purposes for performing DC circuit breaker tests?

Matching

Power Cable Cross Section

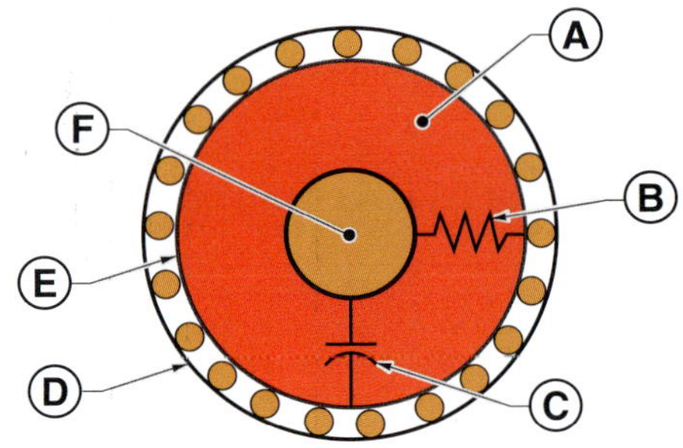

______________ **1.** Conductor

______________ **2.** Capacitor created by proximity to ground

______________ **3.** Insulation

______________ **4.** Resistance

______________ **5.** Semiconductor shielding

______________ **6.** Wire shield

Guard Circuits

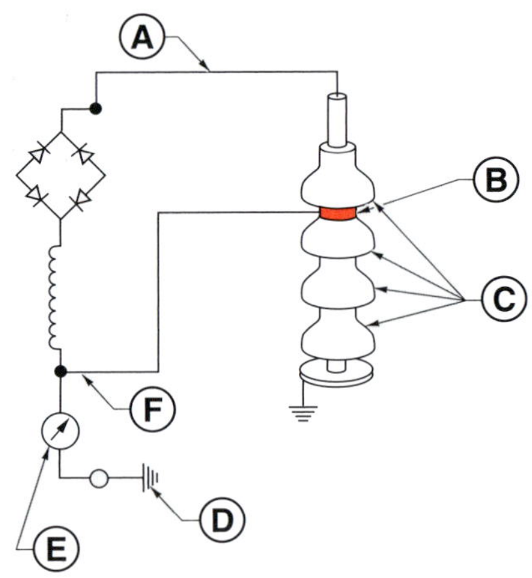

______________ **1.** Earth ground

______________ **2.** Guard

______________ **3.** Guard collar

______________ **4.** Line

______________ **5.** Meter

______________ **6.** Petticoats

Vacuum Bottle Components

Shermco Industries

______________ **1.** Electrical contacts

______________ **2.** Flexible metal bellows

______________ **3.** Metal vapor condensing shield

______________ **4.** Vacuum bottle

chapter **8**

LOW-VOLTAGE OVERCURRENT PROTECTIVE DEVICES

Low-voltage overcurrent protective devices (OCPDs), or trip devices, include thermal-magnetic, series, solid-state, and microprocessor (digital) devices. OCPD functions include long-time delay (LTD), instantaneous (INST), short-time delay (STD), ground-fault, and I^2t/I^4t functions. Proper safety practices must be employed and personal protective equipment (PPE) must be used when performing any type of work on circuit breakers, switchgear, and related components. The manufacturer's instruction books must be consulted, for safety alerts and safe operation procedures. The requirements of NFPA 70E® must be followed when performing electrical testing and maintenance.

OBJECTIVES

- Explain the requirements of NFPA 70E for overcurrent protective devices (OCPDs).
- List the primary types of OCPDs.
- Describe how OCPDs operate.
- List the common functions of OCPDs.
- Explain the purpose of a short-time delay.
- Explain the purpose of the I^2t/I^4t function.

OVERCURRENT PROTECTIVE DEVICES AND SAFETY

Overcurrent protective devices (OCPDs) are important for protecting not only electrical equipment but also people. NFPA 70E®, *Standard for Electrical Safety in Employee Workplaces*, Section 130.5 states, "*The arc flash hazard analysis shall take into consideration the design of the overcurrent protective device and its opening time, including its condition of maintenance.*" The informational note for NFPA 70E Section 130.5 states, "*Improper or inadequate maintenance can result in increased opening time of the overcurrent protective device, thus increasing the incident energy.*" NFPA 70E Section 205.3, *General Maintenance Requirements,* states, "*Electrical equipment shall be maintained in accordance with manufacturers' instructions or industry consensus standards to reduce the risk of failure and the subsequent exposure of employees to electrical hazards.*" Finally, NFPA 70E Section 205.4, *Overcurrent Protective Devices,* states, "*Overcurrent protective devices shall be maintained in accordance with the manufacturers' instructions or industry consensus standards. Maintenance, tests, and inspections shall be documented.*"

The NFPA 70E Technical Committee had concerns that companies and workers did not understand how important their OCPDs were in protecting people. Even though NFPA 70E is not a maintenance standard, these sections were included due to the consequences of not properly maintaining OCPDs.

KEY TERMS

- incident energy
- current sensor

INCIDENT ENERGY

Incident energy is the heat from an electrical arc. Incident energy is proportional to time. For example, if a worker is exposed to an electrical arc that produces 8 cal/cm^2 for 0.1 sec and is wearing arc-rated clothing and PPE that is rated at 8 cal/cm^2 or greater, the chances of injury are small. If that same worker is exposed to that same electrical arc for 0.2 sec, the worker will now be exposed to 16 cal/cm^2. If the time of exposure is tripled, the incident energy received by the worker is also tripled.

OCPDs operate in just a few cycles, typically from two to six cycles. One cycle is 0.0167 sec, so the amount of time is short. The time of operation increases substantially when a circuit breaker misoperates. For example, if a molded-case circuit breaker (MCCB) with a normal operating time of two cycles operates slowly due to a lack of lubrication, the time of operation may possibly increase only two to three cycles.

If the technician were wearing what was thought to be appropriate arc-rated clothing and PPE (8 cal/cm^2), the technician would actually be underprotected. If the same circuit breaker fails to operate, the next device upstream, which is a low-voltage power circuit breaker (LVPCB) would have to open to clear the fault. The upstream circuit breaker may be set so that the short circuit appears to be a large overcurrent. **See Figure 8-1.** That could cause the upstream circuit breaker to operate in 1 sec to 2 sec. The resulting incident energy for a one-second operation would then be 240 cal/cm^2. That is approximately 30 times what the technician would be prepared for, and the chances of injury are almost certain.

SAFETY FACT

As an example of how just a few cycles of time can increase the incident energy, suppose a worker is wearing 8 cal/cm^2 arc-rated clothing and PPE, as suggested by the arc flash warning label on the equipment. The estimated operating time for the OCPD is 2 cycles, or 0.033 seconds. If the circuit breaker opens in 3 cycles (0.5 seconds), that worker would be exposed to 12 cal/cm^2 of incident energy. If the circuit breaker opens in 4 cycles (0.07 seconds), the incident energy would be 16 cal/cm^2.

EFFECTS OF CIRCUIT BREAKER MISOPERATION

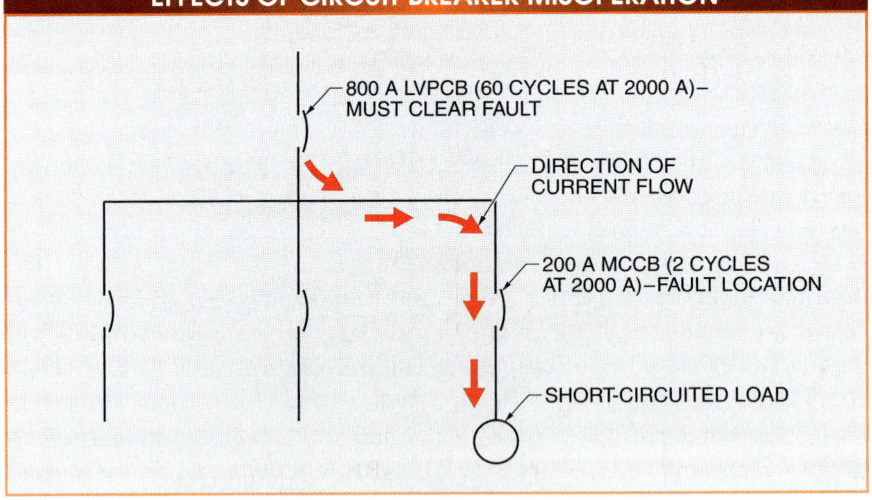

Figure 8-1. If a circuit breaker fails to operate, the next device upstream would have to open to clear the fault.

HISTORY OF OCPDs

As technology has progressed, OCPDs have become more accurate and flexible. Thermal-magnetic OCPDs were among the first to be designed and used, mostly for MCCBs. Being temperature dependent makes them unsuitable for larger industrial applications where the ambient temperature might be elevated.

The first hydraulic-type series OCPD on an LVPCB was manufactured by ITE Circuit Breaker Company in the 1930s. ITE Circuit Breaker Company became ITE-Imperial in the 1960s, which then became ITE/Gould, then ITE Brown-Boveri, then ABB. This early OCPD is known as a sucker disk or adhesion-type device. Two matched, finely-machined surfaces, one convex and the other concave, were placed together with a small layer of oil between them. The interfacial tension of the oil provided the time delay. This device was known as a direct-acting or series OCPD. It was referred to as direct-acting because when it operated, the armature of the OCPD would strike the trip bar of the circuit breaker and directly cause it to open. The "series" identification is due to the fact that load current flowed through the OCPD. **See Figure 8-2.**

ADHESION-TYPE (SUCKER DISK) DEVICES

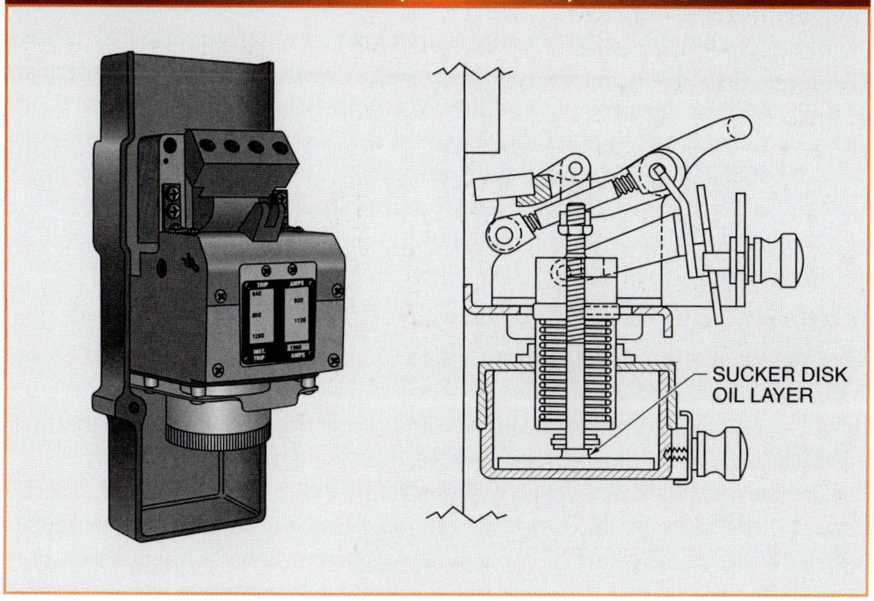

Figure 8-2. Early OCPDs were known as sucker disk or adhesion-type devices, where two matched, finely-machined surfaces (one convex and the other concave) were placed together with a small layer of oil between them.

In the late 1940s, the oil dashpot OCPD was developed and was a large improvement in reliability and accuracy over the adhesion-type series OCPD. Westinghouse Electric developed an air dashpot OCPD that uses an air diaphragm instead of oil to create an adjustable time delay.

The first solid-state OCPDs were introduced by Allis-Chalmers in the 1950s and were not known for being reliable. ABB introduced its first solid-state OCPD in the 1960s, and GE introduced its first solid-state OCPD in 1973. All of the solid-state devices of that era were unreliable.

The difference between a direct acting and indirect acting trip device is that oil and air dashpots that are in series with the load current directly trip the circuit breakers, whereas solid-state and digital OCPDs are indirect acting and use sensors to measure the load current. Solid-state and digital OCPDs are indirect acting because they energize trip coils when they operate and do not trip circuit breakers directly. Each style and type of OCPD has advantages and disadvantages.

The primary disadvantage of series OCPDs is that they contain numerous small linkages plus some type of silicone oil. Over time, the linkages wear and become loose, and the oil used in the dashpots eventually leaks or has to be replaced. Currently there is no company that manufactures oil or air dashpots, and the supply of replacement parts continues to shrink. Solid-state and digital devices offer better accuracy, better repeatability, and lower cost. There are no linkages to wear out, and they offer functions and features that electromechanical OCPDs do not.

One of the disadvantages of solid-state and digital OCPDs is that they can fail without warning. When they fail, all three phases of protection are lost. Currently-manufactured digital OCPDs are reliable and tough.

THERMAL-MAGNETIC OCPDs

Thermal-magnetic OCPDs are the least expensive devices used to protect against damage from overcurrent and short circuits. They are also the least accurate and the most prone to failure. According to some studies, such as the study conducted by Wyle Laboratories for the Nuclear Regulatory Commission (NUREG/CR-5762), the failure rate can be as high as 80% after being in service for three to five years. If they receive minimal maintenance, they are still an acceptable form of protection for noncritical and/or low-power circuits.

Thermal-Magnetic OCPD Construction and Operation

Thermal-magnetic OCPDs are classified as general-purpose OCPDs by the NEC® in Article 100, *Definitions*. A thermal-magnetic overcurrent protection device is an OCPD that provides overload protection by way of the thermal-actuated device and short-circuit protection using a magnetic device.

The thermal portion of the OCPD uses a bimetallic element placed in series with the load current. The bimetallic element is constructed from two metals with differing coefficients of expansion. As current flows through them, they heat up due to I^2R losses, which is heat produced as current passes through resistance.

As the current flow through a conductor increases, the heat given off also increases. This heating causes the bimetal to deflect because the two metals expand at different rates. The metal layer on the right side of the bimetallic element expands at a slower rate than the layer on the left, which forces the bimetal to bend to the right. As the bimetal deflects, it releases a trip latch, which opens the contacts.

The deflection of the bimetal is directly related to the amount of heat created by current flow. As the current flow increases, the heat produced increases, and the bimetal deflects more quickly. The relationship between the heat produced by current flow and the time it takes to trip the circuit breaker is known as an inverse time-current relationship. An increase in current causes a decrease in tripping time.

Thermal-magnetic devices also have an instantaneous trip function. It is used as short-circuit protection. A magnetic coil, or electromagnet, is added to the thermal OCPD. The thermal and magnetic elements are connected in series with each other, so they have the same current flowing through them. During a short circuit, the bimetal will deflect since current is flowing through it. Although thermal elements can react quickly during short-circuit current conditions, they are not fast enough to adequately protect against these conditions. Thermal elements operate in seconds, while short-circuit currents can cause tremendous damage in just cycles. However, if the current exceeds the minimum pickup setting of the magnetic element, the armature of the magnetic unit is attracted to the coil and operates the circuit breaker's trip latch.

The operating time of a magnetic element is referred to as instantaneous. Instantaneous tripping is defined as no intentional time delay. The contacts will open and clear the fault as fast as is mechanically possible. The instantaneous element will typically operate in less than two cycles in an MCCB. **See Figure 8-3.**

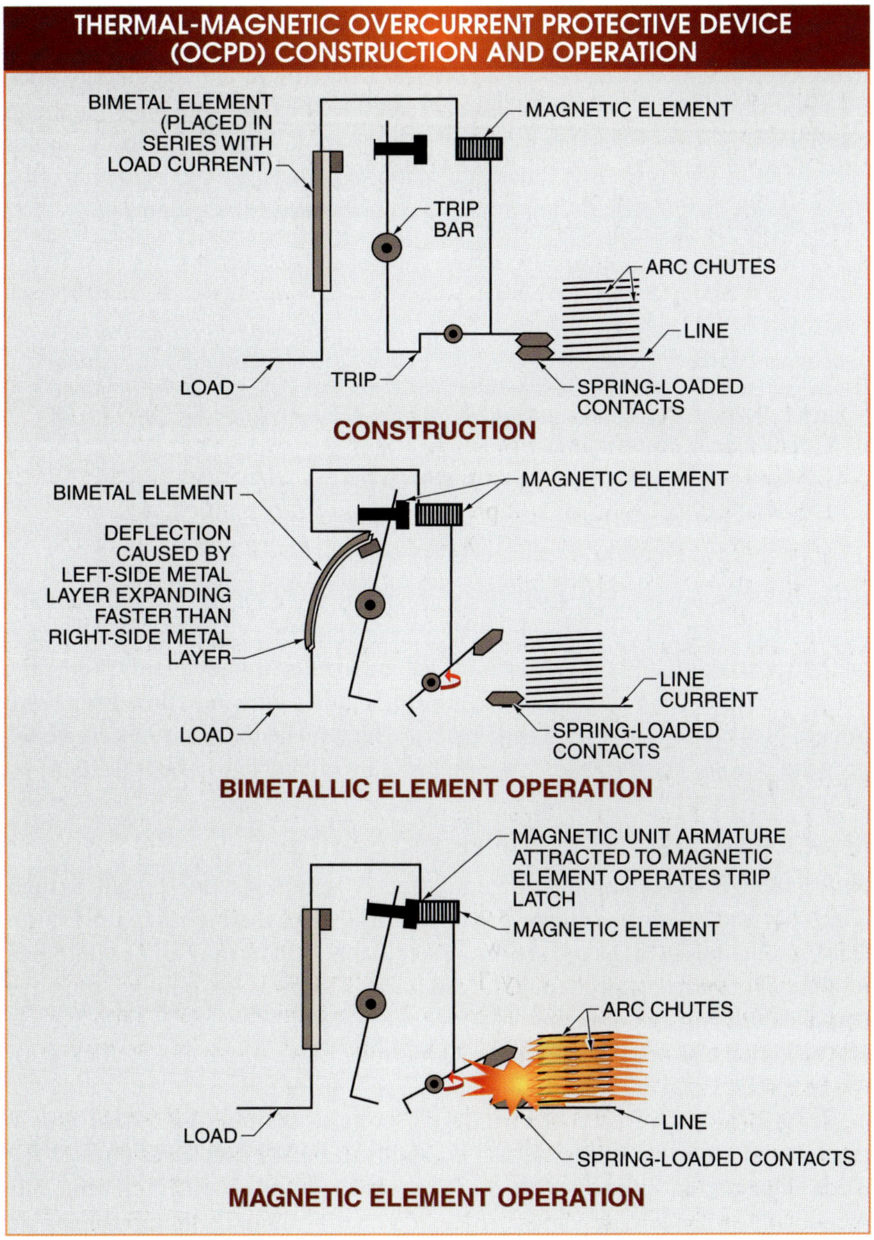

Figure 8-3. Thermal-magnetic OCPDs provide overload protection using a thermal-actuated device and short-circuit protection using a magnetic device.

The minimum operating current value, or pickup of magnetic attraction elements, can be adjusted on interchangeable trip units by changing the tension on the restraining spring. Increasing the spring restraint means more current will be required to overcome it and cause operation. Decreasing spring restraint requires less current to cause it to operate. The minimum pickup of the bimetallic element is factory set and not usually field adjustable.

Ambient Compensation

Ambient-compensated trip elements use a second bimetallic element to compensate for temperature variations between the circuit breaker environment and the equipment it is protecting. An example is when the circuit breaker is indoors where the ambient temperature is high and a motor that is outdoors where ambient temperature is cooler. If a non-ambient-compensated trip element were used in this type of application, the circuit breaker would likely trip too quickly since the circuit breaker is "preheated" as compared to the motor. By using ambient compensation, the circuit breaker will provide protection closer to the motor design limits. The operating times of ambient-compensated trip elements often differ greatly from those of non-ambient-compensated trip elements.

TECH FACT

An attempt to repair series trip devices should not be made. Interchanging components between series trip devices will cause calibration issues because each series trip device was individually calibrated at the factory. Each OEM performed a specific calibration procedure when the devices were manufactured, and it is very difficult to replicate in the field.

SERIES OCPDs

Series OCPDs are used on LVPCBs and have many advantages over thermal-magnetic OCPDs. They are unaffected by temperature, are much more accurate, and can provide multiple operating time bands and more trip functions (short-time delay). Series OCPDs operate on the electromagnetic attraction principle. *Note:* Sucker-disk OCPDs had wide time bands, which made them too inaccurate for growing power systems. They were replaced by oil and air dashpot OCPDs.

Series OCPD Construction and Operation

Series OCPDs were in production until the 1990s, which is a testament to their durability and reliability. However, they require more maintenance than solid-state and digital OCPDs and are not as accurate. As power systems continued to grow, it became apparent that the old technology was not going to meet the demands of ever-tighter power system coordination between OCPDs. Nearly all OEMs have stopped production of series OCPDs and have moved to retrofitting circuit breakers with digital units. **See Figure 8-4.**

On a series OCPD, the electromagnet bolts to the load-side stabs and to the stationary portion of the contact assembly. Load current flows through the electromagnet, and it is sized to carry the full continuous-rated current. The pickup adjustment spring varies the tension on the armature, thereby changing the minimum pickup of the OCPD. When the moving armature is attracted to the electromagnet, the sensitivity adjusting spring puts a restraint on it. The higher the spring tension, the higher the minimum amount of current that will be required to overcome it. **See Figure 8-5.**

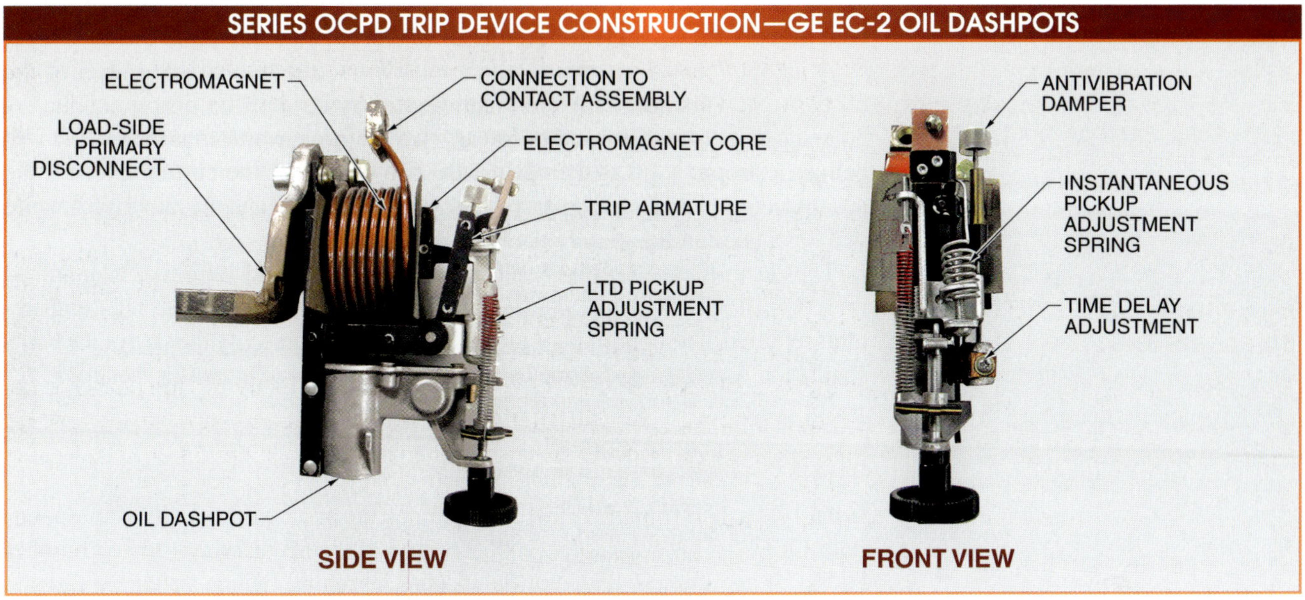

Figure 8-4. Because of their durability and reliability, series OCPDs were in production until the 1990s and have been replaced by more modern solid-state and digital OCPDs.

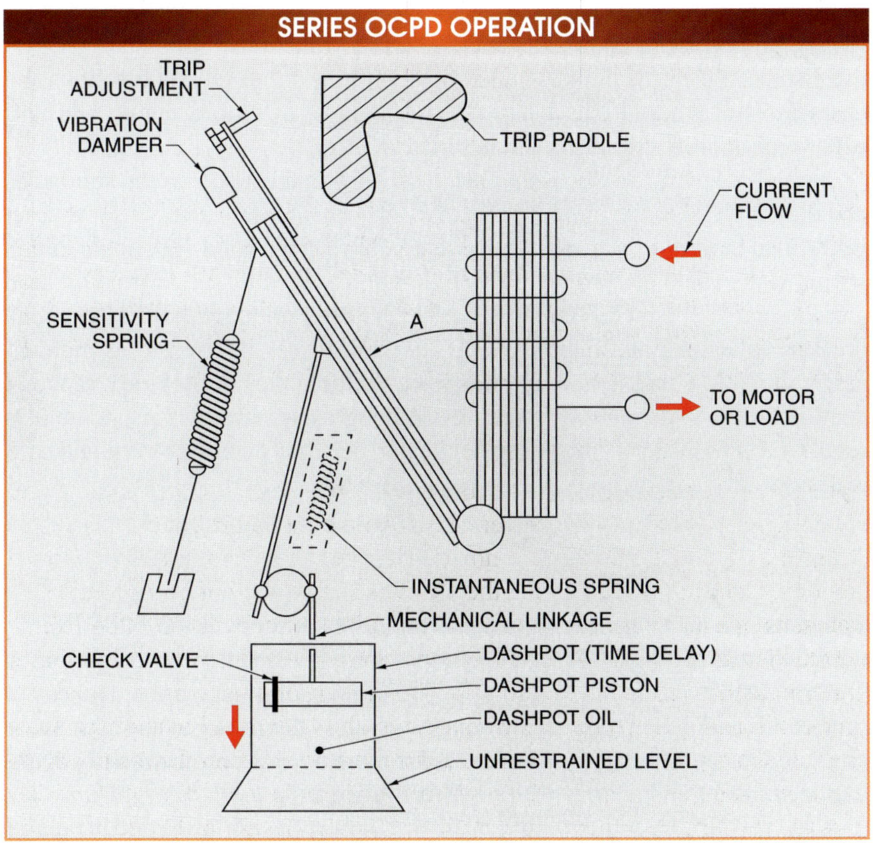

Figure 8-5. With series OCPD operation, when the moving armature is attracted to the electromagnet, the sensitivity adjusting spring puts a restraint on it.

Once the moving armature begins to travel, it must push the piston down through the oil dashpot. The oil must flow from the bottom of the piston to the top, which develops an inverse time-current characteristic. The higher the load current, the more quickly the piston moves through the oil.

At about 75% of piston travel, the piston hits the unrestrained level of the oil dashpot. This area is intentionally cut away so that the piston encounters no resistance from this point or lower. This feature ensures that the OCPD will deliver a strong, solid strike against the trip paddle. If the current exceeds the instantaneous pickup setting, the oil dashpot is bypassed, and the moving armature will have no intentional time delay.

The electromagnetic coil often has one or two turns and surrounds a laminated iron core. The electromagnet can have more than one turn, depending on the ampere-turns required. The continuous current rating is determined by the size of the coil's conductor. As with any conductor, an increase in the size increases the ampacity.

SOLID-STATE OCPDs

In the 1970s, ITE Imperial Electric developed the first commercially viable solid-state OCPD. This was a single-phase time-overcurrent relay (device number 50/51). It had numerous operational issues, especially when voltage transients or power supply spikes hit it. Since that time, solid-state devices have become more accurate and reliable, even to the point where series OCPDs are no longer being sold. Solid-state OCPDs provide three-phase (3ϕ) protection, offer multiple time bands (up to three), and can have long-time delay (LTD), short-time delay (STD), ground fault, and instantaneous (INST) functions. Solid-state OCPDs use discreet electronic circuits and typically mimic series OCPD functions and characteristics. They also have narrower time bands, are more accurate, and have better repeatability capabilities than series OCPDs.

Instead of having an electromagnet in series with the load current, solid-state and digital OCPDs use current "sensors." A *current sensor* is a sensor that reduces load current to a smaller value that is sent to the solid-state logic circuit where, if the value of current and time duration exceed its set points, will send a trip signal to a trip coil. Because the OCPD does not directly trip the circuit breaker, solid-state and digital OCPDs are sometimes referred to as indirect-acting OCPDs. Current sensors often have multiple taps to provide a range of continuous current values and settings, although single-value sensors are also used. Current sensors are not interchangeable between manufacturers, and each is designed for a particular OCPD. **See Figure 8-6.**

MICROPROCESSOR OCPDs

Solid-state OCPDs usually mimic the functions and operation of series trip units, whereas digital OCPDs (sometimes referred to as microprocessor-based OCPDs) can have multiple features not found on solid-state units. Dozens of time bands and pickup points allow nearly endless flexibility in applying these types of devices. The I^2t function allows for more accurate protection of circuits that may have transient overcurrent. **See Figure 8-7.**

CAUTION: Changing overcurrent device trip settings while the circuit breaker is in service may damage the circuit breaker or the equipment being protected by open-circuiting the current sensors.

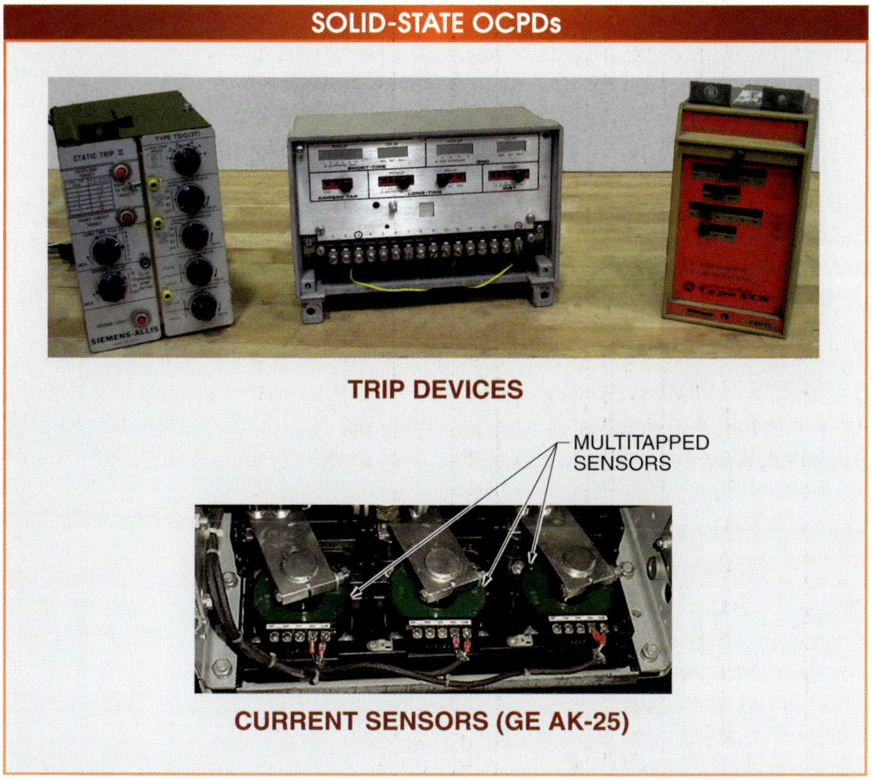

Figure 8-6. Solid-state OCPDs for circuit breakers were first developed by ITE, now ABB, in the 1970s.

Figure 8-7. Digital OCPDs, sometimes referred to as microprocessor-based devices, can have multiple features not found on solid-state units.

OCPD FUNCTIONS

A series OCPD operates individually and may have multiple armatures. This is an advantage and a disadvantage, in that if one OCPD failed, the other two would provide protection. However, since each series OCPD was individually calibrated, it was possible to have them set differently. Some first-generation solid-state OCPDs also had individual phase function settings, but this design quickly ended since all three phases are more than likely set the same. The most common OCPD functions on series trip devices are the following:

- long-time delay (LTD)
- instantaneous (INST)
- short-time delay

Solid-state and digital/microprocessor-based trip devices can also add the following functions:

- ground-fault
- I^2t/I^4t

SAFETY TIP

Series OCPDs are referred to as series because the current flows through the overcurrent device itself. When working on or adjusting them, use extreme caution because the armature and coil are energized.

Long-Time Delay Function

Long-time delay (LTD) provides overload protection, typically two to eight times the continuous current rating of the OCPD. Since it is for overload protection, the long-time delay function usually operates in seconds. All functions are inverse time-current, so if the current is high enough (and no faster function overrides it) the LTD function can operate quickly. LTD can have an adjustable time band and pickup or, what the OEMs would describe as their basic protection, they would have a single time band.

The time band adjustment usually involves a choice of three time bands, which often overlap on series OCPDs (as shown in the blue box). **See Figure 8-8.** Time bands show the minimum and maximum operating times of the OCPD at a given multiple of current. This means that overcurrent devices will not operate at one specific time but will operate within a certain bandwidth.

All series OCPDs must have the delay band and minimum pickup set to the calibration marks scribed on them at the OEM factory. In-between settings throw off the coordination of the power system and cause misoperation and equipment failure. **See Figure 8-9.**

In the example, the lines are individually hand-scribed and do not quite match up with each other from phase to phase. The red line indicates the 80% of continuous current pickup setpoint. Because of manufacturing variations, the 80% setpoint is slightly different on each OCPD. These calibration plates are from a GE AK-2 circuit breaker.

The second scribed mark on each phase is the setpoint for 100% continuous current, the third mark down is for 120% continuous current, and so on. This aspect of the calibration plates is somewhat difficult for less-experienced technicians to interpret since the technicians sometimes believe the mark next to the number is the correct mark. Usually, the scribe marks do not line up with the actual setpoint. Only the 140% setpoint is actually lined up with the factory-marked calibration points (in this example).

LONG-TIME DELAY (LTD) FUNCTION

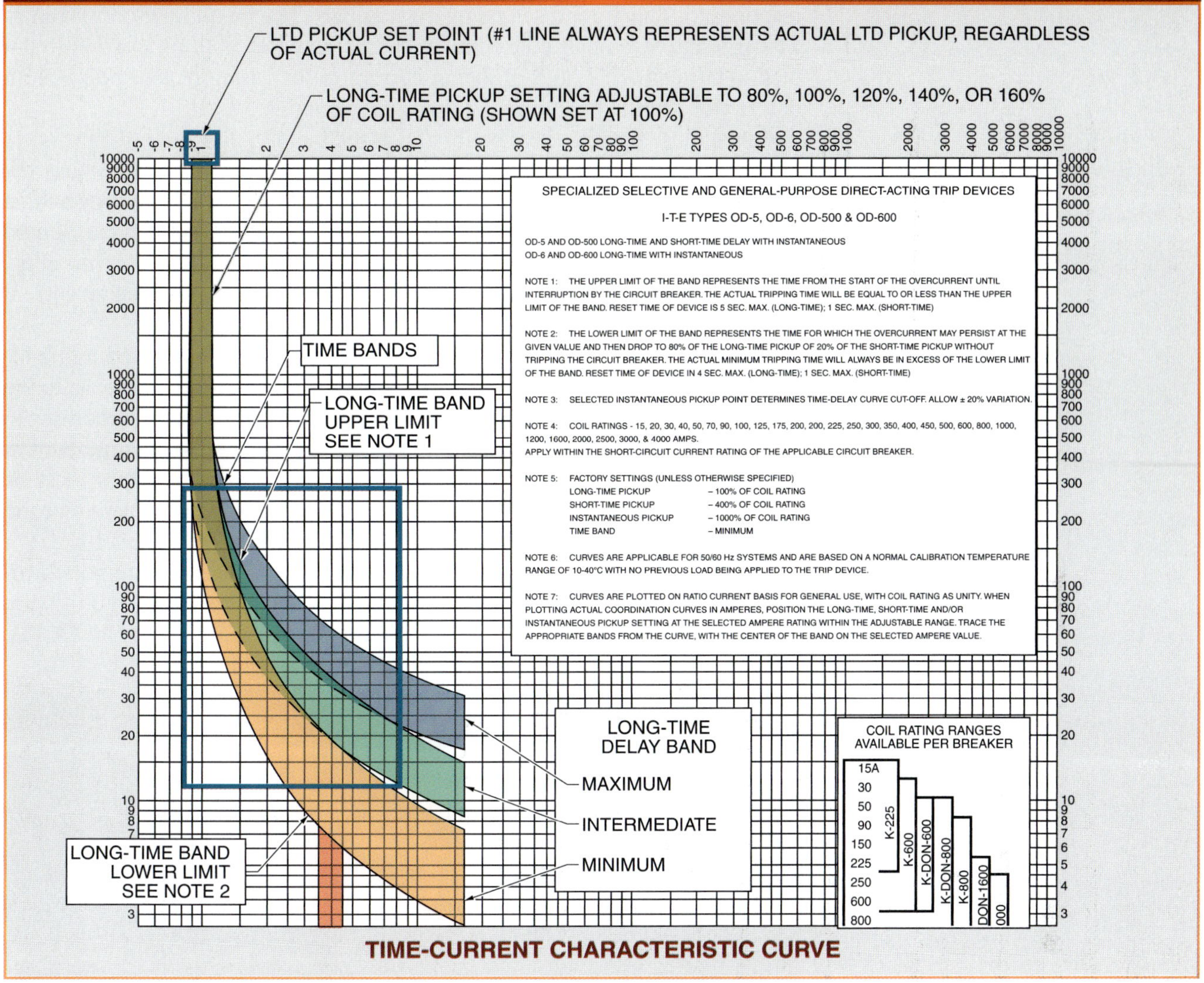

Figure 8-8. The time band adjustment involves a choice of three time bands, which often overlap on series OCPDs. The pickup adjustment sets the minimum value of current that the device responds to.

SERIES OCPDs — ABB/ITE OD TYPE

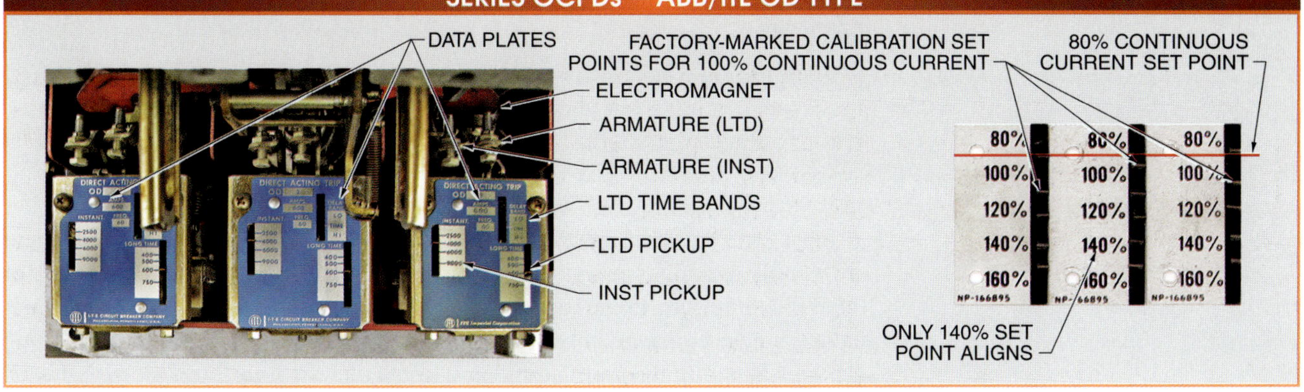

Figure 8-9. All series OCPDs must have the delay band and minimum pickup set to the calibration marks scribed on them at the OEM factory. Note the variation in the individual marks versus the percentages.

The pickup adjustment sets the minimum value of current that the device responds to. This is represented by the number 1 line on all time-current characteristic curves (small blue box). It can be noted as multiples of minimum pickup current or as ratio of actual current to long time pickup point. Both descriptions of the function have the same meaning.

When these devices were calibrated at the factory, they did not have oil in the dashpots. This allowed the minimum pickups to be verified because the armature would respond instantly once the current was high enough to overcome spring restraint. Once oil was added into the dashpots, minimum pickup could no longer be determined, except for the instantaneous function. The time delay was set by the manufacturer by applying multiples of minimum pickup current to the OCPD and monitoring the operating time.

Because of the manner in which they were calibrated compared to the installed condition, both minimum pickup and the time bands must be set to the predetermined marks on the data plates for the devices to function according to OEM specifications. If the OCPDs will no longer calibrate to the time-current characteristic curve, they must be replaced, or the circuit breakers must be retrofitted with digital OCPDs. Changing the settings to any place but the calibration marks renders the OCPDs calibration invalid.

CAUTION: Pickup settings above the continuous current rating are provided for transient overloads only. Circuit breakers will be damaged if they are regularly subjected to current above the continuous current rating of the OCPD.

Shermco Industries

Low- and medium-voltage circuit breakers are commonly used in electric power generating plants.

Series trip devices typically used oil dashpots. On the basic trip device there are usually two pickup adjustments: one for the LTD function and the other for the INST function. These functions are independent of each other. The adjustment spring is connected from the armature to an adjustable screw or tensioning device. Changing the spring tension will change the minimum pickup of each function. If the circuit breaker is equipped with the short-time delay function, there will be an additional adjustment for minimum pickup and time delay.

Westinghouse Electric Corporation went in an entirely different direction from the other circuit breaker OEMs by designing air dashpot OCPDs. With an air dashpot OCPD, an air bellows unit (air dashpot) provided the time delay. When the armature was attracted to the electromagnet, an adjustable orifice metered the air into the chamber below the diaphragm at a given rate and created a time delay. The time delay bands were changed by turning the time-band adjustment screw. Upgraded units that added the short-time delay function featured a black plastic knob on the bottom of the trip unit that was marked with a calibration mark. The trip nut was connected to the armature and physically operated the trip bar on the circuit breaker's operating mechanism. **See Figure 8-10.** The adjustment screw changed the size of the LTD valve or orifice. This changed the rate at which air would enter the bellows unit.

AIR DASHPOT SERIES OCPDs—WESTINGHOUSE

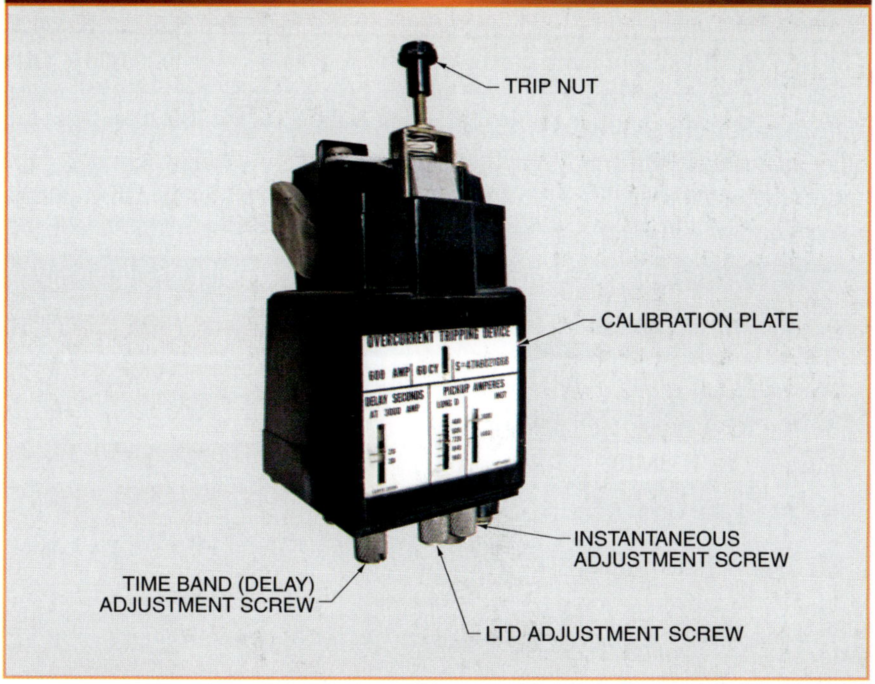

Figure 8-10. With an air dashpot OCPD, an air bellows unit provides the time delay.

As with other types of electromechanical OCPDs, the indicators must be set to the hand-scribed calibration marks for this device to be accurate. Also, parts from series OCPDs cannot be exchanged with parts from other series OCPDs even if they have the same ratings and setpoints. Springs, linkages, and other components will have variations due to wear and manufacturing tolerances. Also, when components from different OCPDs are assembled, the result will not be compatible with the calibration plate.

The LTD function is usually adjustable except for that of low-end, or economy, OCPDs. Most will have three delay bands: minimum, intermediate, and maximum. Westinghouse uses delay seconds such as 20 sec, 30 sec, etc. to indicate time bands. The delay band is adjusted by changing the effective length of the linkage between the armature and dashpot, which causes the piston to travel farther.

Instantaneous Function

The term instantaneous is defined as no intentional time delay. The instantaneous (INST) function is used for short-circuit protection. This does not mean there is no delay at all. Rather, there will be a small but measurable delay due to mechanical friction, travel distance of the armature, etc. With no delay, the INST function uses only a spring to set the minimum pickup current. The INST function is the only function that the actual minimum pickup can measure directly when testing since there is no delay to hamper operation.

Short-Time Delay Function

The short-time delay (STD) function is required to provide proper coordination with other OCPDs that may be in series with each other. **See Figure 8-11.** In the first example, all the OCPDs have only the instantaneous function (no intentional time delay). The main incoming circuit breaker has an instantaneous pickup of 20,000 A. The secondary circuit breaker has an instantaneous pickup of 16,000 A. The load circuit breaker has a pickup of 8000 A. If there were a 22,000 A fault on the load, the circuit breakers most likely to trip would be the main incoming and secondary circuit breakers. All three circuit breakers are in series with each other and, if the fault is large enough, all could trip. This would shut down the entire section of the facility's electrical system and cause production disruptions.

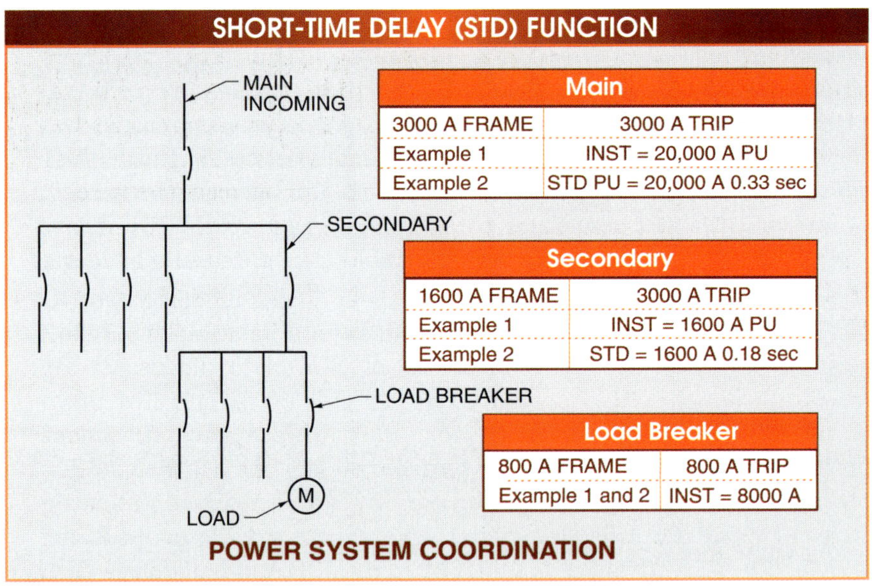

Figure 8-11. STD is required to provide proper coordination on OCPDs that may be in series with each other.

By using an STD function on the main and secondary circuit breakers (example 2), a short delay is required before the upstream circuit breakers trip. As short-circuit current flows through the main and secondary circuit breakers, they begin to respond but cannot trip until the STD function times out. The load circuit breaker, which has an instantaneous trip, operates without any intentional delay and interrupts the short circuit. The main and secondary circuit breakers both reset.

In the event that the load circuit breaker does not operate, the secondary circuit breaker will operate after a 0.18 sec delay of the STD function. If the secondary circuit breaker does not operate, then the main incoming circuit breaker will operate after a longer delay. As the fault "cascades" up into the system, more and more of the power system is shut down. The fault must be interrupted, even if it means shutting down the entire system.

This additional time delay allows the proper coordination of a power system, but it also comes with a price. Because the tripping is delayed for the main incoming and secondary circuit breakers, a fault will last much longer than the anticipated typical 0.07 sec (four cycles). Incident energy is proportional to time, so doubling the operating time doubles the incident energy. This causes safety concerns for personnel working on the part of the power system protected by these devices.

The STD function is independent of the LTD function. Usually, the STD function will have its own minimum pickup adjustment and time delay adjustment. On series OCPDs, STD is provided by using an escarpment (watch-type) timing mechanism in which the timing wheel can only advance one tooth at a time. STD also has inverse time-current characteristics, so increasing the current will cause it to operate more quickly. STD is required to have a 0.5 sec maximum operating time. However, many series OCPDs that have this function do not meet the requirement when set on their maximum time band.

STD also uses time bands, and they are changed by adjusting the effective length of the linkage. As with the LTD function, there are normally three bands that can be selected: minimum, intermediate, and maximum. The exception is the air dashpot OCPD developed by Westinghouse. It uses a second armature with a spring-tensioned adjustment for setting pickup. Time delay is provided by an orifice that is opened when the STD current value is reached. The STD band is changed by adjusting the size of the STD orifice.

The STD and LTD functions must have their pickups set to precalibrated points on the faceplate of the OCPD. **See Figure 8-12.** The INST function is the only function in which the minimum pickup can be set without regard for the calibration marks. This is because the INST function does not have a time delay, so as soon as the current reaches its minimum pickup current value, it will operate.

CAUTION: Never set the minimum pickup or time delay pointers on an OCPD to a position other than one of the factory-marked calibration marks.

OCPD CALIBRATION PLATES

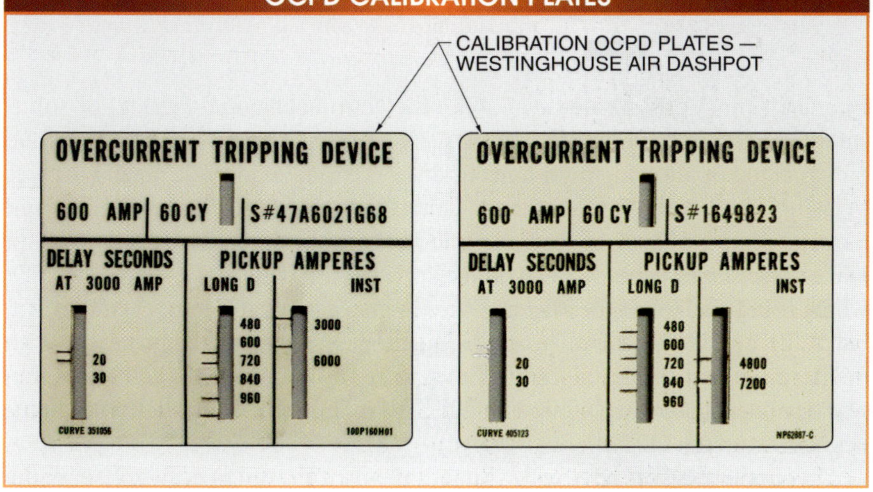

Figure 8-12. On a Westinghouse air dashpot OCPD, all three phases are marked on the calibration plate.

Ground-Fault Function

Per the National Electrical Code® (NEC®) Section 230-95, *Ground Fault Protection of Equipment,* ground-fault protection is required where there is a "*solidly grounded-wye service of more than 150 volts, but not exceeding 600 volts phase-to-phase for each service disconnect rated 1,000 amps or more.*" Ground-fault protection can be provided only when using a solid-state or digital OCPD. Delay times for the ground-fault function are to be 0.5 sec or less. *Note:* The ground-fault function on LVPCBs is not for personnel protection. That is provided by ground-fault circuit interrupters (GFCIs), which operate for an imbalance between the line and neutral when there is more than a 5 mA (±1 mA) difference in current.

The ground-fault function senses unbalanced currents between the three phases. **See Figure 8-13.** The type of ground-fault protection used in most LVPCBs is referred to as residual ground-fault sensing. Any imbalance between the phases will be seen on the wye-connected ground sensor. If that imbalance exceeds a preset amount, the ground-fault function causes a trip. In power systems, the unbalanced currents are kept to a minimum to prevent power quality issues, so the ground-fault pickup is set to a low value of current, sometimes less than 100 A.

GROUND-FAULT FUNCTION

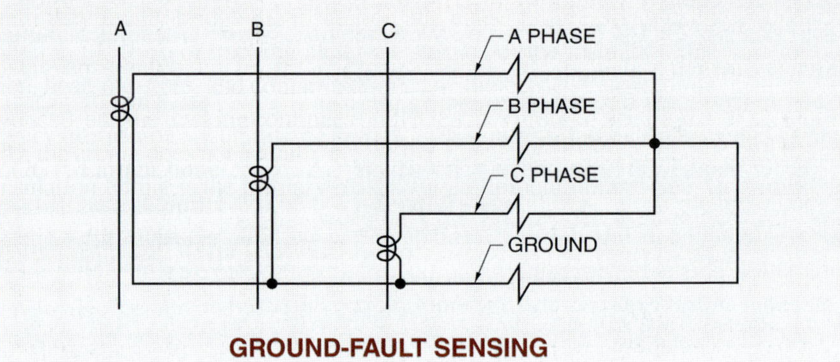

Figure 8-13. The ground-fault function senses unbalanced currents between the three phases.

I^2t/I^4t Functions

I^2t and I^4t functions are not available on electromechanical (series) or solid-state OCPDs, only on digital OCPDs. On solid-state and series OCPDs the only method of allowing high-current-draw equipment to cycle on and off was by using a higher minimum pickup. That is why these OCPDs have minimum pickups of up to 160% of their continuous current rating. This really did not make sense because the OCPD would respond as if it were a much larger OCPD, which could lead to overheating and fires. It also damaged electrical equipment, but at the time it was the only method available to manage the problem. With the development of digital OCPDs that were programmable, the I^2t function was developed and can be found on the LTD, STD, and ground-fault functions, depending on the OCPD.

On circuit breakers having this function, when I^2t (or I^4t) is switched in, the time-current characteristic curve is shifted over from its original position. This shift allows equipment that draws a heavy start-up current to cycle on and off without causing a nuisance trip. I^2t and the newer function I^4t can also be used to provide better coordination between other types of OCPDs and the circuit breaker, which eliminates overlap of the two characteristics. **See Figure 8-14.** By using the I^4t function to offset the characteristic, there is no overlap and less difference between the two characteristic curves.

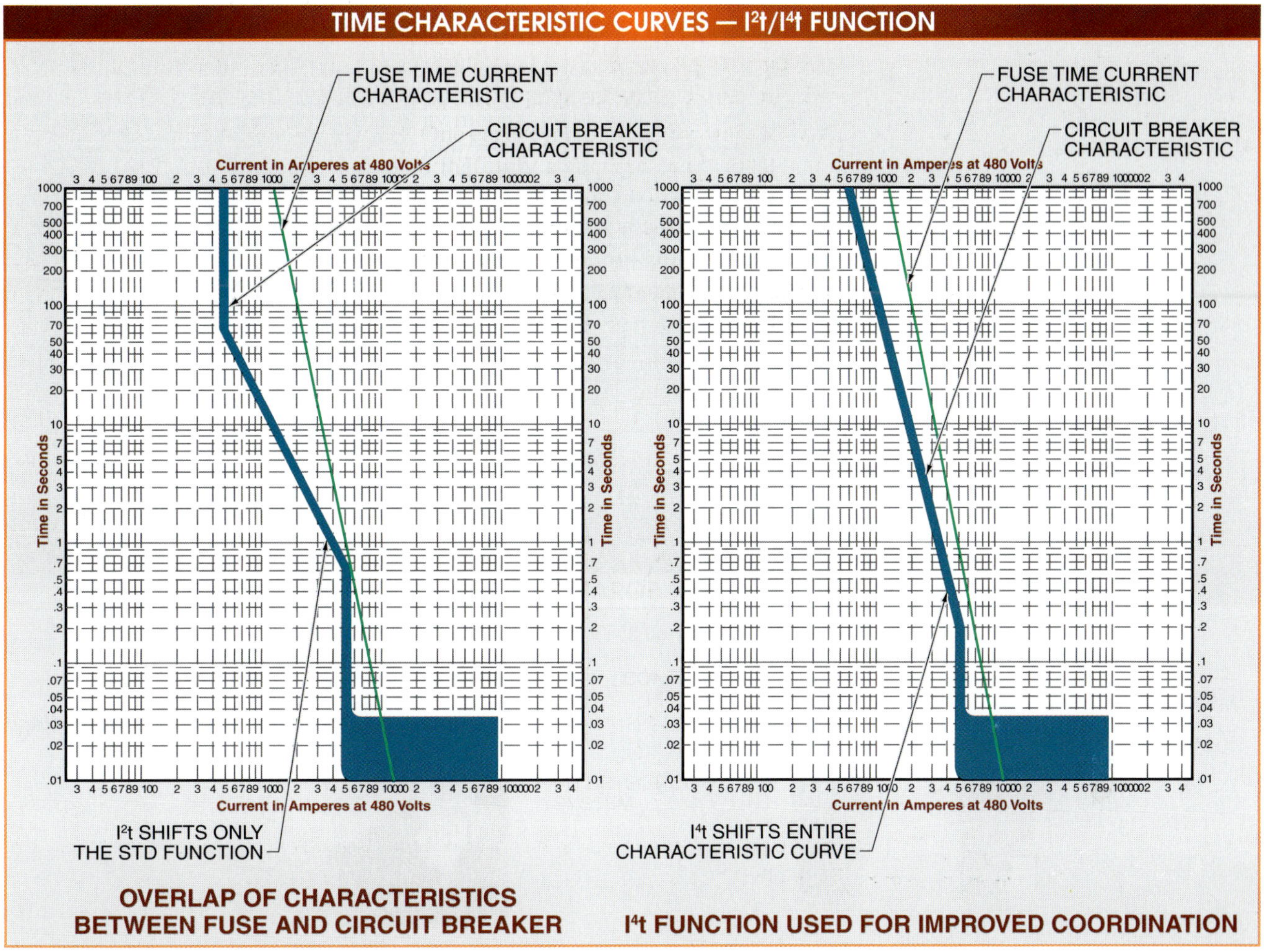

Figure 8-14. On circuit breakers having the I^2t/I^4t function, when I^2t (or I^4t) is switched in, the time-current characteristic curve is shifted over from its original position.

The primary difference between I^2t and I^4t is the shape of the offset. I^4t moves the entire characteristic curve while I^2t shifts only a portion of the characteristic curve. Using a time-current characteristic curve from an Eaton/Hammer DigiTrip OCPD as an example, both curves have an STD function. The curve on the left has the I^2t in, and the curve on the right has I^2t out. **See Figure 8-15.** With I^2t out, the STD characteristic is in its standard configuration. With I^2t out, an air compressor or other high-current-draw device could cause the STD function to operate. This is due to the high starting current of the compressor and the very short operating time of the STD function. With I^2t in, the STD characteristic is shifted over, which prevents the starting current of the compressor from falling into the operating characteristic of the OCPD.

EATON CUTLER-HAMMER DIGITRIP TIME-CURRENT CHARACTERISTIC CURVES

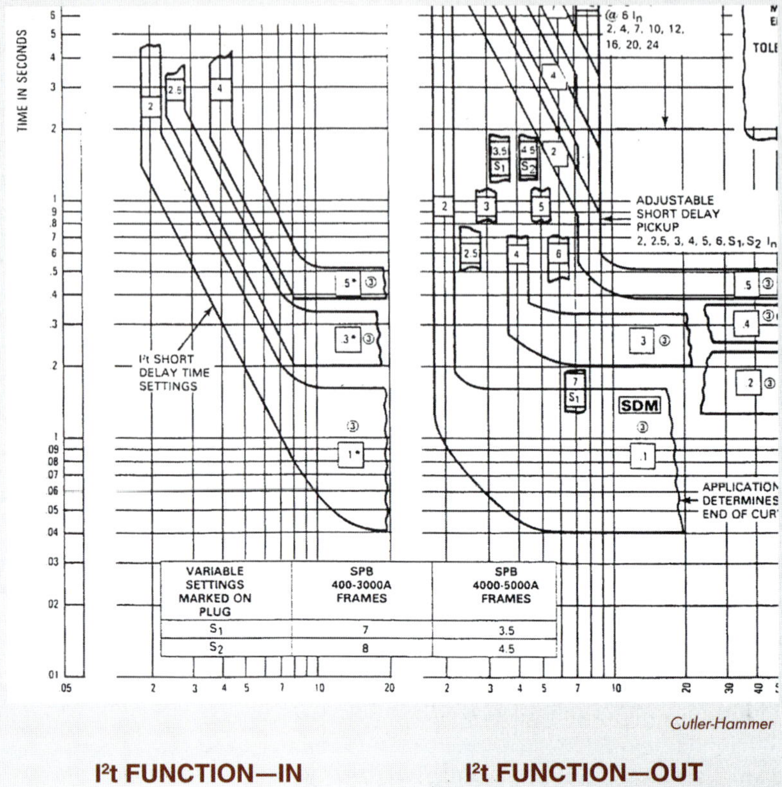

Figure 8-15. With I²t in, the STD characteristic is shifted over, which prevents high transient currents from falling into the operating characteristic of the OCPD.

CASE STUDY—CIRCUIT BREAKER TESTING

An industrial company would have their protective relays on their 13.8 kV power system calibrated each year but would not have the circuit breakers tested or maintained. When it was pointed out that the circuit breakers needed maintenance as well as the protective relays, the company representative responded, "They are like brand new. We never operate them." The implication was that the testing company was just trying to make some extra money by performing work that was not needed. Several months later, an underground feeder cable failed and the fault cascaded through six circuit breakers before it was cleared.

The company was certain that the fault was in the protective relays not being properly calibrated. However, upon inspection, all the protective relays had their operations indicators (flags) dropped, which indicated that they had operated. It was determined that the operating mechanisms of these circuit breakers were so dry from a lack of lubricant that the opening coils burned up on all six circuit breakers.

continued on next page...

CASE STUDY (...continued from previous page)

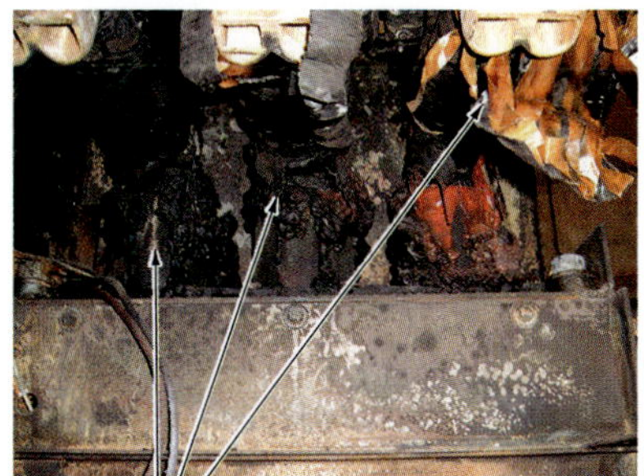

Conclusion: All parts of the protective system must be tested and maintained in order for it to operate properly when a problem occurs.

SUMMARY

Low-voltage OCPDs, or trip devices, include thermal-magnetic, series, solid-state, and microprocessor (digital) devices. OCPD functions include long-time delay (LTD), instantaneous (INST), short-time delay (STD), ground-fault, and I^2t/I^4t functions. OCPDs are undergoing constant evolution. From sucker-disk OCPDs to oil and air dashpots to solid-state devices to digital OCPDs, accuracy, features, and reliability will continue to improve over the previous generation of devices.

DIGITAL RESOURCES
ATPeResources.com/QuickLinks
Access Code: 223436

chapter 8

LOW-VOLTAGE OVERCURRENT PROTECTIVE DEVICES

CHAPTER REVIEW

Name ______________________ Date ____________

True-False

T F 1. I^2t can be used to provide better coordination between other types of OCPDs and the circuit breaker.

T F 2. Incident energy is proportional to time.

T F 3. The STD function is dependent on the LTD function.

T F 4. Ground-fault protection can be provided only when using a series OCPD.

T F 5. Digital OCPDs operate on the electromagnetic attraction principle.

T F 6. Solid-state OCPDs have better repeatability capabilities than series OCPDs.

T F 7. The primary difference between I^2t and I^4t is the shape of the offset.

T F 8. Solid-state OCPDs can have multiple features not found on digital units.

T F 9. Overcurrent protective devices (OCPDs) are important for protecting not only electrical equipment but also people.

T F 10. One OCPD operating cycle is 0.0167 sec.

T F 11. Solid-state OCPDs provide three-phase (3ϕ) protection.

T F 12. Parts from series OCPDs can be exchanged with parts from other series OCPDs if they have the same ratings and setpoints.

T F 13. Nearly all OEMs have stopped production of series OCPDs.

T F 14. Solid-state and digital OCPDs can fail without warning.

T F 15. The operating times of ambient-compensated trip elements do not differ from those of non-ambient-compensated trip elements.

T F 16. Series OCPDs were among the first to be designed and used.

T F 17. The deflection of the bimetal in a thermal-magnetic OCPD is directly related to the amount of heat created by current flow.

T F 18. The higher the spring tension of the sensitivity adjustment spring of a series OCPD, the lower the minimum amount of current that will be required to overcome it.

T F 19. All series OCPDs must have the delay band and minimum pickup set to the calibration marks scribed on them at the OEM factory.

T F **20.** Changing overcurrent device trip settings while the circuit breaker is in service may damage the circuit breaker.

T F **21.** The short-time delay function is required to have a 1.0 sec maximum operating time.

T F **22.** The long-time delay function is usually adjustable except for that of low-end, or economy, OCPDs.

T F **23.** Currently, there is no company that manufactures air or oil dashpots.

T F **24.** Current sensors on solid-state OCPDs are interchangeable between different OEMs.

T F **25.** The NFPA 70E is a maintenance standard.

Multiple Choice

______ **1.** Since it is for overload protection, the ___ function usually operates in seconds.
A. I^2t/I^4t
B. ground-fault
C. short-time delay
D. long-time delay

______ **2.** The short-time delay function is required to ___.
A. protect against overloads
B. protect against short circuits
C. provide proper coordination with all OCPDs that may be in series with each other
D. allow equipment that draws a heavy start-up current to cycle on an off without causing a nuisance trip

______ **3.** ___ OCPDs are the least accurate devices used to protect against damage from overcurrent and are the most prone to failure.
A. Solid-state
B. Digital
C. Series
D. Thermal-magnetic

______ **4.** Ambient-compensated trip elements use a second ___ to compensate for temperature variations between the circuit breaker environment and the equipment it is protecting.
A. trip armature
B. bimetallic element
C. vibration damper
D. current sensor

______ **5.** The ___ function is the only function that the actual minimum pickup can measure directly when testing since there is no delay to hamper operation.
A. instantaneous
B. ground-fault
C. long-time delay
D. short-time delay

______________ **6.** Instead of having an electromagnet in series with the load current, ___ OCPDs use current sensors.
- A. thermal-magnetic and digital
- B. general-purpose and series
- C. solid-state and digital
- D. solid-state and series

______________ **7.** OCPDs operate in just a few cycles, typically from ___ cycles.
- A. two to six
- B. four to eight
- C. six to ten
- D. eight to twelve

______________ **8.** The pickup adjustment sets the ___ that the OCPD responds to.
- A. minimum value of current
- B. maximum value of current
- C. minimum value of heat
- D. maximum value of heat

______________ **9.** The thermal portion of the thermal-magnetic OCPD uses a ___ placed in series with the load current.
- A. trip armature
- B. sensitivity spring
- C. magnetic element
- D. bimetallic element

______________ **10.** Being temperature dependent makes ___ OCPDs unsuitable for larger industrial applications where the ambient temperature might be elevated.
- A. series
- B. thermal-magnetic
- C. solid-state
- D. digital

______________ **11.** The ___ function is the only function in which the minimum pickup can be set without regard for the calibration marks.
- A. long-time delay
- B. short-time delay
- C. instantaneous
- D. ground-fault

______________ **12.** The primary disadvantage of ___ OCPDs is that they contain numerous small linkages plus some type of silicone oil.
- A. series
- B. microprocessor
- C. solid-state
- D. thermal-magnetic

______________ **13.** Series OCPDs ___.
- A. are less accurate than thermal-magnetic OCPDs
- B. can provide less trip functions than thermal-magnetic OCPDs
- C. can provide multiple operating time bands
- D. are affected by temperature

______ **14.** Delay times for the ground-fault function are to be ___ sec or less.
- A. 0.5
- B. 0.8
- C. 1.0
- D. 1.5

______ **15.** There are normally ___ time bands that can be selected with the STD and LTD functions.
- A. two
- B. three
- C. four
- D. six

Completion

______ **1.** I^2t and I^4t functions are only available on ___ OCPDs.

______ **2.** Pickup settings above the continuous current rating are provided for ___ only.

______ **3.** ___ is the heat from an electrical arc.

______ **4.** ___ tripping is defined as no intentional time delay.

______ **5.** A(n) ___ overcurrent protection device is an OCPD that provides overload protection by way of the thermal-actuated device and short-circuit protection using a magnetic device.

______ **6.** A(n) ___ is a sensor that reduces load current to a smaller value that is sent to the solid-state logic circuit where, if the value of current and time duration exceed its set points, will send a trip signal to a trip coil.

______ **7.** The ___ function provides overload protection, typically two to eight times the continuous current rating of the OCPD.

______ **8.** Solid-state OCPDs usually mimic the functions and operation of ___ trip units.

______ **9.** The instantaneous (INST) function is used for ___ protection.

______ **10.** The first hydraulic-type series OCPD was referred to as ___-acting because when it operated, the armature of the OCPD would strike the trip bar of the circuit breaker and directly cause it to open.

______ **11.** Series trip devices typically used ___ dashpots.

______ **12.** The electromagnetic coil of a series OCPD often has one or two turns and surrounds a laminated ___ core.

______ **13.** The ___ of a thermal-magnetic OCPD is constructed from two metals with differing coefficients of expansion.

______ **14.** The ___ function senses unbalanced currents between the three phases.

______ **15.** The relationship between the heat produced by current flow and the time it takes to trip the circuit breaker is known as a(n) ___ relationship.

______ **16.** The first ___ OCPDs were introduced by Allis-Chalmers in the 1950s and were not known for being reliable.

______ **17.** ___ OCPDs are the least expensive devices used to protect against damage from overcurrent and short circuits.

______________ **18.** Because the OCPD does not directly trip the circuit breaker, solid-state and digital OCPDs are sometimes referred to as ___ OCPDs.

______________ **19.** Circuit breakers will be damaged if they are regularly subjected to current above the continuous current rating of the ___.

______________ **20.** The type of ground-fault protection used in most LVPCBs is referred to as ___.

Short Answer

1. What is a problem caused by the additional time delay created by the short-time delay function?

2. What is an advantage and a disadvantage of series OCPDs operating individually and having multiple armatures?

3. Explain how the tension on the restraining spring affects the minimum operating current value of an interchangeable trip unit.

4. What are advantages of the shift of the time-current characteristic curve caused by the I^2t and I^4t functions?

5. How is the LTD delay band adjusted?

Matching

Thermal-Magnetic OCPD Construction

_______________ **1.** Load

_______________ **2.** Arc chutes

_______________ **3.** Bimetal element

_______________ **4.** Trip

_______________ **5.** Magnetic element

_______________ **6.** Spring-loaded contacts

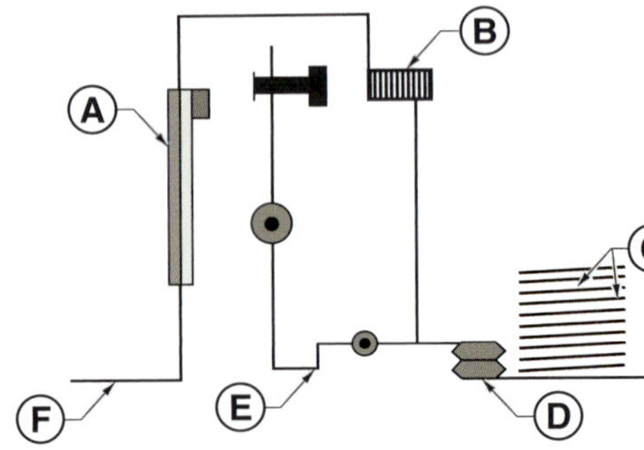

Series OCPD Construction

_______________ **1.** Oil dashpot

_______________ **2.** Antivibration damper

_______________ **3.** Connection to contact assembly

_______________ **4.** Time delay adjustment

_______________ **5.** Electromagnet

_______________ **6.** Long-time delay adjustment spring

_______________ **7.** Trip armature

_______________ **8.** Load-side primary disconnect

_______________ **9.** Instantaneous pickup adjustment spring

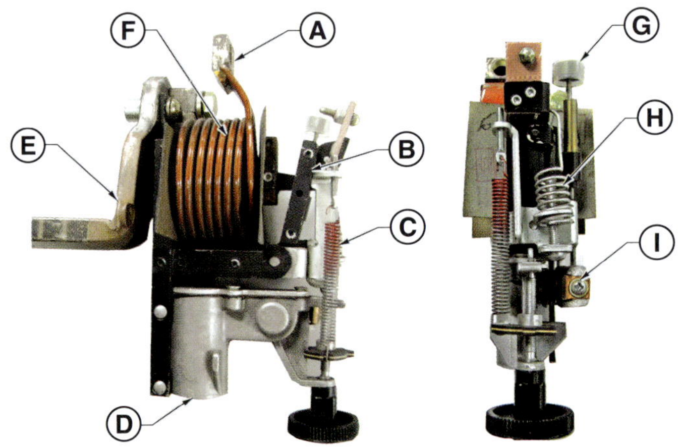

Air Dashpot Series OCPD Construction

_______________ **1.** Long-time delay adjustment screw

_______________ **2.** Instantaneous adjustment screw

_______________ **3.** Trip nut

_______________ **4.** Time band (delay) adjustment screw

_______________ **5.** Calibration plate

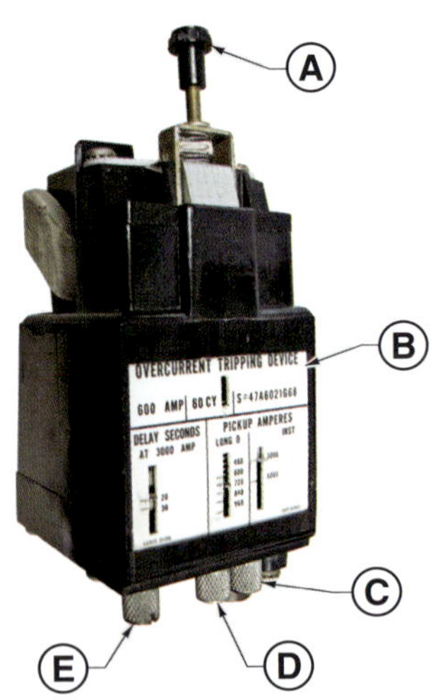

chapter 9

INTERPRETING TIME-CURRENT CHARACTERISTIC CURVES

Each original equipment manufacturer (OEM) produces time-current characteristic curves for its overcurrent protective devices (OCPDs). These curves are specific to each type of device, and the devices that the curves may be used for are listed on the curves. Time-current characteristic curves are not interchangeable, even within the same manufacturer. Sometimes an OCPD may have more than one curve, depending on its continuous current rating. It should always be verified that the curve being used is appropriate for the OCPD being tested.

Time-current characteristic curves are used by maintenance personnel to determine whether the OCPD meets the specifications of the OEM. As they age in service, OCPDs tend to require calibration. Series OCPDs may leak dashpot oil, or linkages will begin to wear and bind. Air dashpot devices can develop cracks in the flexible air bellows unit. Solid-state and digital OCPDs often fail without warning while in service, or the components begin to drift from their original specifications.

Electrical engineers use time-current characteristic curves to select the correct circuit breakers, OCPDs, and settings for power system coordination (engineering) studies. During power system coordination studies, the required protection setpoints are determined, and the devices are set so that selective tripping will occur. Selective tripping means that the device closest to the load will be the first one to operate and, if it does not operate, then the next device upstream will operate. If circuit breakers are not set according to the engineering study, false trips and nuisance tripping will occur.

Personnel safety is another factor that must be considered. If settings from the engineering study are changed, the device may not trip properly or take too much time to trip. Personnel working on or near the device could be exposed to much higher incident energies than expected, which could lead to serious injury or death.

OBJECTIVES

- Explain the purpose of the number 1 line.
- Describe how to interpret time-current characteristic curves that have multiple scales.
- Determine the operating time for the long-time delay function using the manufacturer's time-current characteristic curves.
- Determine the operating time for the short-time delay function using the manufacturer's time-current characteristic curves.
- Define the band limits on a time-current characteristic curve.
- Determine the tolerance for the instantaneous function using the manufacturer's time-current characteristic curves.
- Describe how to defeat the ground-fault function on overcurrent protective devices.
- List the advantages of performing secondary injection tests on overcurrent protective devices.

KEY TERMS

- maximum total clearing time line
- upper band time limit
- lower band time limit
- potentiometer
- secondary injection test

> **WARNING**
>
> *Never change any settings on any OCPD without an engineering study.*

THERMAL-MAGNETIC OCPD CURVES

Thermal-magnetic overcurrent protective devices (OCPDs), or trip devices, are used primarily on molded-case circuit breakers. The continuous current rating is stamped on the handle or frame of most molded-case circuit breakers. If the circuit breaker is of the interchangeable-trip type, the continuous current rating will be located on the OCPD. The continuous current rating of an interchangeable-trip molded-case circuit breaker is equal to or less than the frame rating. For example, the frame rating may be 400 A, but the OCPD may be only 60 A.

One of the first items that should be reviewed is the information box on the time-current characteristic curve. **See Figure 9-1.** The information box located at the bottom of the curve indicates that the curve applies to GE E-100 Line TEF and THEF molded-case circuit breakers. The information box also indicates the model number as well as the available continuous current, voltage, and frequency ratings that the circuit breaker can be used for. On the right side of the model number, it is stated that there are no adjustments for long-time delay or instantaneous trips.

The horizontal lines of the graph represent time, usually in seconds, while the vertical lines represent multiples of current rating. These lines also represent multiples of the number 1.

THERMAL-MAGNETIC DEVICE CURVE INFORMATION

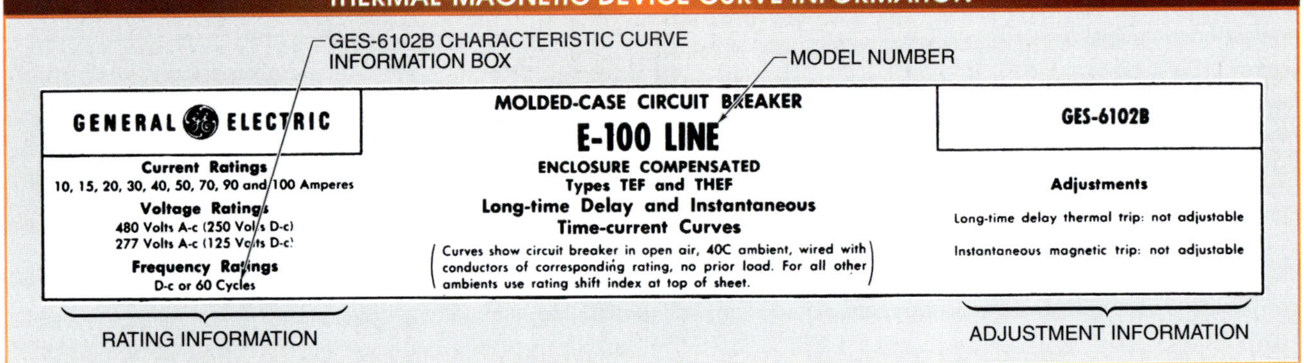

General Electric Company

Figure 9-1. One of the first pieces of information that should be reviewed is the information box on the time-current characteristic curve.

Number 1 Line

The number 1 (#1) line on a time-current characteristic curve may be referred to as the minimum pickup setpoint, 100% of continuous current rating, coil rating, or ampere tap, depending on the type of circuit breaker OCPD. Typically, the #1 line is the long-time delay minimum pickup setpoint. For example, if the long-time delay minimum pickup is 100 A, the #1 line would represent 100 A, the #2 line would represent 200 A, the #3 line would represent 300 A, and so on. For different OCPD minimum pickup setpoints, the #1 line represents that specific value. The example time characteristic has two multiple scales as well. The multiple scale to the left is used with the curve on the left, while the scale on the right is used with the curve on the right. There can be several curves on one sheet. **See Figure 9-2.**

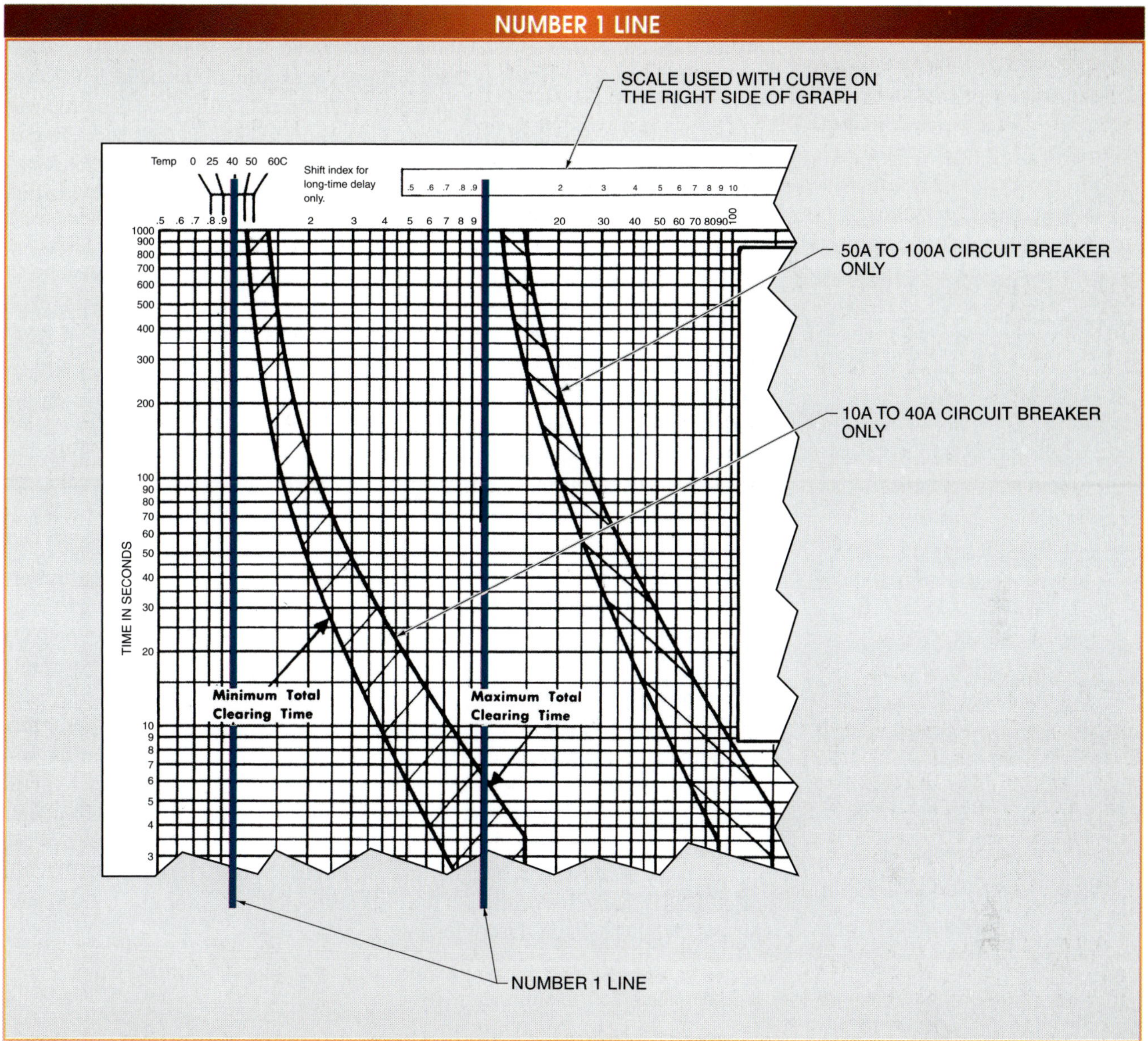

Figure 9-2. The number 1 (#1) line on a time-current characteristic curve may be referenced as the minimum pickup setpoint, 100% of continuous current rating, coil rating, or ampere tap, depending on the type of circuit breaker OCPD.

Long-Time Delay Function

The band at the top left side of the time-current characteristic curve is the long-time delay (LTD) band. There are two lines that make up this band: the upper time limit (maximum total clearing time) and the lower time limit (minimum total clearing time). On molded-case circuit breakers, the #1 line does not actually cross the lower band time limit as it does on power circuit breakers. This indicates that the circuit breaker should not trip at rated continuous current within the time specifications of the curve (in this case 1000 sec).

The upper time limit line represents the maximum time from the start of an overcurrent until interruption by the circuit breaker. The amount of time delay will always be equal to or less than the upper time limit of the band. The lower time limit line functions like the upper time limit, except that the trip time will be equal to or greater than the lower time limit. This means that, according to OEM specifications, no circuit breaker will operate at a specific time. A circuit breaker will operate within a bandwidth, and the time bands provide that specification. The standard test point for the LTD function is three times the continuous current rating. With an E-100 Line OCPD with a minimum pickup of 100 A, 300 A will be applied to it for the timing test.

Note: There are two curves on the page. The time curve on the left can only be used for circuit breakers with continuous current ratings from 10 A to 40 A. The time curve on the right is to be used for circuit breakers rated from 50 A to 100 A. Since a 100 A circuit breaker is used as an example, the curve on the right is used to determine operating time.

The maximum amount of time that the example circuit breaker should take to trip is 85 sec at a test value of three times (300%) the minimum pickup setpoint. This time value is found by reading the vertical #3 line to the upper limit and then reading left or right on the horizontal line to the trip time of 85 sec.

The lower time limit is read in the same manner. The #3 line (multiplier) is read upward until it intersects with the lower limit and then is read over horizontally to the time delay. Using the LTD test value of three times the LTD continuous current rating (300 A), the minimum operating time of the example circuit breaker is 35 sec.

The example OCPD will operate anywhere between 35 sec and 85 sec when three times its continuous current flows through it. For any given minimum pickup of the OCPD, if three times its continuous current is applied to it, it will always trip in 35 sec to 85 sec. The actual continuous current does not matter as long as three times the continuous current is applied. If the device trips the circuit breaker, there will be a definite amount of time before the OCPD will reset. This is due to the heating of the bimetallic element. Time must be allowed for the OCPDs to reset during testing.

Low- and medium-voltage circuit breakers are typically found in large commercial buildings.

Instantaneous Function

The instantaneous (INST) function time-current characteristic curve is located at the bottom of the curve for the thermal-magnetic OCPD. The INST device is identified as adjustable or nonadjustable according to the note to the right of the model number in the information box.

The INST curve lower pickup limit is drawn on the #9 line. The upper pickup limit is drawn on the #15 line. The INST setpoint (pickup) is in the center of these two limits, or the #12 line. This means that the INST pickup will operate somewhere between 9 times the setpoint to as much as 15 times the setpoint. The tolerance of the INST OCPD is then ±25% of the setpoint. **See Figure 9-3.**

INSTANTANEOUS (INST) FUNCTION TIME-CURRENT CHARACTERISTIC CURVE

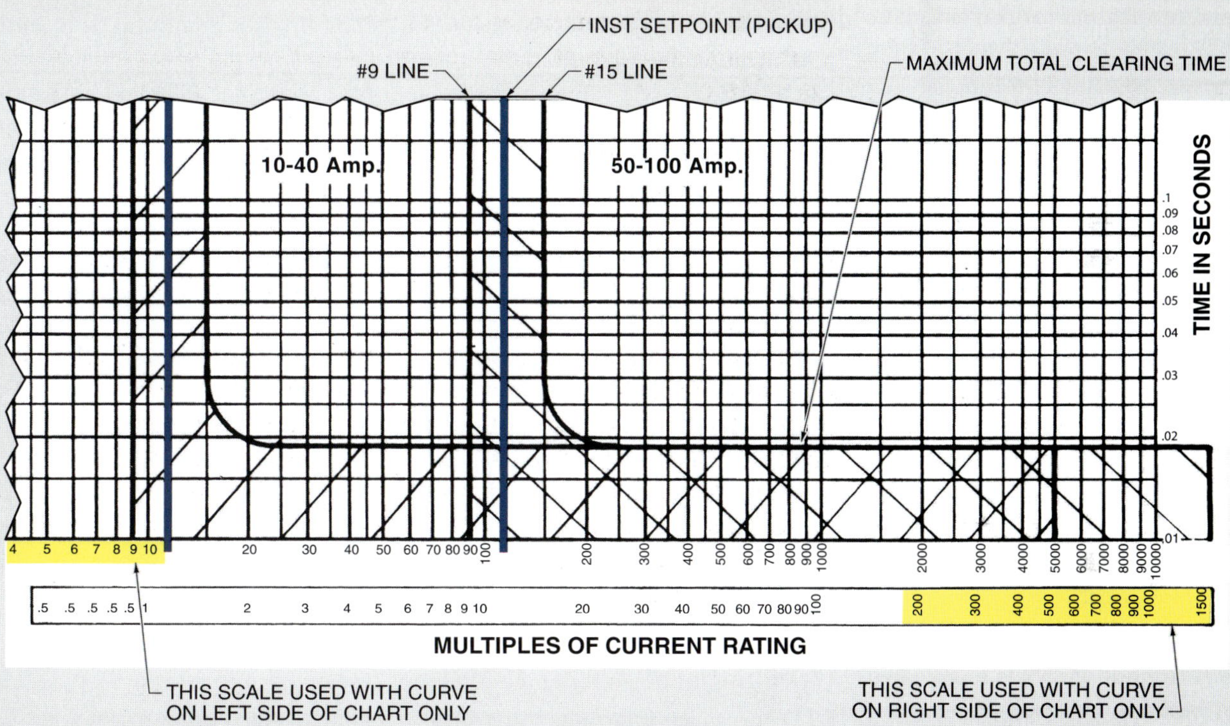

Figure 9-3. The INST function time-current characteristic curve is located at the bottom of the characteristic curve for the thermal-magnetic OCPD.

The line for the maximum total clearing time is drawn at the top of the INST curve. The *maximum total clearing time line* is the line on an INST curve that represents the maximum amount of time for circuit breaker contacts to open and extinguish an arc. The actual clearing time should never exceed the maximum total clearing time.

Generally, molded-case circuit breakers are considered two-cycle devices, which means that they should clear a fault in no more than two cycles. In this case, the circuit breaker will clear a fault much quicker, 0.018 sec. There are 0.0167 seconds per cycle, so this circuit breaker can clear a fault in just over one cycle. If the unit has an adjustable INST pickup, the curve is read in a similar manner as that of the nonadjustable units.

The next time-characteristic curve is for a GE J-600 Line molded-case circuit breaker. Below the model number, it is stated that the circuit breaker is ambient compensated and that the curve is also used for circuit breaker types TJJ (up to 400 A), TJK, and THJK. **See Figure 9-4.** This curve has two time curves on one page, according to the continuous current rating of the OCPD. The curve on the left-hand side is used for circuit breakers having a continuous current rating of 125 A to 400 A. The curve on the right-hand side is used for circuit breakers having a continuous current rating of 500 A to 600 A. Both curves have adjustable INST (magnetic) OCPDs. They are adjustable from 3 to 10 times the continuous current rating. The pickup, or setpoint, is in the center of the pickup limits. The INST unit may trip anywhere from 9 to 11 times the continuous current rating on the high setpoint (±10% tolerance).

Since this is an ambient-compensated circuit breaker, the temperature around the circuit breaker (ambient temperature) can be compensated for. The higher the ambient temperature, the faster the circuit breaker could trip. It is similar to preheating the trip unit. If the ambient temperature is known, the time curve can be moved to the left to compensate for the abnormal ambient temperature. If the ambient temperature is between 10°C and 50°C, no compensation is needed. If the ambient temperature is above 50°C or below 10°C, the curve is to be shifted, and the new trip times are used.

During field testing, the circuit breaker will not be tested outside normal room temperature, and ambient temperature compensation is not necessary. However, when engineers coordinate the system protection, this aspect of the curve can be important.

SERIES OCPD CURVES

Series OCPD time-characteristic curves include curves for the LTD function, multiple time bands, short-time delay (STD) function, and INST function. Series OCPDs are in series with the flow of current through the circuit breaker. This means that, when in service, these devices are energized at full voltage and the actual current flowing through the circuit breaker. This needs to be considered whenever troubleshooting these devices as the device itself is energized.

LTD Function—Series OCPDs

LTD curves are interpreted in a similar manner as the curves for molded-case circuit breakers. There are some differences, most having to do with terminology. One major difference is that the #1 line never intersects the lower band time limit for molded-case circuit breakers, whereas it does for series OCPDs. This means that series OCPDs can trip circuit breakers, even though only 100% of their continuous current is applied to them.

The first band at the top left side of the OD-3 curve is the LTD band. As with the examples in the molded-case section, the time band consists of the upper band time limit and the lower band time limit. The #1 line is centered between the upper time limit and lower time limit at the top of the curve. **See Figure 9-5.**

Note: The upper limit represents the maximum time from the start of an overcurrent until interruption by the circuit breaker. The amount of time delay will always be equal to or less than the upper limit of the band. The lower band time limit functions like the upper band time limit, except that the trip time will be equal to or greater than the lower band time limit. Refer to Note 2 on the time-current characteristic curve.

GE J-600 LINE TIME-CURRENT CHARACTERISTIC CURVE—AMBIENT COMPENSATED

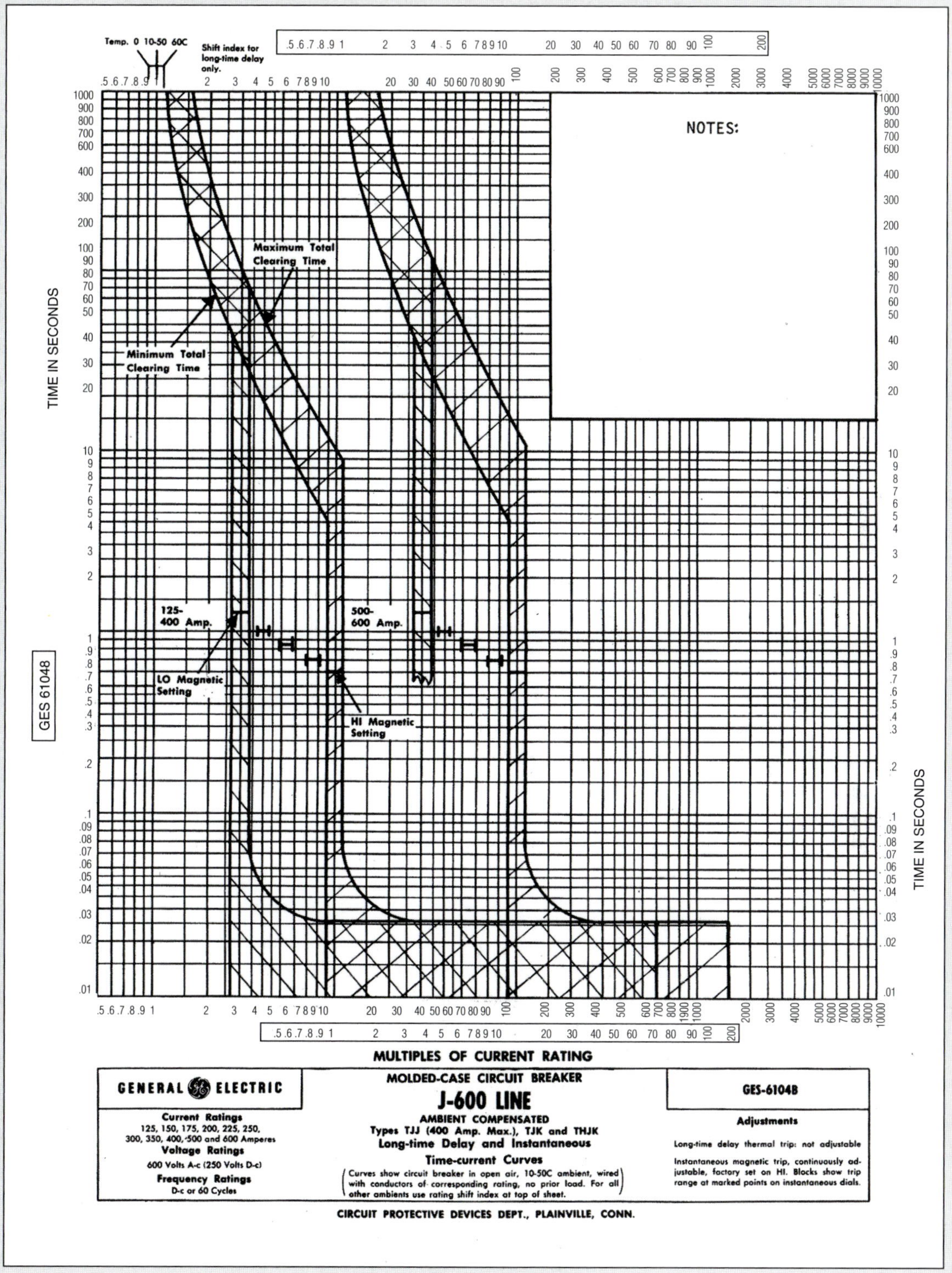

General Electric Company

Figure 9-4. The GE J-600 Line molded-case circuit breaker is ambient compensated. The time-current characteristic curve can also be used for circuit breaker types TJJ (up to 400 A), TJK, and THJK.

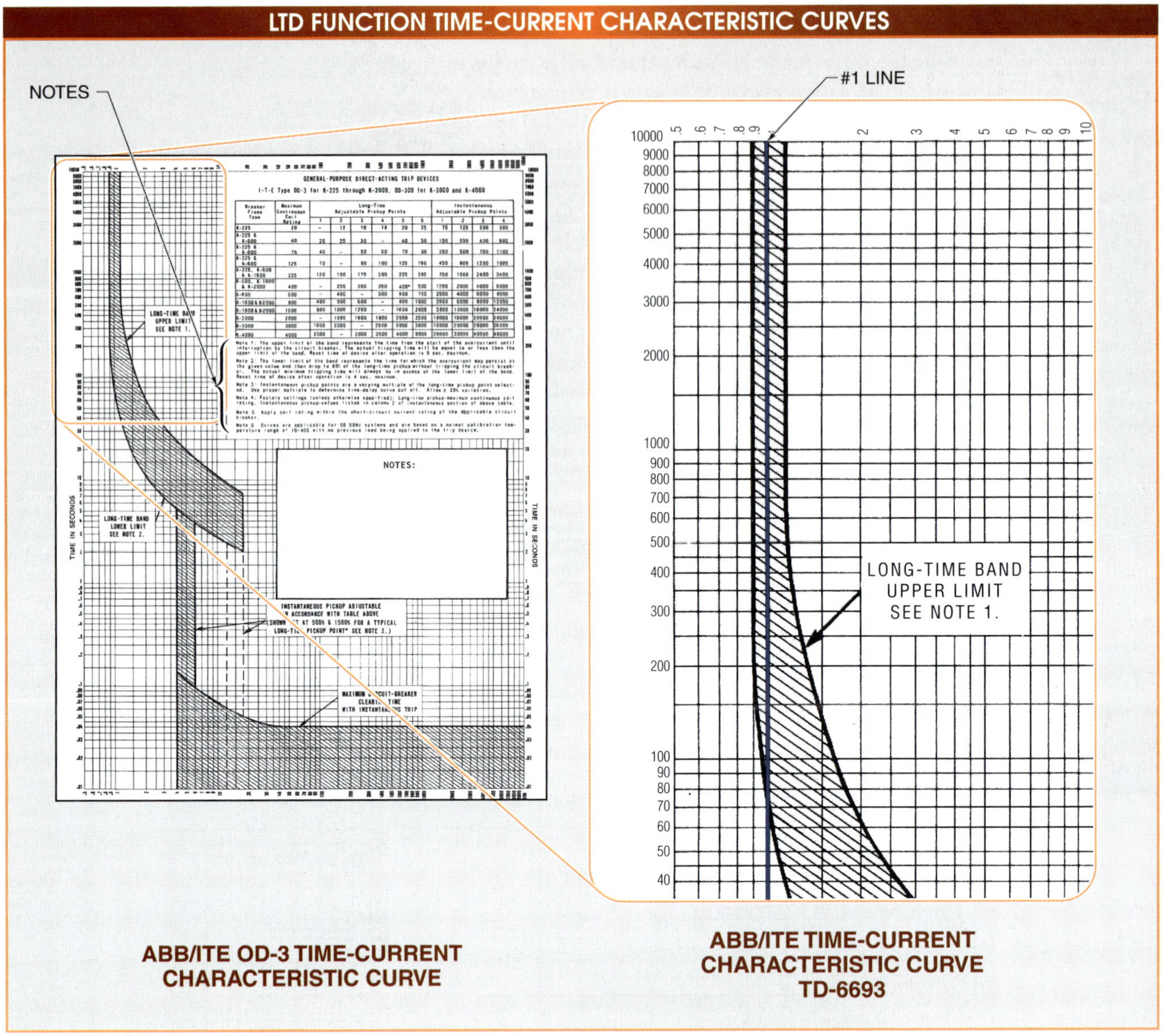

ABB Inc.

Figure 9-5. With LTD function time-current characteristic curves, the #1 line does not intersect the lower band time limit for molded-case circuit breakers, whereas it does for series OCPDs.

The standard test point for the LTD function is three times the minimum pickup current it is set to. With an OD-3 OCPD set to 1200 A, 3600 A will be applied to it for the timing test. The maximum amount of time that the OCPD should take to trip is 35 sec. This time value is found by reading the vertical #3 line to the upper band time limit and then reading left or right on the horizontal line to the trip time of 35 sec.

The lower band time limit is read in the same manner as the upper band time limit. The #3 line (multiplier) is read upward until it intersects with the lower band time limit and then over horizontally to the time delay. Using the LTD test value of three times the LTD setpoint (3600 A), it is determined that the minimum trip time is 6.5 sec. **See Figure 9-6.**

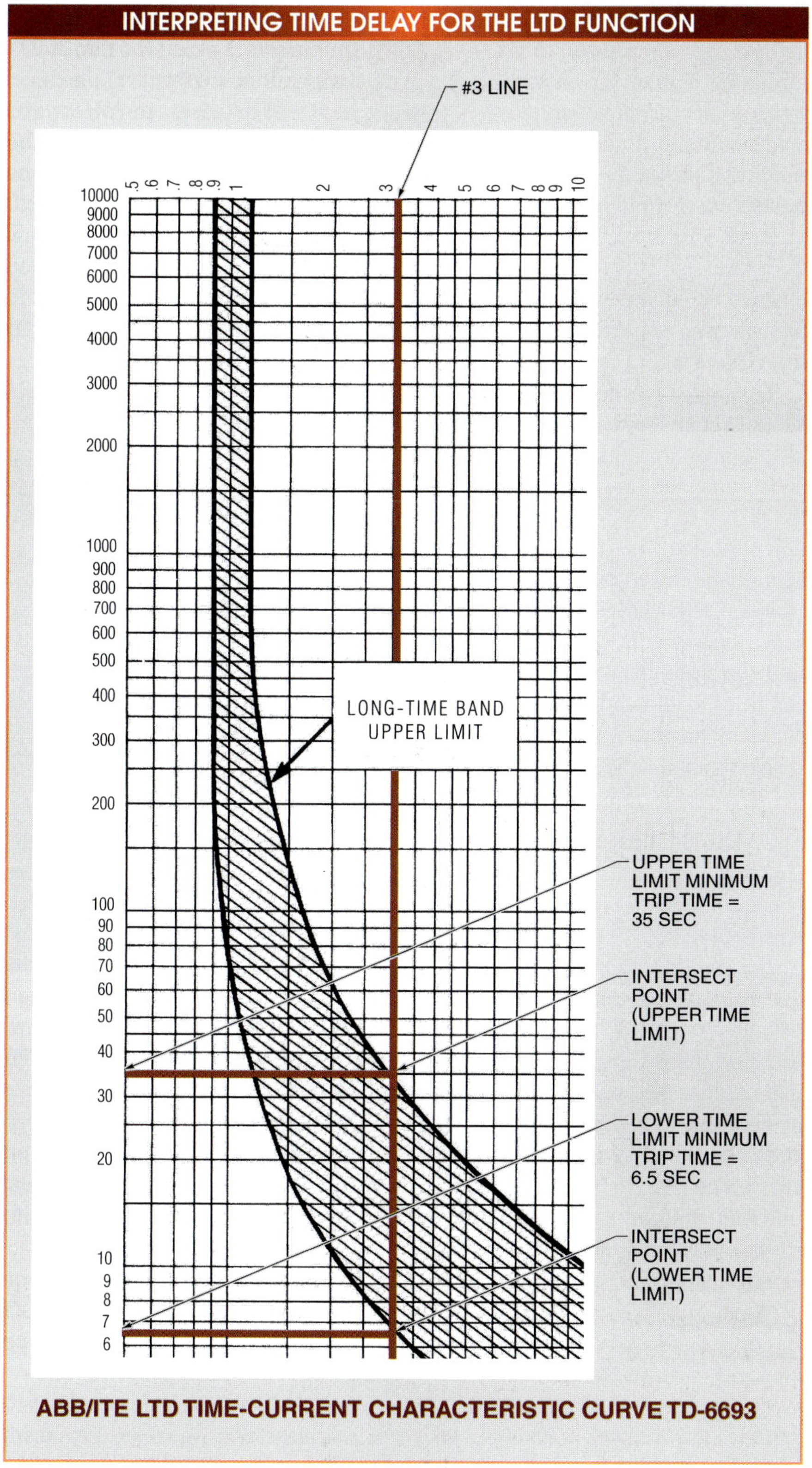

ABB Inc.

Figure 9-6. The values of the upper and lower band time limits are found by reading the vertical #3 line to the upper/lower band time limits and then reading left or right on the horizontal line to the trip time readings.

The example OCPD in Figure 9-6 will operate anywhere between 6.5 sec and 35 sec when three times its setpoint (minimum pickup) current flows through it. For any given minimum pickup setting of the OCPD, if three times its setpoint current is applied to it, it will always trip in 6.5 sec to 35 sec. The actual minimum pickup current does not matter as long as three times the minimum pickup current is applied during testing. It is important to note that the OCPD will not operate at a specific time, only within its time delay band.

If the circuit breaker trips, it will not reset immediately. It takes time for the piston to travel through the oil to its original start position. If the trip time reaches the upper limit, it will take a maximum of 5 sec to reset (Note 1). If the trip time only reaches the lower limit, it will take a maximum of 4 sec to reset (Note 2).

Time must be allowed for the OCPDs to reset during testing. This can affect testing in two ways. First, if the OCPD has not completely reset during a timing test, then an inaccurate (short) trip time will occur. Second, if the test current is being jogged up when doing an INST pickup test, time must be allowed for the piston to reset before doing the actual pickup test. If the piston is drawn down through the oil dashpot while the current is being set, the piston may actually be faster than the INST unit.

Note: A dashed, or ghost, line shown on a time-current characteristic curve only illustrates the adjustable limits of that characteristic and cannot be used to determine operating time. Only the solid time-current characteristic curve can be used to determine operating time. This is true for all OEM time-current characteristic curves.

Multiple Time Bands

The ABB/ITE OD-5 OCPD curve has three LTD bands: minimum, intermediate, and maximum. This allows for more flexibility in system coordination, but it complicates interpretation of its time-current characteristic curve. **See Appendix.**

For the OD-5 OCPD, it must be noted that the lower band limits for the intermediate and maximum bands are dashed lines. Care must be exercised when reading these time bands to make sure that the correct time band is used. The minimum (bottom) time band is shaded, so it is easy to determine. The intermediate time band is defined by the next dashed line and the solid line above it. The maximum time band is defined by the second dashed line and the top solid line. The same situation occurs with the STD time-current characteristic curve. **See Figure 9-7.** Operating times at the standard timing point of three times the minimum pickup setpoint for the three time bands are the following:

- minimum—10 sec to 29 sec
- intermediate—27 sec to 50 sec
- maximum—46 sec to 90 sec

STD Function—Series OCPDs

The STD time band is located on the OD-5 curve directly below the lower band time limit of the LTD minimum time band limit. *Note:* The LTD setting has no effect on the STD setting or the INST setting, since each function has its own armature. The adjustable current range for each trip function is strictly related to the coil rating.

LONG-TIME DELAY CHARACTERISTIC CURVE INTERPRETATION

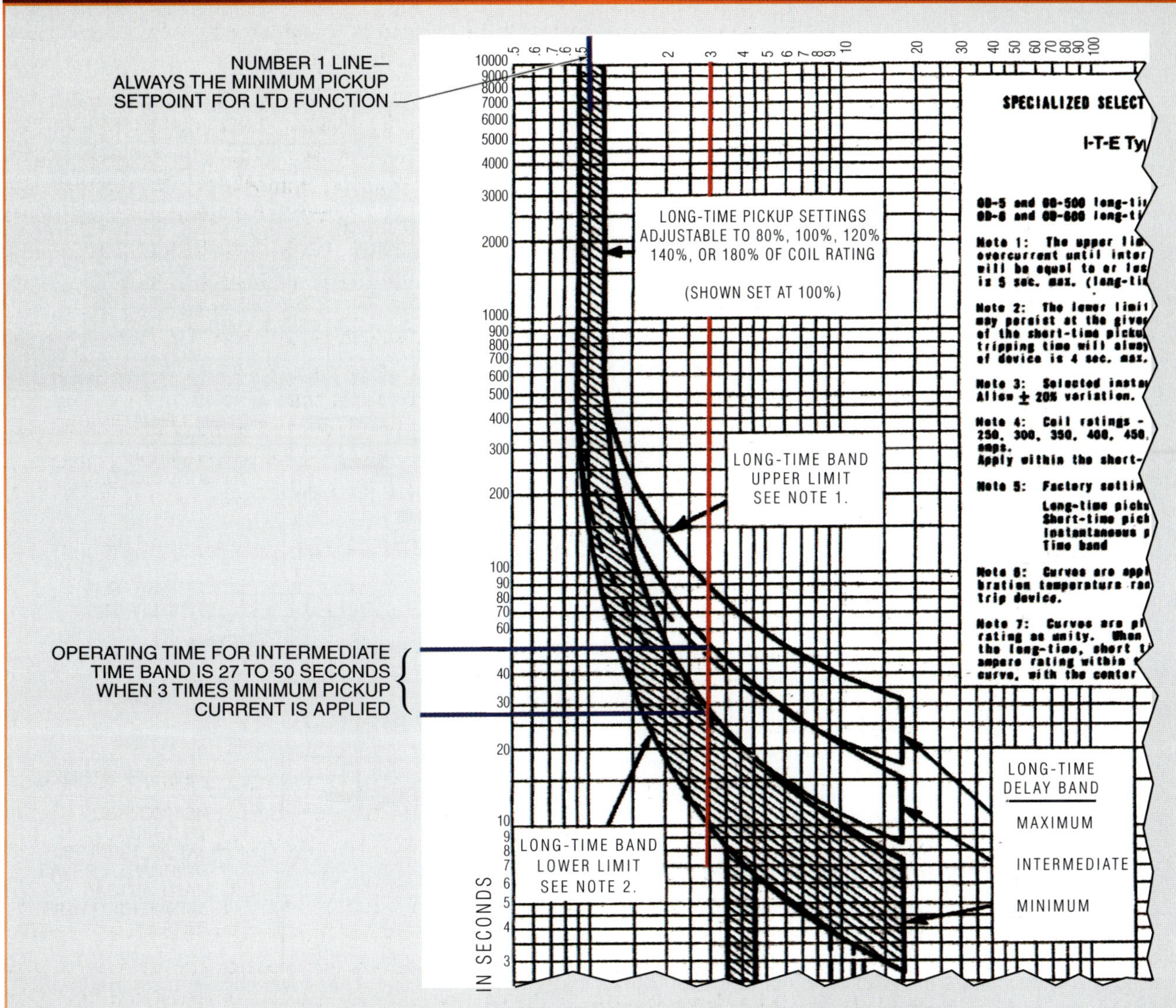

ABB/ITE STD TIME-CURRENT CHARACTERISTIC CURVE TD-6695

ABB Inc.

Figure 9-7. With an ABB/ITE TD-6695 curve, the minimum (bottom) time band is shaded. The intermediate time band is defined by the next dashed line and the solid line above it. The maximum time band is defined by the second dashed line and the top solid line.

The note just to the right of the STD time band states, "Short-time delay pickup setting is adjustable to 400%, 700% and 1000% (shown set at 400% and 1000%)." The center of the STD band is found on the #4 line between the upper and lower band time limits. There is also a dashed curve outlined for the STD band that is drawn with its center on the #10 line. These two curves show that the STD device can be adjusted between 4 and 10 times the OCPD's coil rating. **See Figure 9-8.** The purpose for showing the two delay curves is to illustrate the adjustable range. They cannot be used to determine operating time. *Note:* The solid STD curve must only be used for determining trip times. The dashed curve is only for illustrative purposes.

SHORT-TIME DELAY (STD) CHARACTERISTIC CURVE INTERPRETATION

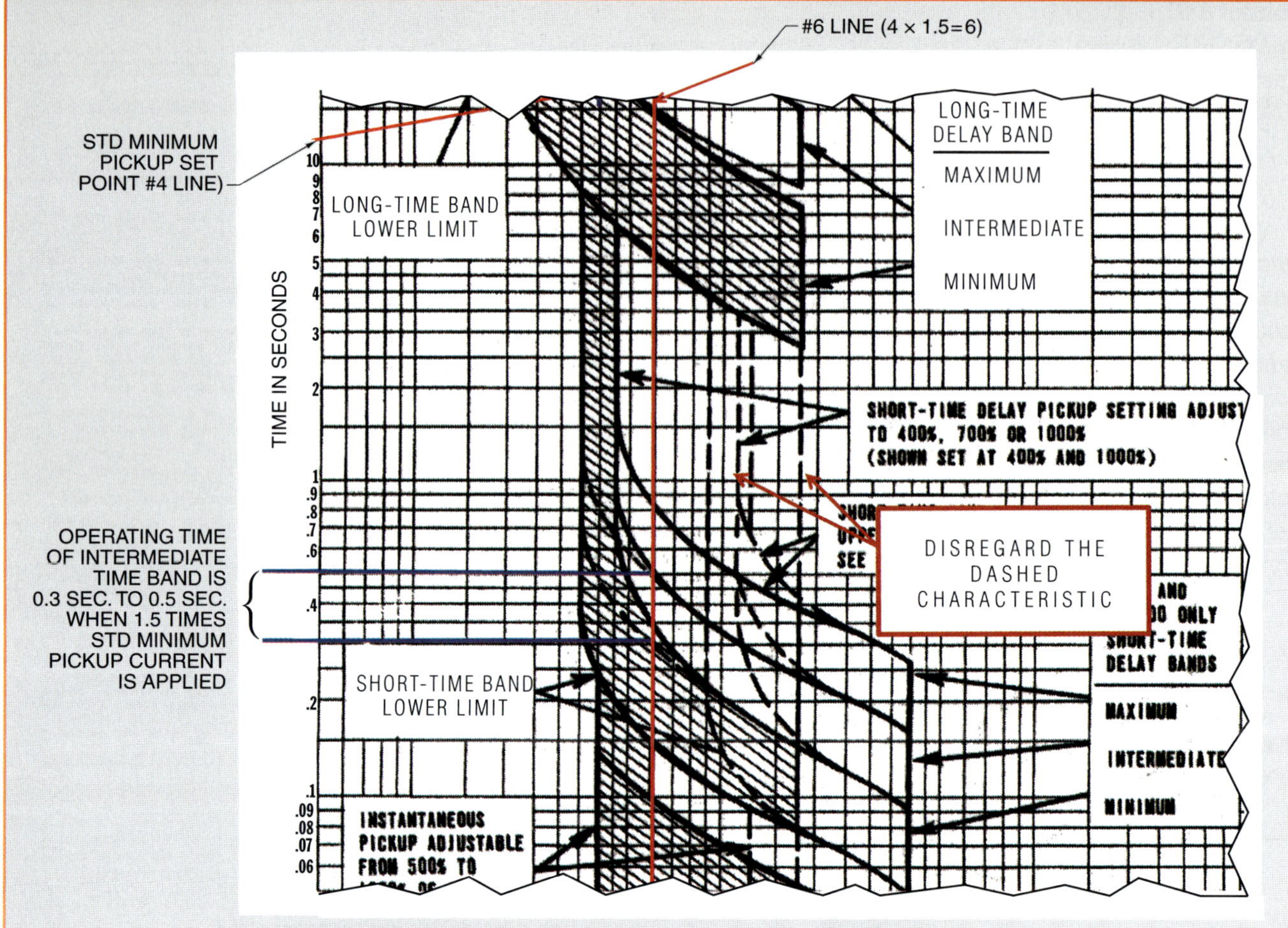

ABB/ITE STD TIME-CURRENT CHARACTERISTIC CURVE—TD-6695

ABB Inc.

Figure 9-8. The center of the STD band is found on the #4 line between the upper and lower band time limits. There is a dashed curve outlined for the STD band that is drawn with its center on the #10 line. These two curves show that the STD device can be adjusted between 4 and 10 times the OCPD's coil rating.

The center of the STD band, the #4 line in this case, is used the same way as the #1 line for the LTD function. The center of the characteristic curve represents the minimum pickup setpoint of the STD function. Each STD band has an upper and lower limit. An *upper band time limit* is the time that represents the maximum operating time on a time-current characteristic curve. A *lower band time limit* is the time that represents the minimum operating time on a time-current characteristic curve. The upper limit and lower limit of the bands are used in the same manner as the LTD bands, since they represent the maximum and minimum operating time. The following procedure is used to determine the operating times for the STD function:

1. Find the center of the STD time-current characteristic curve. On the OD-5 OCPD, it is the #4 line. *Note:* This will probably be different for other OEM OCPDs.
2. Multiply the number that the STD curve is drawn on by the multiple of test current that is going to be used. For example, the standard timing test point

for the STD function is 1.5 times the STD minimum pickup setpoint. As an example, if the STD minimum pickup setpoint is 6000 A and the time band is set to minimum, 9000 A will be injected into the circuit breaker (6000 A × 1.5 = 9000 A). The same ratio will be applied to the time curve itself. The #4 line representing the setpoint will be multiplied by 1.5 with a result of 6. The #6 line, the multiplier line, will be used to determine operating times for the ITE OD-5 OCPD.

3. Read up the appropriate vertical multiplier line, which is the #6 line in this case, until it intersects the upper and lower band limits to determine the operating time. Using the OD-5 curve, follow the #6 line vertically to the lower band time limit, then read either left or right on the horizontal line to the operating time of 0.15 sec. Continue reading up the vertical #6 line to the upper band time limit, and follow the horizontal line left or right. The upper band time limit is 0.35 sec.

The STD function will operate between 0.15 sec and 0.35 sec when 1.5 times the STD minimum pickup setpoint current is applied to the OCPD. Other operating times for the OD-5 STD function are the following:

- minimum—0.15 sec to 0.35 sec
- intermediate—0.3 sec to 0.5 sec
- maximum—0.5 sec to 0.85 sec

Low- and medium-voltage circuit breakers are typically installed in institutional applications where many buildings are located in a small area, such as college campuses.

INST Function—Series OCPDs

The INST device is adjustable in accordance with Note 3 and states, "Selected instantaneous pickup point determines time-delay curve cutoff. Allow ±20% variation." The note just to the right of the INST band shows that this function is adjustable from 500% to 1500% of the coil rating and is shown set at 500% and 1500%. The center of the INST time-current characteristic curve represents the minimum pickup setpoint for the INST function.

The INST minimum pickup is shown on the #5 line and has lines drawn to its left and right. These lines represent the ±20% tolerance. The maximum clearing time line is drawn at the top of this curve. The maximum clearing time line represents the maximum amount of time for the circuit breaker to open and quench an arc. The actual clearing time should never exceed the maximum clearing time limit. **See Figure 9-9.**

INST FUNCTION TIME-CURRENT CHARACTERISTIC CURVE INTERPRETATION

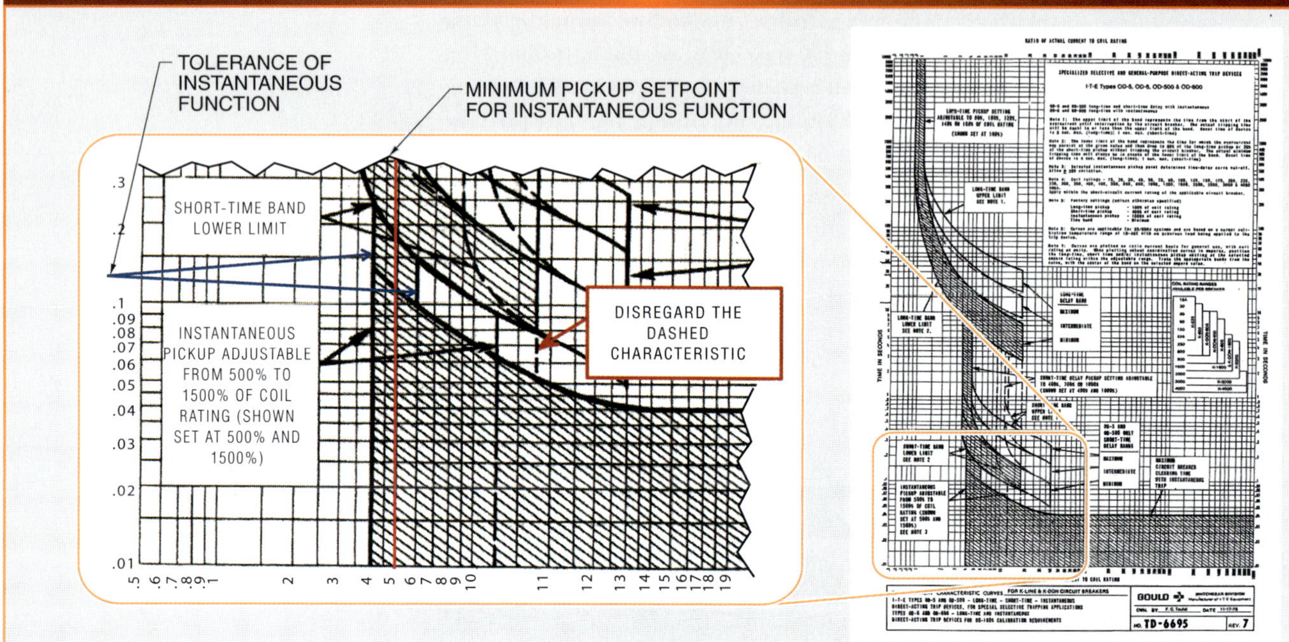

NOTE: ACTUAL OPERATING CURRENT WILL BE BETWEEN 4 TIMES AND 6 TIMES INSTANTANEOUS MINIMUM PICKUP CURRENT

ABB Inc.

Figure 9-9. The note just to the left of the INST band states that this function is adjustable from 500% to 1500% of the coil rating and is shown set at 500% and 1500%. The center of the INST curve represents the minimum pickup setpoint for the INST function.

SOLID-STATE AND DIGITAL OCPD CURVES

Solid-state OCPD curves are interpreted in a similar manner as series OCPD curves. Solid-state and digital curves are easier to read, since the time bands do not overlap. The time bands are also smaller and tighter, and the devices are more accurate.

Setting Device Functions

Potentiometers were sometimes used on early solid-state OCPDs to adjust the minimum pickup and time bands for the various functions for each phase. A *potentiometer* is a variable-resistance electric device that divides voltage proportionally between two circuits. Because of this, the devices had to be calibrated to specific setpoints that were unique to each device, much like series OCPDs. **See Figure 9-10.** On OCPDs like the Static Trip II, the function can only be set to the specific calibration marks. Rather than specific calibration marks, ITE solid-state OCPDs have tap plugs that are inserted into tap settings.

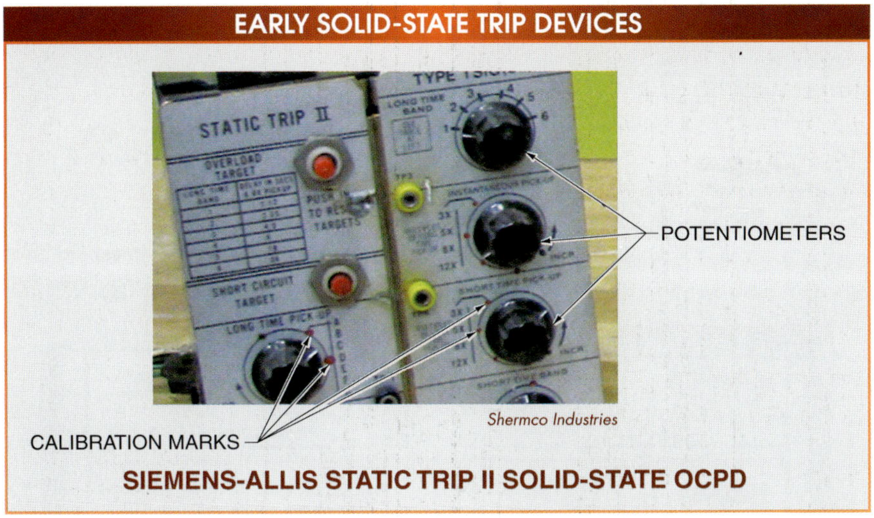

Figure 9-10. Early solid-state OCPDs sometimes used potentiometers and calibration marks to adjust the minimum pickup and time bands for various functions.

Note 5 reads, "The AMPERE TAP setting and all pickup and time band settings are determined by placement of plug into appropriate tap block on front of device." This is for the ITE Power Shield only. Other OEMs have different methods of setting and calculating their pickups.

LTD Function—Solid-State and Digital OCPDs

The LTD function has three time delay bands. All bands are clear and easy to use. The #1 line represents the minimum pickup setpoint of the LTD function. Whatever the actual minimum pickup current is, it is represented by the #1 line. Using the ITE type SS-5 curve as an example, the LTD function would be tested at three times the LTD minimum pickup current. As an example, the ampere tap is set to 400 A. Inserting the plug into this tap setting connects the OCPD to the 400 A winding of the sensor/current transformer (CT).

WARNING

Since the tap connects directly to the sensor/CT, removing the tap plug when the device is energized would cause an open-circuit condition on the sensor/CT, could create voltages that will damage the OCPD, and could possibly cause injury or death.

The ampere tap is multiplied by the LTD minimum pickup setting. For example, with a 400 A sensor and OCPD set to the LTD 1.0 tap, the minimum pickup is 400 A (400 A × 1.0 = 400 A). The LTD is set to the maximum time band setting. **See Figure 9-11.**

To perform the timing test, 1200 A is applied to the circuit breaker (400 A × 3 = 1200 A). The #3 line on the time-current characteristic curve is followed up until it intersects with the maximum upper and lower band time limits. The operating time is determined by reading the appropriate time, in seconds. In this case, the OCPD would operate between 55 sec and 95 sec. In reality, whenever three times the LTD minimum pickup current is applied to this OCPD, it will always operate between 55 sec and 95 sec, regardless of the actual current applied during the test.

SOLID-STATE AND DIGITAL OCPD CURVES—LTD FUNCTION

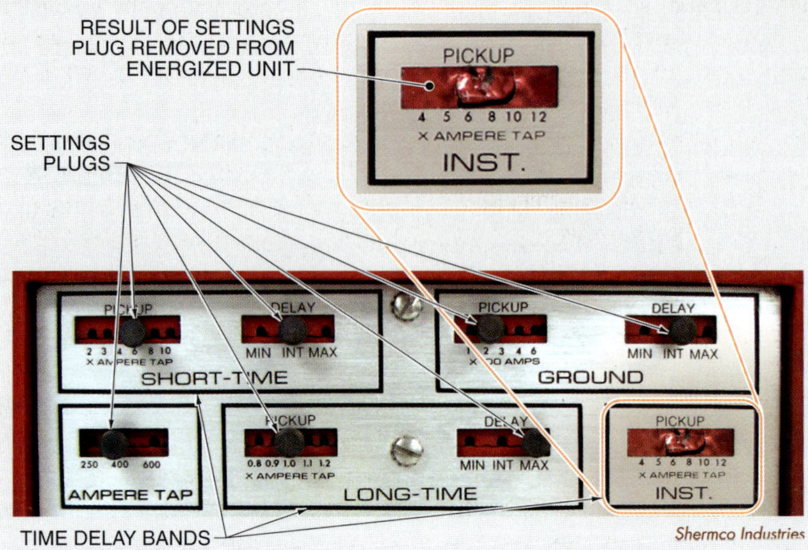

ITE POWER SHIELD TYPE SS SOLID-STATE OCPD

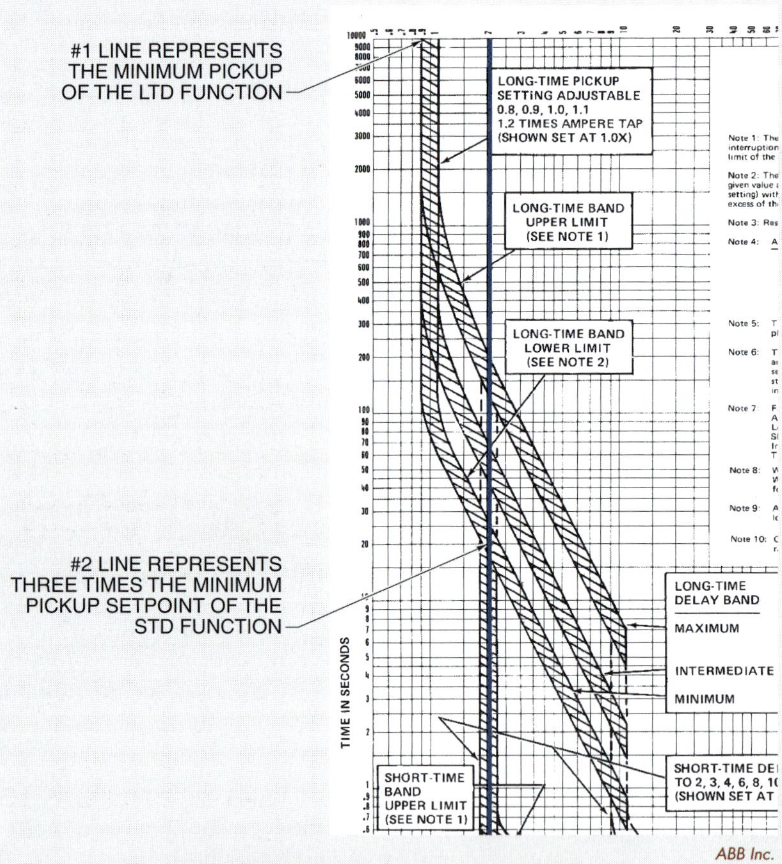

LTD SOLID-STATE TIME-CURRENT CHARACTERISTIC CURVE INTERPRETATION

Figure 9-11. The LTD function has three time delay bands, which are clear and easy to use. The #1 line represents the minimum pickup setpoint of the LTD function, while the #3 line represents three times the LTD minimum pickup.

STD Function—Solid-State and Digital OCPDs

The STD band is located on the SS-5 curve directly below the lower band time limit of the LTD minimum band limit. **See Appendix.** *Note:* The LTD setting has no effect on the STD or INST settings, since each function is an independent electronic feature.

The note just to the right of the STD band states, "Short-time delay pickup setting is adjustable to 2, 3, 4, 6, 8, 10 times ampere tap (shown set at 2x and 10x)." The center of the STD band is found on the #2 line between the upper and lower pickup limits. There is also a dashed curve outlined for the STD band that is drawn with its center on the #10 line. These two curves show that the STD device can be adjusted between 4 and 10 times the OCPD's tap setting. The purpose for showing the two delay curves is to illustrate the adjustable range. *Note:* Only the solid-line STD curve should be used for determining trip times. The dashed curve is only for illustrative purposes to indicate that the characteristic is adjustable.

In this example, the center of the STD band (#2 line) is used in the same manner as the #1 line for the LTD function. The center of the curve represents the minimum pickup setpoint of the STD function. **See Figure 9-12.**

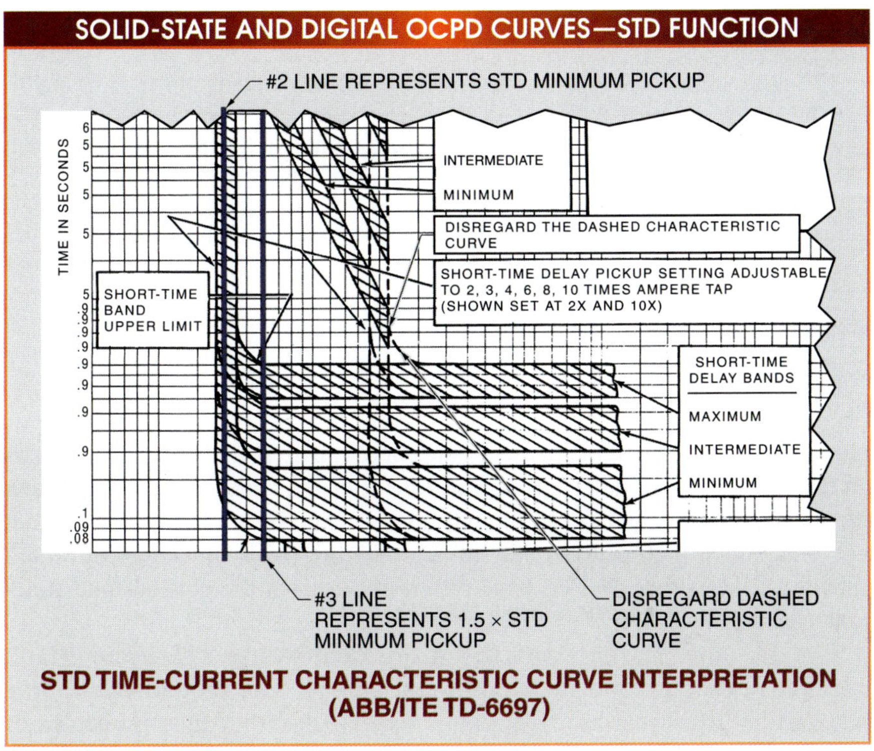

ABB Inc.

Figure 9-12. The STD band is located on the SS-5 curve directly below the lower band time limit of the LTD minimum band time limit.

Each STD band has an upper and lower limit. The upper and lower limits of the time bands are used in the same manner as the LTD bands, since they represent the maximum and minimum operating time. The following procedure is used to determine the operating times for the STD function:

1. Find the center of the STD time-current characteristic curve. For the SS-4 or SS-5 OCPD, it is the #2 line. The center of the STD curve will probably be different for other OEM OCPDs.
2. Multiply the number that the STD curve is drawn on by the multiple of test current that is going to be used. For example, the standard timing test point for STD is 1.5 times the STD minimum pickup setpoint. The STD minimum pickup setpoint is six times the ampere tap setting, which is 400 A. The minimum pickup setpoint for the STD function is 2400 A (400 A × 6 = 2400 A).

   The STD time band in this example is set to the intermediate time band. Test current applied to the circuit breaker will be 3600 A and will be injected into the circuit breaker (2400 A × 1.5 = 3600 A). The same ratio of minimum pickup to test current will be applied to the curve itself. The #2 line (representing the minimum pickup setpoint) will be multiplied by 1.5, with a result of 3. The #3 line (the multiplier line) will be used to determine operating times.
3. Read up the appropriate vertical multiplier line until it intersects the upper and lower STD band limits to determine the operating time. In the example SST-5 curve, follow the #3 line vertically to the lower band limit, then read either left or right on the horizontal line to the trip time of 0.07 seconds. Continue reading up the vertical #3 line to the upper band time limit, and follow the horizontal line left or right. The upper band time limit is 0.175 seconds. The STD function will operate between 0.07 and 0.175 seconds when 1.5 times the minimum pickup setpoint current is applied to the OCPD. Other operating times for the SS-4 and SS-5 STD function are the following:

- minimum—0.07 sec to 0.175 sec
- intermediate—0.2 sec to 0.35 sec
- maximum—0.4 sec to 0.5 sec

Ground-Fault Function—Solid-State and Digital OCPDs

Ground-fault time-current characteristic curves are read in the same manner as STD curves. Ground-fault tripping times are often the same as the STD function for the same time band. Even though this is generally true, the ground-fault curve should still be used to determine operating times for the ground-fault function, since some OCPDs have different times for the ground-fault function. **See Figure 9-13.**

For example, multiples of 100 A are used on the ITE ground-fault curve to set the ground-fault minimum pickup. The ground-fault curve has vertical lines that are in amperes times 100 rather than multiples of the setting. *Note:* On other OEM curves, the multiples may be multiplied by the ampere tap, sensor rating, or some other value. The actual methods used for interpreting these curves must be verified with the manufacturer notes on each curve.

For circuit breakers that have been retrofitted from series OCPDs to digital OCPDs, care must be taken to ensure that the proper curve is used. It is not uncommon to have a circuit breaker from one OEM and a retrofit kit and OCPD from another. The curve from the OEM of the retrofit kit must be used in this case.

SOLID-STATE AND DIGITAL OCPD CURVES—GROUND FAULT FUNCTION

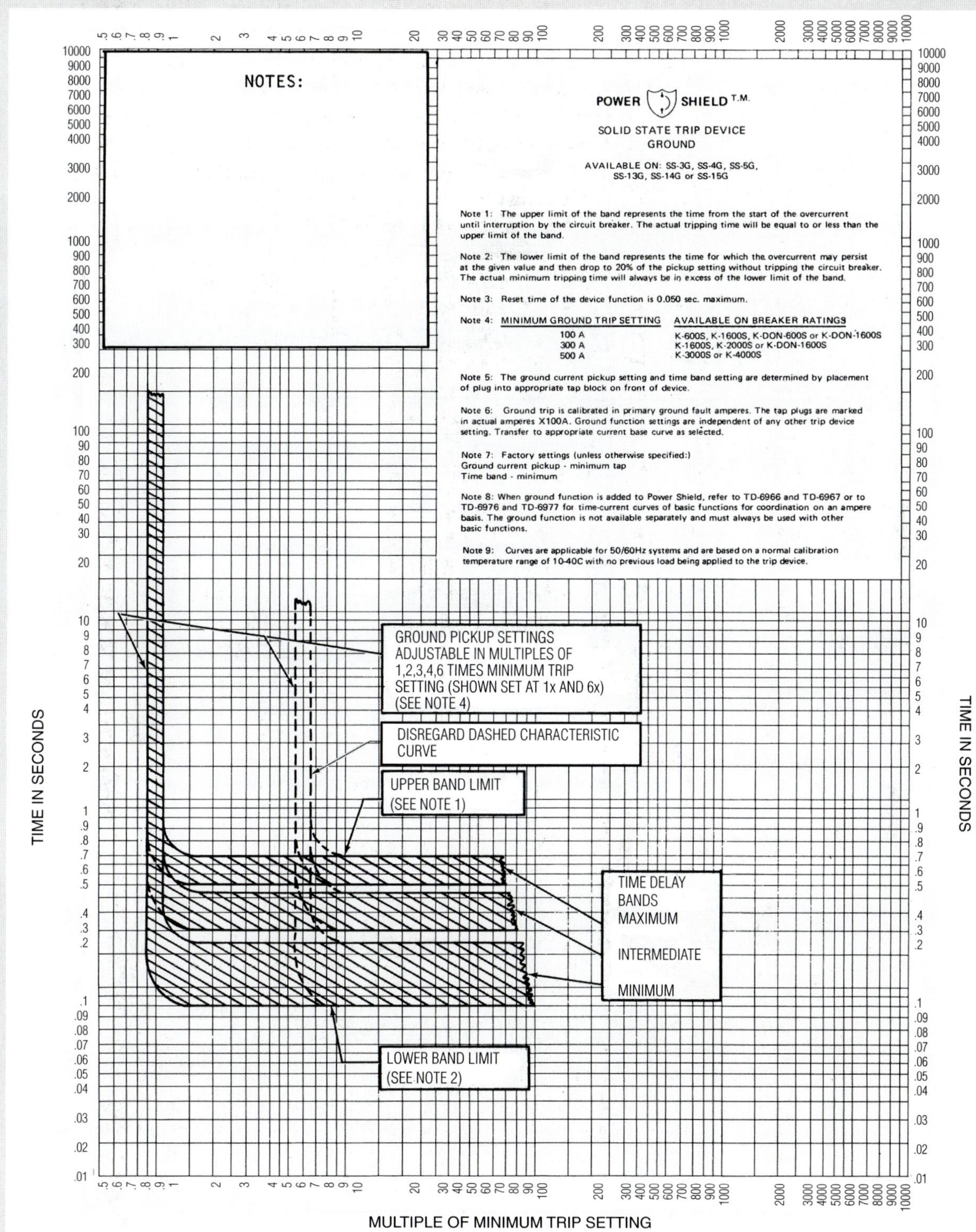

ABB/ITE GROUND-FAULT FUNCTION CHARACTERISTIC CURVE

ABB Inc.

Figure 9-13. Ground-fault curves are read in the same manner as STD curves. Ground-fault tripping times are often the same as the STD function for the same band.

I^2t Function—Solid-State and Digital OCPDs

The I^2t function is used to shift the STD curve to allow for transients, such as motor starting, without having to increase the STD pickup. This provides the most sensitive protection while preventing nuisance trips. The I^2t function can also be found with the ground-fault function, although it is not used as often as it is with the STD function. I^2t is only found on solid-state and digital OCPDs. Each OEM provides different methods of setting its OCPDs. **See Figure 9-14.**

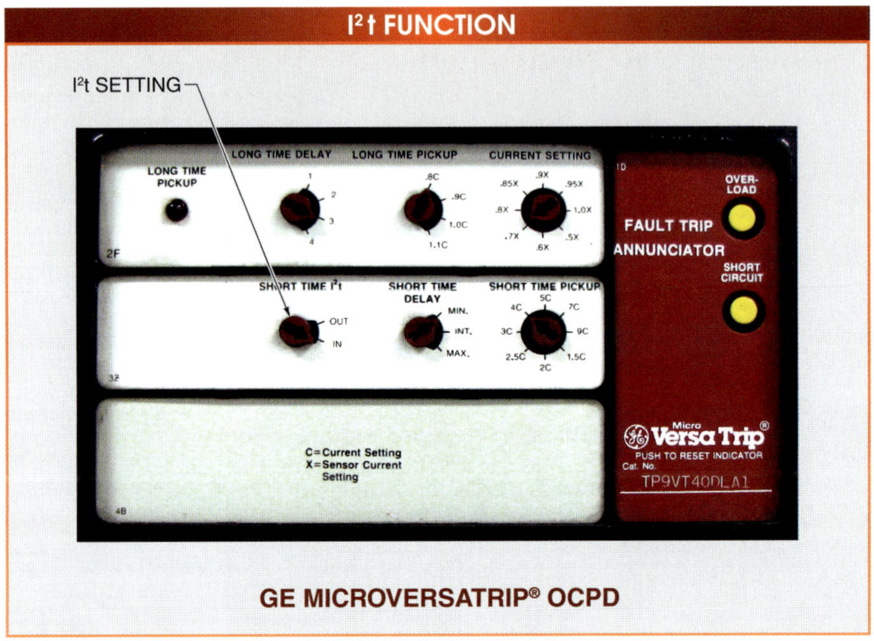

Shermco Industries

Figure 9-14. The I^2t function is only found on solid-state and digital OCPDs.

In this example, the STD function shows the I^2t function as being in use. On the GE MicroVersaTrip® OCPD, *C* represents the minimum pickup current setting, while *X* represents the sensor rating. This can be confusing when first learning to use the GE time-current characteristic curves. STD pickup is set to 4*C*, (four times the current setting), which is 840 A (1200 A sensor × 0.7 = 840 A). Therefore, the STD pickup is 3360 A (840 A × 4 = 3360 A). The time delay band is set to minimum. The GE MicroVersaTrip® STD function is shown with the minimum pickup on the #1.5 line. The standard test point, or test current, for the STD function is 1.5 times the STD minimum pickup, which is 5040 A (1.5 × 3360 A = 5040 A).

On the example OCPD, the LTD minimum pickup setting is 0.8C. Since C is set to 0.7 times 1200 A for the sensor times X, the minimum pickup setting for the LTD function is 960 A (0.7 × 1200 A × 0.8 = 672 A). This gives the coordination engineer great flexibility in setting the minimum pickup to the exact value needed.

The STD minimum pickup is on the #1.5 line. The I^2t function of this circuit breaker is in, and the operating time of the STD function is determined differently than when the I^2t function is out. The STD test current is divided by the sensor rating, plug rating, or other specified value. The result determines the line used for the multiple line on the time-current characteristic curve.

$$line = \frac{test\ current}{sensor\ rating}$$

$$line = \frac{5040\ \text{A}}{1200}$$

$$line = \mathbf{4.2}$$

The #4.2 line is followed vertically until it intersects the lower and upper band time limits. The operating time would then be between 0.75 sec and 1.25 sec for all bands. **See Figure 9-15.**

TIME-CURRENT CHARACTERISTIC CURVE WITH I^2t FUNCTION

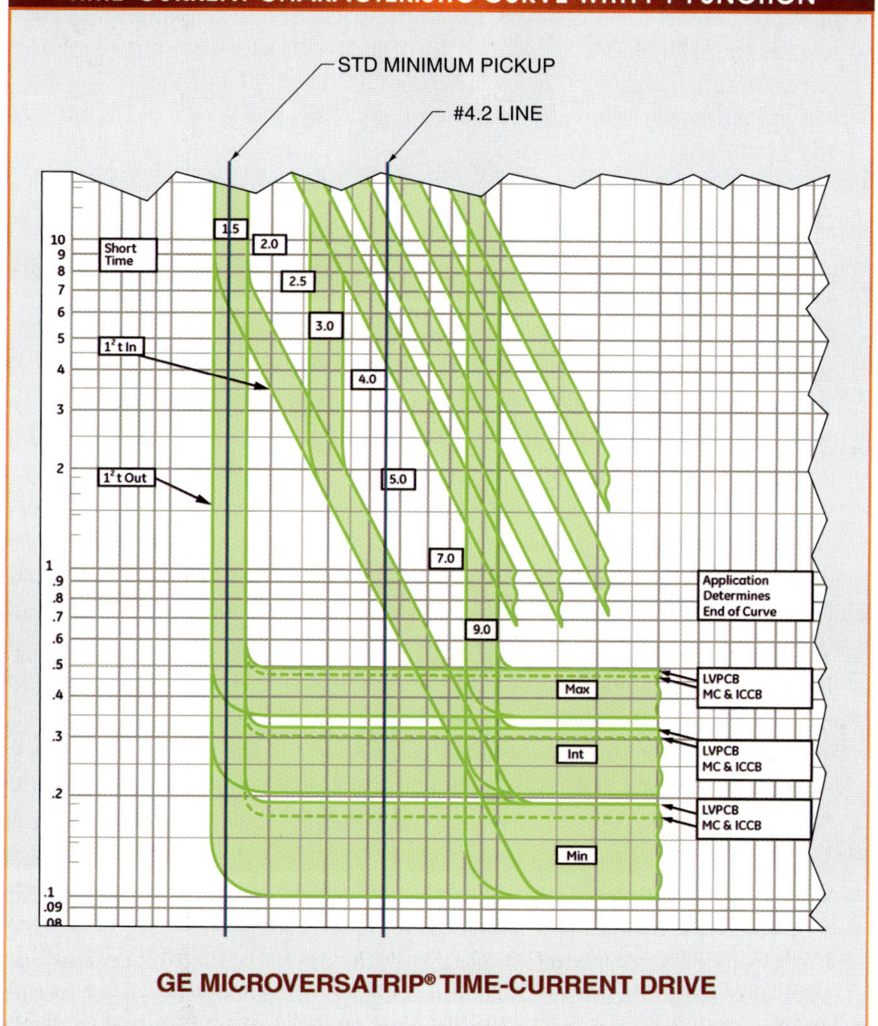

Figure 9-15. The #4.2 line is followed vertically until it intersects the lower and upper time band limits. The operating time is figured to be between 0.75 sec and 1.25 sec for all time bands.

OCPD TESTING

All OCPDs, as any other mechanical or electronic device, must be tested periodically. Should an OCPD fail to operate, the equipment the circuit breaker is protecting, the switchgear, or even the circuit breaker itself may be damaged. Personal injury or a fatality could also result from a defective OCPD. There are two methods used to test OCPDs used on circuit breakers: primary injection testing and secondary injection testing.

- Primary injection testing: Primary injection testing is the preferred method of testing OCPDs used on circuit breakers. Current at primary (actual) current values is injected directly through the primary stabs of the circuit breaker. The primary injection test will test all components in the circuit, including the power supply and current transformer circuits. Primary injection test sets can be used on all types of OCPDs.
- Secondary injection testing: A *secondary injection test* is a test used only on solid-state and digital OCPDs in which test current is injected directly into the logic and timing circuits. Secondary test currents and circuits differ between OEMs. It is necessary to purchase a different secondary injection test set for each OEM, and the OEM often specifies a different secondary injection test set for each of its different OCPDs. With secondary injection testing, the power supply, interconnecting wiring, and sensors/CTs are usually not tested.

Primary Injection Testing Method

When testing the OCPDs on a low-voltage power circuit breaker (LVPCB), several points must be considered:

- High-current test sets output high current and low voltage. The impedance and length of the connections have a limiting effect on the output ability of the test set. Larger circuit breaker test sets use adjustable stabs, which allow connection directly to the circuit breaker primary disconnects (stabs). This provides the least impedance and the highest output current.
- Rated voltage and current must be available for the test set to operate properly. If sufficient voltage is not available, the current requirements of the test set increase rapidly. As a result, the test set circuit breaker could trip when testing higher-rated circuit breakers.
- If the circuit breaker being tested has current-limiting fuses, they must be removed. Links must be used to bypass them. When testing the STD or INST functions, it is possible to cause these fuses to operate.
- When using stabs to connect the circuit breaker to the test set, the correct stabs must be used. The stabs attached to the test set should be clean and tight. The primary fingers on the circuit breaker should also be clean.
- The next best method for connecting the circuit breaker to the test set is to use busbar bent to fit the circuit breaker connection points. The current output will be decreased, but there will be minimal loss.
- If it is not possible to connect directly to the circuit breaker disconnects or use busbars, welding cable can be used. This is often necessary when testing molded-case circuit breakers. The output of the test set will be greatly reduced due to the impedance of the cables. If using multiple cables, twisting the cables will help reduce the overall impedance and increase the output of the set. On smaller, molded-case circuit breakers with high internal impedance,

it may be necessary to use a "Z-link." The Z-link connects the output transformer of the primary injection test set into a series configuration doubling the output voltage. The higher voltage will overcome the impedance and allow a test to be performed. The current output is cut in half, so larger circuit breakers cannot be tested using the Z-link. **See Figure 9-16.**

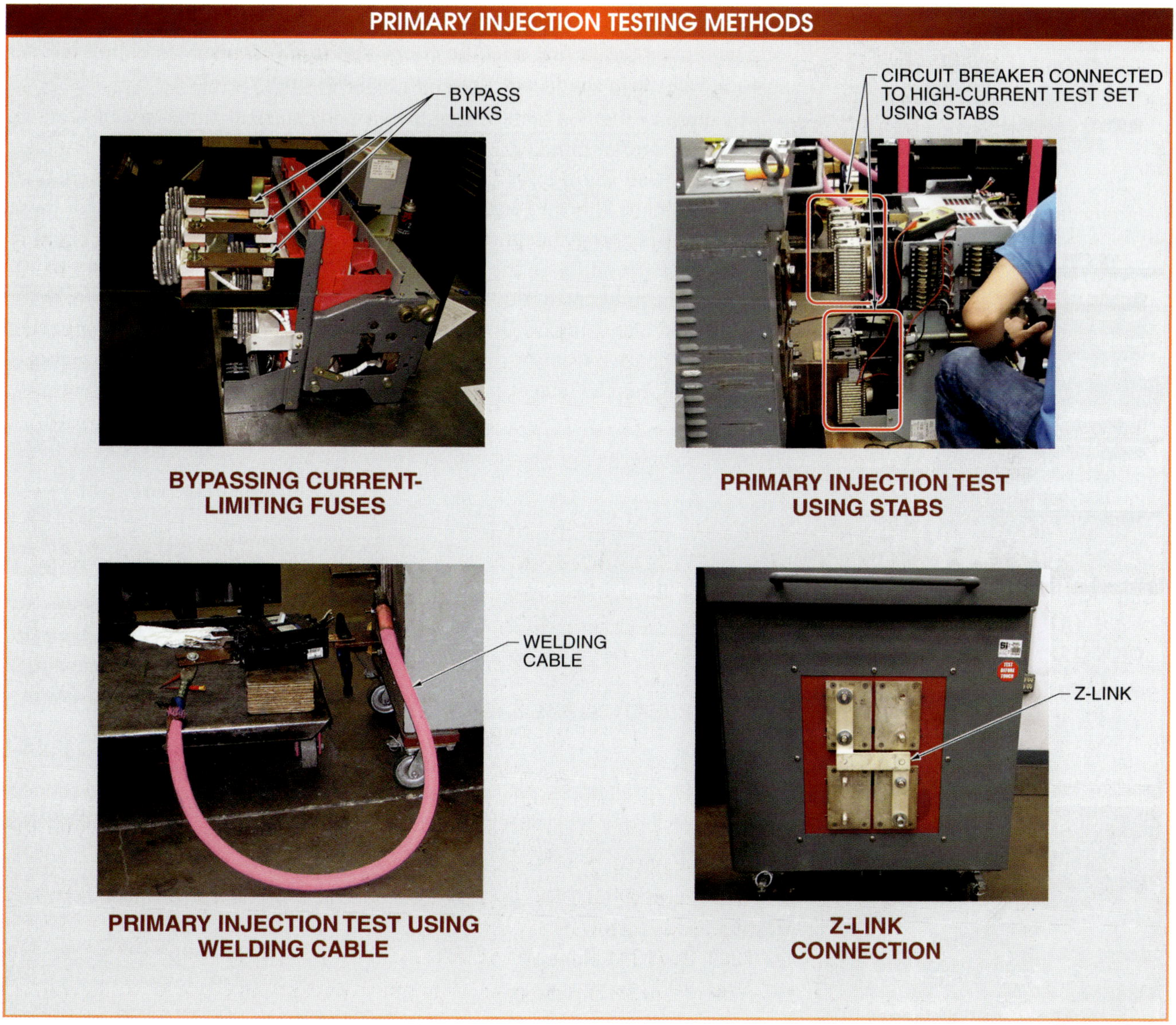

Shermco Industries

Figure 9-16. Primary injection testing can be performed using bypass links, high-current test sets using stabs, or welding cables. Test sets are connected in series with Z-links.

- OCPDs must be allowed to fully reset before tests are performed and in between tests. After setting the current on the test set to the appropriate value, allow a reasonable amount of time before conducting the test and between tests. Solid-state and digital OCPDs reset almost immediately, so there is no need to wait. Molded-case circuit breakers using thermal-magnetic OCPDs need at least 30 sec, while series trip OCPDs used on LVPCBs only need 4 or 5 to 10 sec.

TECH TIP

When OCPDs are being tested using the primary injection test method, components in the current path, including the test stabs, can get hot enough to cause burns to fingers.

- Circuit breakers are tested one phase at a time.
- The ground-fault function must be bypassed or short circuited to perform the long-time, short-time, and instantaneous tests. Since the ground-fault function is so sensitive and fast, it will operate before the other functions. Table 1, *Circuit Breaker Ground-Fault Defeat,* must be used for defeating the ground-fault function on some types of LVPCB OCPDs. The instruction manual can be consulted on how to bypass the ground-fault function. Sometimes the device must be connected to a secondary injection test set to defeat the ground-fault function. **See Appendix.**

An alternate method of defeating the ground-fault function is to connect two poles of the circuit breaker in series, such as A-phase to B-phase or B-phase to C-phase. **See Figure 9-17.** Connecting the circuit breaker in this manner causes current to flow into one phase and out the other. The vector sum of these two currents is zero, since they are 180° apart in phase angle and are equal in magnitude. Depending on the test current involved, it may be necessary to use busbars as the jumpers to provide a low-impedance path. This method works for all circuit breakers, regardless of manufacturer. A jumper could be connected to all three phases, if desired. The primary drawback of this method is that it is more complicated to set up.

PRIMARY INJECTION TESTING—ALTERNATE METHOD

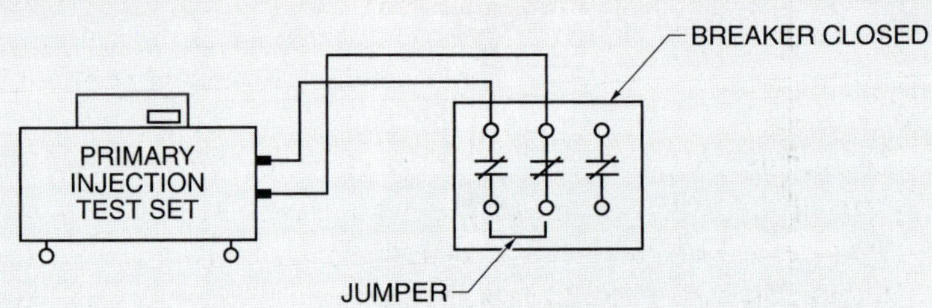

DEFEATING THE GROUND-FAULT FUNCTION BY USING A SERIES CONNECTION

Figure 9-17. An alternate method of defeating the ground-fault function is to connect two poles of the circuit breaker in series, such as A-phase to B-phase or B-phase to C-phase.

When testing OCPDs, the recommended test points for primary injection testing are as follows:

- LTD: 3 × LTD pickup
- STD: 1.5 × STD pickup
- ground fault: 1.5 × ground fault pickup
- INST: pickup only at set point

Secondary Injection Testing Method

The secondary injection testing method can be used with solid-state and digital OCPDs. Series OCPDs cannot be tested using the secondary injection testing method. Advantages of the secondary injection testing method are the following:

- The test sets are small and portable.
- Secondary injection test sets cost a fraction of primary injection test sets.
- Most test sets are straightforward and have detailed instructions.

- There are lower human resource requirements. It is not necessary to remove the circuit breaker from its cubicle, so one technician can conduct the test. The circuit breaker should be disconnected from its load and racked into the disconnected position. The trip coil will be activated once the unit times out.

The main disadvantages of secondary injection testing are the following:

- Test sets are often specific for each type of trip unit, so multiple test sets may be required. Often, a different secondary injection tester is required for each generation of a single OEM's OCPD. If an end user facility has circuit breakers of several different OEMs in service, the price advantage is quickly nullified.
- The circuit breaker is subject to incomplete testing. This is because the power supply, CTs, and interconnecting wiring are not tested.
- Tests are usually performed at specific settings, not on the as-found settings. Even though the OCPD should be linear, there is a small possibility that a defect may not be found. With primary injection testing, the OCPD is tested on its as-found settings.

A small current is injected directly into the logic and timing circuits of the OCPD. The minimum pickup and timing are verified against OEM specifications. The test set supplies power to the logic circuits, so the power supply is not checked. Since the test set supplies the test current directly into the OCPD, the CTs and the interconnecting wiring are bypassed. Secondary injection test sets have low output current, typically 5 A or less. Each type of secondary injection test set has specific operating procedures, so general guidelines cannot be provided.

Secondary injection testing is useful for checking the calibration of the OCPD and ensuring that the timing and logic functions are operating properly. Therefore, secondary injection testing should be used for routine maintenance. Because it does not test all the components, a primary injection test should be performed every three to five years. **See Figure 9-18.**

SECONDARY INJECTION TESTING

USING SECONDARY INJECTION TEST SETS

Shermco Industries

Figure 9-18. Secondary injection testing is useful for checking the calibration of an OCPD and ensuring that the timing and logic functions are operating properly. Secondary injection testing should be used for routine maintenance.

SUMMARY

The maintenance and testing of circuit breakers are critical for power system reliability and worker safety. Misoperation or nonoperation of power system OCPDs can produce devastating results on the equipment and anyone nearby. All OEMs recommend regular maintenance and testing of their circuit breakers, typically on a three-to-five year schedule. Heat, moisture, and mechanical, electrical, and chemical stresses deteriorate insulation, lubricants, and current-carrying components. It is only by testing and maintaining these critical devices that their condition can be verified.

DIGITAL RESOURCES
ATPeResources.com/QuickLinks
Access Code: 223436

chapter 9

INTERPRETING TIME-CURRENT CHARACTERISTIC CURVES

CHAPTER REVIEW

Name ______________________________ Date ______________

True-False

T F **1.** The LTD setting has no effect on the STD or INST settings.

T F **2.** The information box on a time-current characteristic curve indicates the model number as well as the available continuous current, voltage, and frequency ratings that the circuit breaker can be used for.

T F **3.** The vertical lines of the graph on a time-current characteristic curve represent time.

T F **4.** Molded-case circuit breakers using thermal-magnetic OCPDs need at least 30 sec to fully reset before tests are performed.

T F **5.** Secondary injection test sets can be used on all types of OCPDs.

T F **6.** If the circuit breaker has current-limiting fuses, they must be removed for primary injection testing.

T F **7.** The band at the top left side of the time-current characteristic curve is the LTD band.

T F **8.** The time bands of a series OCPD are smaller and tighter than those of a solid-state OCPD.

T F **9.** The continuous current rating of an interchangeable-trip molded-case circuit breaker is equal to or less than the frame rating.

T F **10.** If a circuit breaker has been retrofitted from a series OCPD to a digital OCPD, the curve from the original series OCPD must be used.

T F **11.** The lower band time limit is read in the same manner as the upper band time limit.

T F **12.** On molded-case circuit breakers, the #1 line does not actually cross the lower band time limit.

T F **13.** Each STD band has an upper and lower limit.

T F **14.** The actual clearing time for a circuit breaker should never exceed the maximum clearing time limit.

T F **15.** For an ambient-compensated circuit breaker, the higher the ambient temperature, the slower the circuit breaker could trip.

Multiple Choice

______ 1. ___ OCPD curves are interpreted in a similar manner as series OCPD curves.
- A. Parallel
- B. Series-parallel
- C. Vacuum
- D. Solid-state

______ 2. Only the ___ time-current characteristic curve can be used to determine operating time.
- A. ghost
- B. lower limit
- C. solid
- D. upper limit

______ 3. Typically, the #___ line is the long-time delay minimum pickup setpoint.
- A. 1
- B. 2
- C. 3
- D. 4

______ 4. The LTD function has ___ time delay bands.
- A. two
- B. three
- C. four
- D. five

______ 5. The instantaneous (INST) function time-current characteristic curve is located at the ___ of the curve for a thermal-magnetic OCPD.
- A. top right
- B. top left
- C. center
- D. bottom

______ 6. The amount of time delay will always be ___ than the upper limit of the band.
- A. less
- B. greater
- C. equal to or less
- D. equal to or greater

______ 7. Generally, molded-case circuit breakers are considered ___-cycle devices, which means that they should clear a fault in no more than this number of cycles.
- A. one
- B. two
- C. three
- D. four

______ 8. ___ is only found on solid-state and digital OCPDs.
- A. STD
- B. LTD
- C. I^2t
- D. INST

_______________ **9.** Ground-fault tripping times are often the same as the ___ function for the same time band.
A. INST
B. STD
C. LTD
D. I^2t

_______________ **10.** The actual minimum pickup current of a series OCPD does not matter as long as ___ times the minimum pickup current is applied during testing.
A. three
B. five
C. seven
D. nine

Completion

_______________ **1.** The ___ line is the line on an INST curve that represents the maximum amount of time for circuit breaker contacts to open and extinguish an arc.

_______________ **2.** A(n) ___ is a variable-resistance electric device that divides voltage proportionally between two circuits.

_______________ **3.** The ___ function is used to shift the STD curve to allow for transients, such as motor starting, without having to increase the STD pickup.

_______________ **4.** ___ injection testing is the preferred method of testing OCPDs used on circuit breakers.

_______________ **5.** A(n) ___ is the time that represents the maximum operating time on a time-current characteristic curve.

_______________ **6.** The ___ function must be bypassed or short circuited to perform tests on the other functions because it is so sensitive and fast that it will operate before the other functions.

_______________ **7.** During a secondary injection test, a small current is injected directly into the logic and ___ circuits of the OCPD.

_______________ **8.** A(n) ___ is the time that represents the minimum operating time on a time-current characteristic curve.

_______________ **9.** A(n) ___ test is a test used only on solid-state and digital OCPDs in which test current is injected directly into the logic and timing circuits.

_______________ **10.** During the primary injection testing of smaller, molded-case circuit breakers with high internal impedance, it may be necessary to use a "___," which connects the output transformer of the primary injection test set into a series configuration doubling the output voltage.

Short Answer

1. List the procedure for determining the operating times for the STD function.

2. What are three methods used to connect a circuit breaker to a primary injection test set?

3. Why must time be allowed for an OCPD to reset during testing?

4. What functions are included on a series OCPD time-current characteristic curve?

5. List two advantages and two disadvantages of secondary injection tests.

chapter **10**

MAINTENANCE AND SAFETY ISSUES

The National Fire Protection Association (NFPA) 70E® 2012(1) contains a number of requirements for maintaining overcurrent protective devices. Chapter 1, Section 130.5 Informational Note (IN) No. 1 states, "Improper or inadequate maintenance can result in increased opening time of the overcurrent protective device, thus increasing the incident energy." Section 130.5 IN No. 4 refers the reader to Chapter 2 for more safety-related maintenance requirements in NFPA 70E®. Chapter 2, Section 205.3 NFPA 70E® states, "Electrical equipment shall be maintained in accordance with manufacturers' instructions or industry consensus standards to reduce the risk of failure and the subsequent exposure of employees to electrical hazards." Section 205.4 states, "Overcurrent protective devices shall be maintained in accordance with the manufacturers' instructions or industry consensus standards. Maintenance, tests, and inspections shall be documented."

To understand the reason for preventive maintenance and the causes for problems, an understanding of the construction and operation of the typical stored-energy type circuit breaker is needed. These issues are common to all low-voltage power, medium-voltage power, and medium-voltage vacuum circuit breakers.

OBJECTIVES

- Identify the vulnerable points of a circuit breaker.
- List the most common causes of circuit breaker failure.
- Describe the problems that are associated with a lack of lubrication.
- List the different types of contact adjustments.
- Describe the problems that can occur from the abuse and neglect of equipment, electrical problems, plastic subcomponent failure, and extreme operating environments.

KEY TERMS

- organic lubricant

VULNERABLE CIRCUIT BREAKER POINTS

The left side of Figure 10-1 shows a simple prop-and-roller operating mechanism using accordion-style linkage that is similar to that used on early circuit breakers. However, it is not used on modern circuit breakers due to its inherent weaknesses. The circuit breaker contacts are held in the closed position by a mechanical prop-and-roller mechanism. The centerline of the prop is in line with the centerline of the main roller. The opening springs (not shown) are pulling the contact assemblies rearward, exerting pressure on the prop and roller. As long as the prop and roller remain aligned, the circuit breaker will stay in the contact-closed position. To trip the circuit breaker, the trip bar pushes the prop off the roller, which collapses the linkage and opens the contacts. The right side of Figure 10-1 shows the collapsed linkage. **See Figure 10-1.**

This same principle is used on modern circuit breakers, although it is greatly modified. **See Figure 10-2.** Figure 10-2 illustrates one style of a modern prop-and-roller mechanism. Without the extending linkages, it is much more rugged and has fewer issues from misalignment and breakage (the top left portion of Figure 10-2 shows the operating mechanism in the closed position). The insulated coupling (12) holds the contacts closed due to parts 2, 5, 6, 11, and 14 being placed in an interference fit. The trip latch (11) holds the secondary latch (14) from rotating clockwise. The secondary latch is positioned against the secondary latch roller (6; yellow), which in turn extends the cam (main) roller (5; red) against the prop (2). The opening spring (15) is not shown in this view but exerts pressure on the contacts to open.

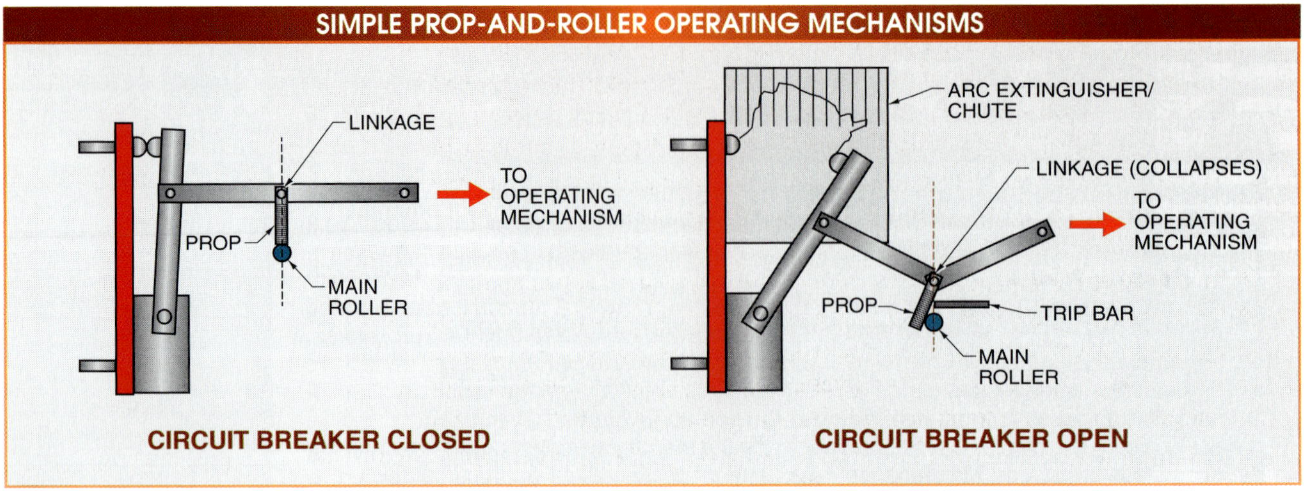

Figure 10-1. A simple prop-and-roller operating mechanism using accordion-style linkage is similar to that used on early circuit breakers. It is not used on modern circuit breakers due to its inherent weaknesses.

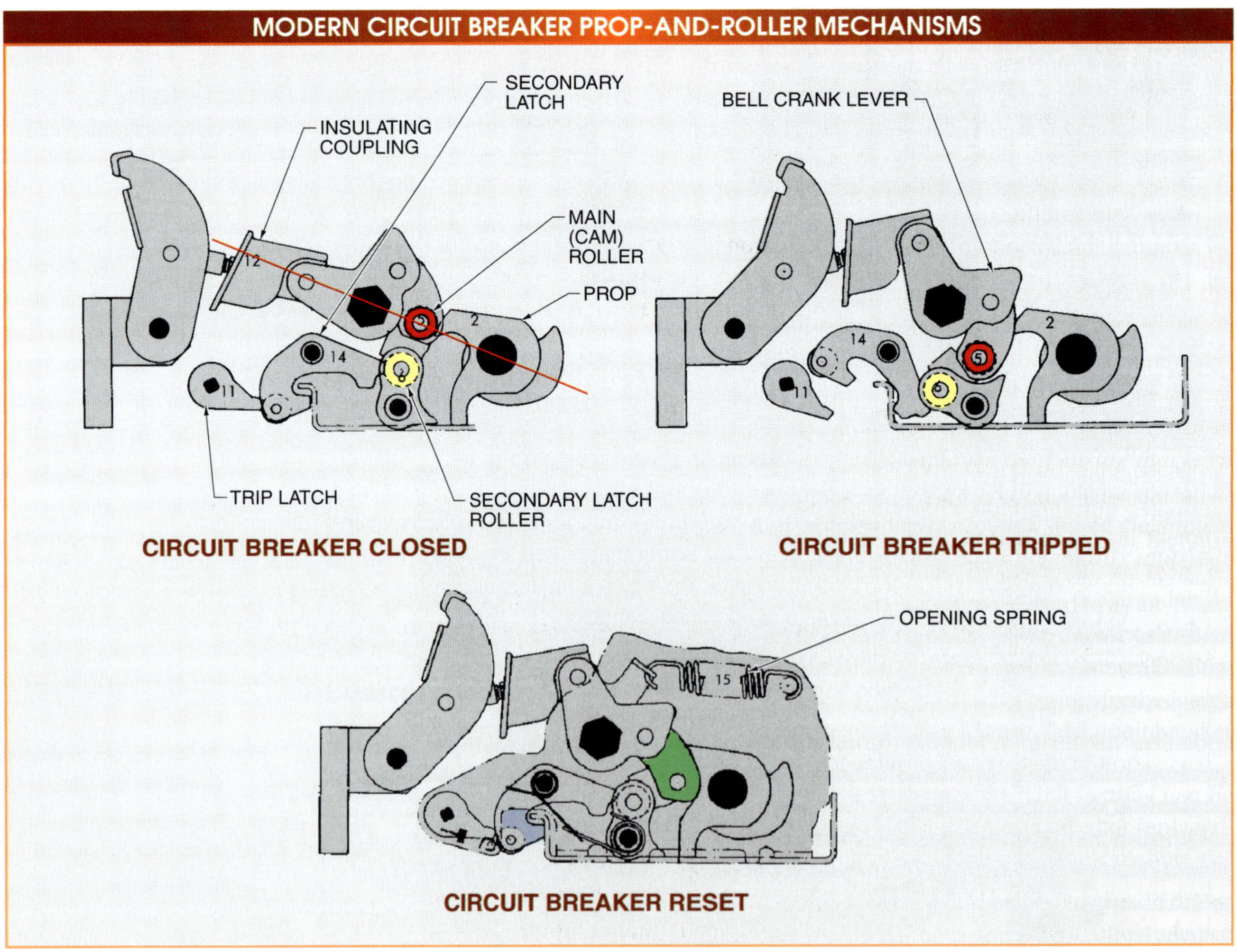

Figure 10-2. Modern prop-and-roller mechanisms, without the extending linkages, are much more rugged and have fewer issues from misalignment and breakage than older mechanisms.

With the circuit breaker closed, the secondary latch (14) is held against the secondary latch roller (6), which pushes the main roller (5) and its linkage into a vertical position, as seen in the middle section of Figure 10-2. The main roller is then held against the prop (2), which prevents it from overextending. The centerline of the insulating coupling pin is in a straight line with the main roller through the camshaft. In this position, the contacts cannot open until the linkage collapses, which cannot occur until the trip latch (11) releases the secondary latch (14).

The top right portion of Figure 10-2 shows the operating mechanism in the tripped position. To open the circuit breaker, the trip latch is rotated clockwise, allowing the secondary latch to rotate clockwise (circuit breaker is in the tripped position). When the secondary latch rotates clockwise, the main roller (5) and the secondary latch roller (6) collapse. This allows the opening springs to pull the contacts open. Part 7 is a bell-crank lever, which is used to change motion in one direction to another direction, usually at a 90° angle.

The top right portion of Figure 10-2 shows the operating mechanism in the reset position. This position is the condition of the mechanism just before closing. With the circuit breaker in the reset position, the trip latch (11) and the secondary latch (14) are reset to the same position as when the circuit breaker is in the closed position. To place the mechanism in this position, the cam (3) has to be rotated slightly counterclockwise until the prop (2) is lifted up. This allows the main roller (5) and linkage (green) to slip into the crook of the prop.

By doing this, the linkage is slightly extended, and the secondary latch (14) engages the front of the frame. As the linkage collapses, the bell-crank rotates and allows the contacts to open, shown (in blue), which allows a gap between the trip latch and the secondary latch. The prop and cam reset to their original position and, if the circuit breaker close button is pressed, the closing springs will accelerate the contacts closed. The cam and prop rotate, extending the linkage and forcing the components into the same positions, as seen in the top portion of Figure 10-2.

It is easy to see that if these components begin to gum up and slow down, the circuit breaker will not operate properly. Even an idle circuit breaker can become gummy and sluggish. All lubricants dry out over time. An *organic lubricant* is a lubricant composed of emulsifiers and oil. Over time, the oil dries out and leaves the emulsifiers, which thicken. When a circuit breaker is exercised, it helps to keep the lubricants as pliable as possible. But they still dry out over time. As the circuit breaker carries load current, heat is produced by the I^2R losses. The higher the load current, the greater the I^2R losses and heat produced. This heat increases the rate at which the lubricants dry out, even if the circuit breakers are exercised regularly.

MOST COMMON CAUSES OF CIRCUIT BREAKER FAILURE

Circuit breakers fail for many reasons. However, the most common reasons for circuit breaker failure include a lack of lubrication, improper circuit breaker adjustment, abuse and neglect of equipment, electrical problems, plastic subcomponent failure, and the type of operating environment where it is used. These problems can cause the circuit breaker to malfunction, overheat, and ultimately fail.

Lack of Lubrication

A lack of lubrication is the most common problem that can cause circuit breakers to fail. Typically, one out of every four circuit breakers has lubrication issues when maintenance is performed. This lack of lubrication results in circuit breakers that will not open, will not close, or open after an inadmissable extended delay.

All rollers and latch surfaces in a circuit breaker require lubrication. The old lubricant must be cleaned off and new lubricant applied. Each OEM recommends a specific lubricant for its circuit breakers: one for the current path and one for the operating mechanism. Nearly all OEMs now recommend, or at least allow the use of, a product equal to or similar to Mobilgrease 28 HT (High Temperature).

Note: Mobilgrease 28 is a red synthetic-based lubricant sold in tubes or cans that can be used in both the circuit breaker's current path and the operating mechanism.

It is not just the operating mechanism that requires lubrication. The contact pivot points of all circuit breakers are lubricated at the factory. During the course of time, this lubricant dries out, becomes gummy, dries further, and then flakes off, resulting in metal-to-metal wear. When metal-to-metal wear takes place, problems occur. Wear of the parts is greatly accelerated, parts become misaligned, and the operation of the circuit breaker slows down.

Current Paths. The points that wear takes place should be noted. In the example circuit breaker contact assembly, the primary points of wear are at the contact roller/tensioner contact (1) and the arcing contact/tensioner contact (2). A lack of lubricant at point 1 causes wear across both contact faces. A lack of lubricant at point 2 causes the circuit breaker to slow down and grooving to be cut into both the roller and tensioner contact. Eventually, the circuit breaker will seize if the components are not replaced and new lubricant is not applied. **See Figure 10-3.**

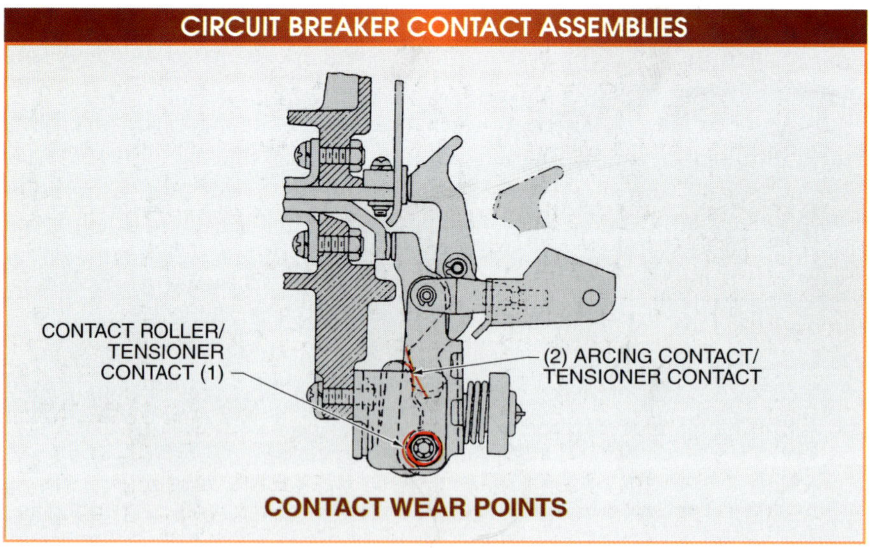

Figure 10-3. In the example circuit breaker contact assembly, the primary points of wear are at the contact roller/tensioner contact (1) and the arcing contact/tensioner contact (2). Other circuit breaker's contact assemblies may have different wear points.

Some technicians spray lubricant into the contact pivot point or the operating mechanism. To the technician using it, it appears that the spray lubricant has solved the problem. The circuit breaker functions as it should, opening and closing on repeated cycling. However, any remaining lubricant has now been flushed out of the pivot points and replaced with a petroleum-based lubricant. More than likely, when the circuit breaker heats up due to current flow through it, the lubricant will quickly dissipate and significant wear will begin. Other technicians use white lithium grease or some unapproved lubricant. None of these lubricants are acceptable, even if they seem to work at the time of application.

One of the reasons OEMs have been particular about the types and brands of lubricants used on their circuit breakers is that the circuit breakers must perform under a variety of conditions, often in adverse conditions. Using lubricants not designed for or proven to function well in adverse conditions is shortsighted and can possibly ruin an expensive circuit breaker or worse, cause it to seize during a fault condition. **See Figure 10-4.**

CONTACT ASSEMBLY WEAR

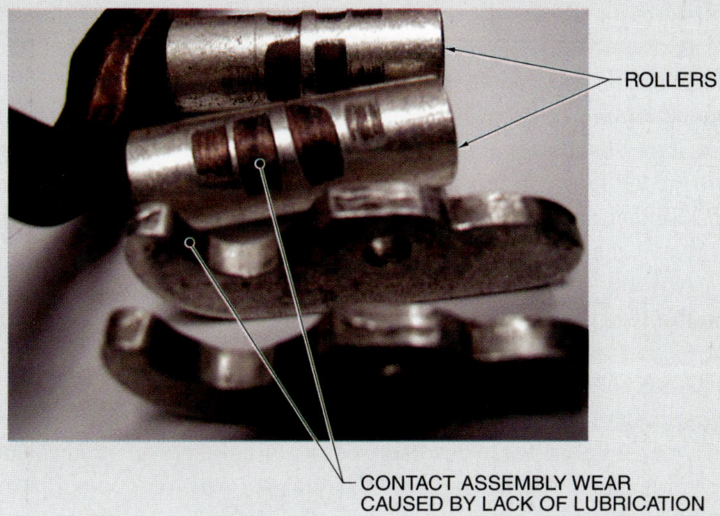

Shermco Industries

Figure 10-4. Using lubricants not designed for or proven to function well in adverse operating conditions can possibly ruin an expensive circuit breaker or worse, cause it to seize during a fault condition.

Note: It is not just one particular manufacturer or model of circuit breaker that a lack of lubrication will affect. Lack of lubrication is a major problem for all circuit breakers.

Besides circuit breaker damage, there are also issues involving equipment the circuit breaker protects as well as possible injury to personnel working on the circuit breaker when it fails to clear a fault. A delayed trip, or no trip, could push the incident energy well above the ratings of the PPE the worker may be wearing.

Contact Resistance Testing. The corrective action for increased contact resistance sometimes is not straightforward. There is no way to see inside the contact pivot points, and it is too expensive to take the circuit breaker apart just to see what is going on inside. To determine the condition of the contact pivot point area, a micro-ohmmeter test, often referred to as a contact resistance test, can be performed. **See Figure 10-5.**

CONTACT RESISTANCE TESTING

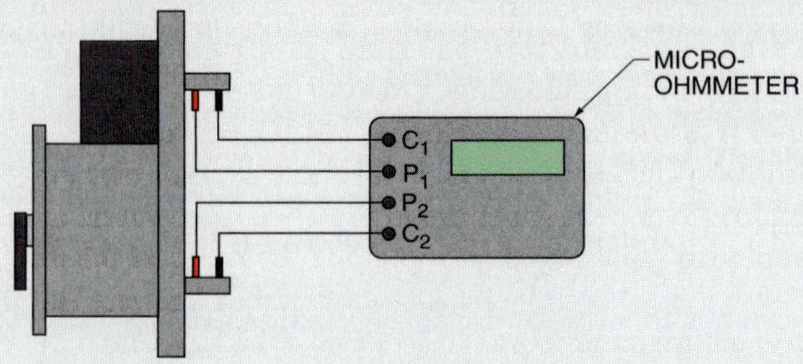

CONTACT RESISTANCE TEST USING A MICRO-OHMMETER

Figure 10-5. To determine the condition of the contact pivot point area, a micro-ohmmeter test, often referred to as a contact resistance test, can be performed.

By measuring the resistance of the entire current path, any increase in resistance can be determined. In low- and medium-voltage power circuit breakers, a part of the increased resistance is due to lubricant drying out. By measuring the resistance each year, a trend can be established. When the resistance becomes too great, the circuit breaker is due for a tear down and relubrication. In general, the contact resistance should be as low and consistent as possible between the contacts of all three phases. However, this does not help to diagnose problems.

A 4000 A, 480 V circuit breaker should have a reading of less than 30 μΩ if it is in good condition. Smaller circuit breakers may have readings of 60 μΩ to 80 μΩ. Molded-case, 480 V circuit breakers can have contact resistance readings in the hundreds of micro-ohms.

Note: Occasionally, an OEM will recommend values but not often.

One guideline for low-voltage power circuit breakers that can be used is that readings up to 100 μΩ indicate a circuit breaker in good operating condition, while readings that exceed 300 μΩ indicate problems. For medium-voltage circuit breakers, the values need to be somewhat lower. Readings above 200 μΩ indicate the need to investigate for problems, and a circuit breaker in good operating condition should have readings less than 60 μΩ. Any value in between these values may indicate that the lubricant is beginning to dry out but has not yet reached the problem plateau. The important part of this process is to perform the test annually and trend the changes in readings.

Operating Mechanisms. Operating the circuit breaker on a regular basis is good for the mechanism. It loosens up the grease that is still pliable and spreads it throughout the moving components. Listening to the circuit breaker and watching it when it operates may reveal a malfunction. Circuit breakers should have a clear, solid sound. If operating five circuit breakers, it can be obvious if one of them is operating slower than the others.

The lubricant on the latch faces and around rollers should be visually inspected. If the lubricant is dirty, dusty, or thick, it must be replaced. If a circuit breaker is operated and it will not reset until it has been cycled a few times, there is a good chance that the mechanism is gummed up. In general, the operating mechanism will need to be cleaned at about the same time as the current path.

If the main roller in the prop-and-roller mechanism gums up, the circuit breaker will not reset properly. Any accessible latch and roller surface should be relubricated. A roller that is sticky or hard to rotate may need to be removed and lubricated. This can be time consuming, especially if the roller is hard to reach. **See Figure 10-6.** Also, some rollers may be under tension from latches or springs. These are spring-charged mechanisms that move very fast and hard.

PROP-AND-ROLLER MECHANISM LOCATION

PROP-AND-ROLLER MECHANISM

Shermco Industries

Figure 10-6. A roller that is sticky or hard to rotate may need to be removed and lubricated.

Before starting any maintenance procedure on a circuit breaker, the circuit breaker must always be cycled to ensure that the springs are discharged and the contacts are open. **See Figure 10-7.** The bearing shown is from a GE AK-75 operating mechanism and is often found sticking, which causes the circuit breaker to slow down, reset improperly, or not open. The trip paddle also has bearings that, when sticky, cause the circuit breaker to seem as though it is in a trip-free condition.

GE AK-75 OPERATING MECHANISM

Shermco Industries

Figure 10-7. Before starting any maintenance procedure on a circuit breaker, the circuit breaker must always be cycled to ensure that the springs are discharged and the contacts are open. The bearing of a GE AK-75 operating mechanism is often found sticking, which causes the circuit breaker to slow down, reset improperly, or remain closed.

TECH FACT

Most issues with circuit breakers involve their contact structures. Although the operating mechanism can cause problems, contact wear and dried lubrication account for the majority of circuit breaker failures.

Improper Adjustment

The second most common problem that can cause circuit breaker failure is improper adjustment. The adjustments on the operating mechanism are usually close to specifications. If they were not close to being correctly adjusted, the circuit breaker would not operate. However, misadjustment of the contact assemblies causes the majority of the problems. The three primary and common adjustments on low- and medium-voltage air circuit breakers are as follows:

- contact pressure or contact wipe
- contact sequence or contact make
- contact gap

Contact Pressure Adjustment. Too much contact pressure leads to a number of problems, including cracked contact faces, excessive wear, and the bending of components. A properly adjusted circuit breaker has enough contact pressure to reduce contact resistance to specifications but not so much that it damages the circuit breaker. To determine contact pressure, most OEMs require that a measurement be taken from a stationary part of the contact assembly to the contact face, with the circuit breaker open. The same measurement is repeated after the circuit breaker has been slow closed. Slow, or maintenance, closing allows the circuit breaker contacts to be closed slowly by bypassing the spring-charging mechanism. **See Figure 10-8.**

CONTACT PRESSURE ADJUSTMENT

USING THE SLOW CLOSE FUNCTION

Shermco Industries

Figure 10-8. To determine contact pressure, most OEMs require that a measurement be taken from a stationary part of the contact assembly to the contact face, with the circuit breaker open. The same measurement is repeated after the circuit breaker has been slow closed.

CASE STUDY—IMPROPER CONTACT PRESSURE

Technicians should be aware that opening springs are fully charged when slow closing a circuit breaker, and the circuit breaker is very dangerous. When taking measurements, it is necessary to place the hands around the contacts where they could be injured if the circuit breaker were to open, so caution is needed. Each OEM has a specific method of checking contact pressure and for slow closing the circuit breaker.

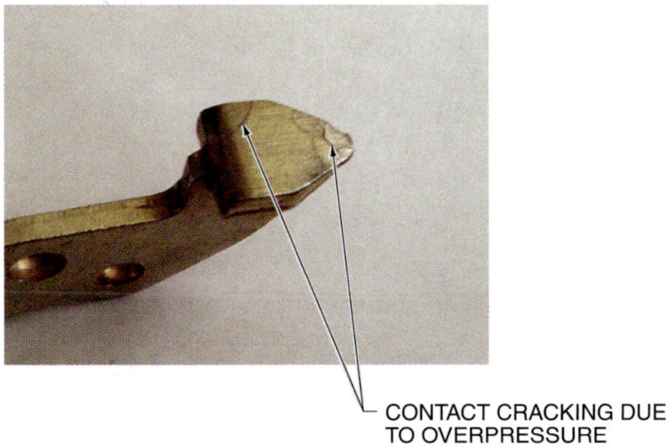

The images shown are the result of too much contact pressure. The excessive contact pressure caused the contact face to crack. Then, during a fault, pieces of the face blew off the contact, and the entire contact melted. This is because the arcing contact face is made of an alloy of zinc, copper, and silver, whereas the portion under the face is made of silver-plated copper. Damage to the contact face was not immediately apparent when an inspection was first made. Only after cleaning the contact surfaces was the damage revealed.

Contact Sequence Adjustment. When a circuit breaker closes, all three contacts should make at about the same time. If they are too far off, the arcing contacts will suffer accelerated wear, which leads to failure of the main contacts. Most OEMs of low-voltage power circuit breakers recommend that all contacts touch within 0.125″ (1⁄8″) of each other. The adjustment point for contact sequence is the same as for contact pressure, which is the stud between the insulating link and the contact assembly. **See Figure 10-9.**

CONTACT SEQUENCE ADJUSTMENT

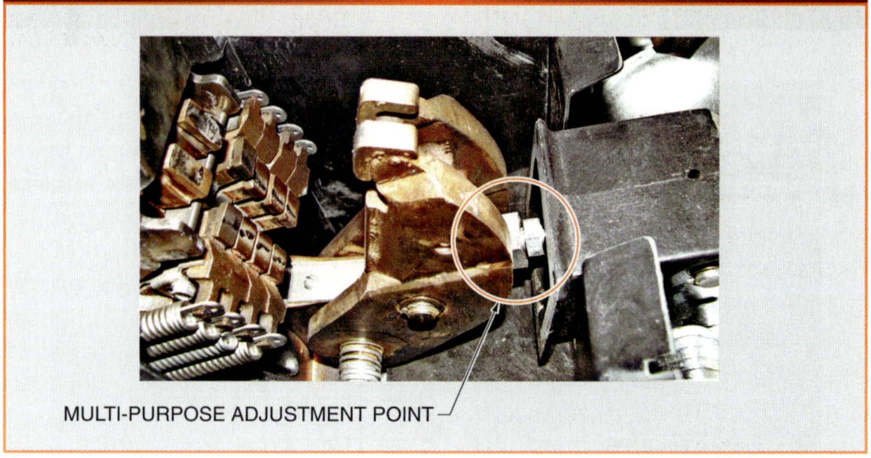

Shermco Industries

Figure 10-9. The adjustment point for contact sequence is the stud between the insulating link and the contact assembly.

Contact Gap Adjustment. The contact gap adjustment is not quite as critical as the contact sequence adjustments. Many times, a specific gap is not provided in OEM instruction books for low-voltage power circuit breakers. However, a contact gap is usually specified for medium-voltage air-magnetic circuit breakers. If the contacts do not open far enough, a circuit breaker will suffer a restrike condition in which it cannot interrupt a fault and is destroyed.

Air is a good dielectric when dry. Air is not a good dielectric when contaminated with moisture, carbon, or other conductive particles or if there is an insufficient air gap between the contacts. These conditions can cause a restrike.

The contact gap is measured, with the circuit breaker open, from the closest points between the stationary and moving contact assemblies. **See Figure 10-10.** If the contact gap is too small, it can be adjusted using the same stud that is used for the other adjustments on most circuit breakers. Again, a properly adjusted circuit breaker is a compromise between all three specifications. A circuit breaker with good contacts should fall within the OEM-specified tolerances for all three adjustments.

CONTACT GAP ADJUSTMENT

Shermco Industries

Figure 10-10. The contact gap is measured, with the circuit breaker open, from the closest points between the stationary and moving contact assemblies.

CASE STUDY—WILDLIFE ACCESS TO ELECTRICAL EQUIPMENT

A Shermco Industries customer was having normal preventive maintenance performed when one of its technicians racked out one of the 4.16 kV circuit breakers. As the circuit breaker was being racked out, the technician noticed something unusual under one of the contact assemblies. Upon closer inspection, it was discovered that two rats had made the circuit breaker their home. One of them got just a little too close to the energized contact and fell victim to an electrical discharge—something just short of a fault.

CIRCUIT BREAKER—RACKED OUT

CONTACT ASSEMBLIES

continued on next page...

CASE STUDY (...continued from previous page)

Conclusion: Small mammals, reptiles, and insects use switchgear for a home. It is warm, clean, and provides protection from predators and the elements. Any individual in the area at that time could have been injured, depending on the settings of the upstream circuit breaker. To prevent this type of accident, all openings in switchgear must be verified to be completely closed and protected from wildlife access.

Abuse and Neglect of Equipment

The third most common problem that can cause circuit breaker failure is the abuse and neglect of equipment. Workers seem to take out their frustrations and anger on circuit breakers and other inanimate objects. This can lead to injury and/or damage that can be dangerous to the power system. Many times, the abuse is due to a lack of training, not maliciousness. **See Figure 10-11.**

ABUSE AND NEGLECT OF EQUIPMENT

Shermco Industries

Figure 10-11. Damage to circuit breaker components can be dangerous to the power system and individuals in the immediate area.

Note: The racking gear in Figure 10-11 came out of a low-voltage circuit breaker that had been racked-in so hard that it bent the hardened stop pin (inset circle). If the stop pin were to shear, the circuit breaker would rack in, then when it reaches the end of its travel, rack back out. The damage was caused by overtorquing the racking handle. As the lubricant gets gummy, it will take more effort to rack in the circuit breaker. If the stabs were not lubricated, the force required to make the primary stabs connect will increase.

ABUSE AND NEGLECT OF EQUIPMENT CAUSED BY POOR MANAGEMENT PRACTICES

The abuse and neglect of equipment is not limited to careless technicians. Equipment damage also can be caused by managers who consider the maintenance of electrical protective devices to be unnecessary overhead, or another cost they do not need. This attitude is shortsighted and dangerous. Cobwebs in electrical equipment are not uncommon, since insects and spiders use electrical equipment as nesting places. Cobwebs in equipment indicate improper maintenance. **See Figure 10-12.**

There is no doubt that the circuit breaker has been neglected for many years. The chances of it actually opening during a fault are small, which means that a lot of expensive switchgear will be damaged and someone may be seriously injured or even killed. The cost of a life cannot be compared to the cost of maintaining switchgear. Managers do not often look at these issues in that perspective. Often, it is a matter of spending the money budgeted for maintenance on something considered more pressing.

H. Landis Floyd, in a presentation at the 13th Annual IEEE Electrical Safety Workshop (in 2006 dollars), estimated the costs of a serious electrical injury to be 23 million dollars. Mr. Floyd used the Consumer Price Index to adjust his number for inflation based on the referenced paper.

POORLY MAINTAINED EQUIPMENT

SEVERELY-NEGLECTED OPERATING MECHANISM WITH COBWEBS

Shermco Industries

Figure 10-12. Cobwebs on equipment and rust indicate improper maintenance.

Electrical Problems

Electrical problems represent about 25% of the problems that can cause circuit breaker failure. The electrical problems usually include pinched wiring, bad connectors, or faulty components, such as motor cutoff switches, spring charging motors, or closing coils. **See Figure 10-13.**

COMMON ELECTRICAL PROBLEMS

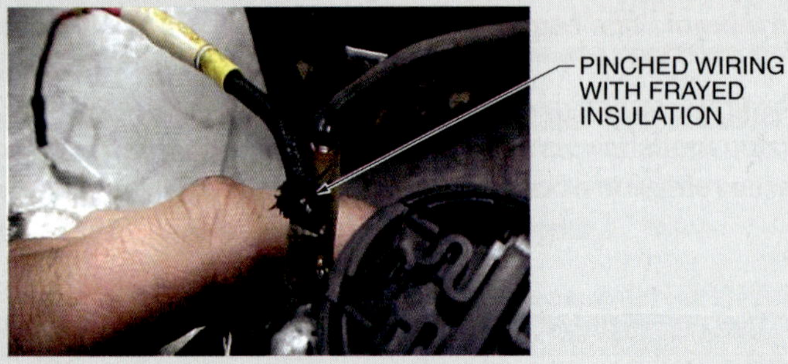

Shermco Industries

Figure 10-13. Electrical problems that can cause circuit breaker failure include pinched wiring, bad connectors, and faulty components, such as motor cutoff switches, spring charging motors, and closing coils. Electrical problems represent about 25% of problems with circuit breakers.

Shermco Industries

Even though they require less in the way of maintenance and servicing, vacuum circuit breakers will fail if not properly maintained.

Plastic Subcomponent Failure

Plastic subcomponents represent about 10% of the problems that can cause circuit breaker failure. As circuit breakers age, the plastic subcomponents, such as secondary contact blocks, relay housings, and terminal blocks, become brittle due to being heated by the current passing through the circuit breaker. Circuit breakers were never designed to be in use for 30, 40, or even 50 years, although some circuit breakers are still in service beyond that point. Some circuit breaker parts cannot be purchased new and the only source is the used parts market.

Operating Environments

Some circuit breakers are used in environments they were never originally designed to be used in. Facilities such as wastewater treatment plants, pulp and paper mills, and other structures in oil and gas fields may have environments that are corrosive or react with the silver plating on the current-carrying parts of circuit breakers. Pulp and paper mills often have sulfur fumes present due to the manufacturing process. Sulfur corrodes the silver plating on contacts and other current-carrying components. Hydrogen sulfide or sulfur dioxide exposure can cause serious damage to circuit breakers. **See Figure 10-14.**

Figure 10-14. Some circuit breakers become damaged because they are used in environments they were never intended to be used in.

Note: The motor starter shown in Figure 10-14 was in service at an oil field. The silver whiskers shown grow out of the silver plating due to the chemical reaction to the gases. These are conductive and must be cleared on a regular basis to prevent shorting.

The general environment also can play an important role in circuit breaker longevity. Circuit breakers located outdoors, even though they are in "weather-tight" enclosures, do not fare as well as ones located indoors. Humidity, heat, dust, and dirty air filters can impact their performance in a relatively short period of time. Circuit breakers located outdoors probably require maintenance more frequently than ones located indoors, especially if the indoor circuit breakers are in a climate-controlled, positive-pressure switchgear room.

There is no "one size fits all" for maintaining circuit breakers and their associated switchgear. Each unit must be evaluated individually to ensure that it stays functional and safe.

SUMMARY

The mechanical operating mechanisms found in low- and medium-voltage power circuit breakers share certain construction characteristics and have the same types of operation and maintenance issues. Age and environment play large roles in how well circuit breakers hold up in service and must be major considerations when planning any maintenance program. Circuit breakers in harsh environments can be expected to age faster than those in clean environments. Also, the operating mechanisms and the current path require disassembly, cleaning, and lubrication every three to five years. These requirements may be extended if the circuit breakers are lightly loaded and in a clean, air-conditioned environment. The monitoring of contact resistance will help determine a schedule for maintenance frequency.

DIGITAL RESOURCES
ATPeResources.com/QuickLinks
Access Code: 223436

chapter 10

MAINTENANCE AND SAFETY ISSUES

CHAPTER REVIEW

Name ______________________________ Date ______________

True-False

T F 1. All lubricants dry out over time.

T F 2. The second most common problem that can cause circuit breaker failure is improper adjustment.

T F 3. A circuit breaker with good contacts should fall within the OEM-specified tolerances for two of the three adjustments.

T F 4. Lack of lubrication results in circuit breakers that will not open, will not close, or open after a delay.

T F 5. Circuit breakers located outdoors, even though they are in "weather-tight" enclosures, do not fare as well as ones located indoors.

T F 6. Operating the circuit breaker on a regular basis is bad for the mechanism because it increases wear.

T F 7. The contact gap adjustment is more critical than the contact sequence adjustments.

T F 8. A simple prop-and-roller operating mechanism using accordion-style linkage is not used on modern circuit breakers due to its inherent weaknesses.

T F 9. Circuit breakers should have a clear, solid sound when they operate.

T F 10. It is important to perform the contact resistance test weekly.

T F 11. The third most common problem that can cause circuit breaker failure is the abuse and neglect of equipment.

T F 12. If lubricant is dirty, dusty, or thick, it must be replaced.

T F 13. If the contact gap is too small, it can be adjusted using the same stud that is used for the other adjustments on most circuit breakers.

T F 14. Some circuit breakers are used in environments they were never originally designed to be used in.

T F 15. Mobilgrease 28 is a red synthetic-based lubricant sold in tubes or cans that can be used in only the circuit breaker's operating mechanism.

Multiple Choice

____________________ **1.** ___ is the most common problem that can cause circuit breakers to fail.

A. Improper adjustment
B. Abuse of equipment
C. An electrical problem
D. A lack of lubrication

____________________ **2.** Without the ___, the modern prop-and-roller mechanism is much more rugged and has fewer issues from misalignment and breakage than the older, simple style.

A. extending linkages
B. insulated couplings
C. main roller
D. secondary latch roller

____________________ **3.** Too ___ leads to a number of problems, including cracked contact faces, excessive wear, and the bending of components.

A. little contact pressure
B. much contact pressure
C. small of a contact gap
D. large of a contact gap

____________________ **4.** To determine the condition of the contact pivot point area, a(n) ___ test, often referred to as a contact resistance test, can be performed.

A. micro-ohmmeter
B. voltmeter
C. ammeter
D. insulation

____________________ **5.** The contact ___ is measured, with the circuit breaker open, from the closest points between the stationary and moving contact assemblies.

A. pressure
B. sequence
C. gap
D. wipe

____________________ **6.** As circuit breakers age, the ___ subcomponents, such as secondary contact blocks, relay housings, and terminal blocks, become brittle due to being heated by the current passing through the circuit breaker.

A. silver
B. steel
C. copper
D. plastic

____________________ **7.** One guideline for low-voltage power circuit breakers that can be used is that contact resistance test readings up to ___ μΩ indicate a circuit breaker in good operating condition.

A. 100
B. 150
C. 300
D. 450

______________ **8.** Corrosive environments are bad for circuit breakers because sulfur corrodes the ___ plating on contacts and other current-carrying components.
A. cadmium
B. nickel
C. silver
D. copper

______________ **9.** Most OEMs of low-voltage power circuit breakers recommend that all contacts touch within ___″ of each other.
A. 0.125
B. 0.25
C. 0.375
D. 0.75

______________ **10.** In general, the contact resistance should be as ___ as possible between the contacts of all three phases. However, this does not help to diagnose problems.
A. low and consistent
B. low and different
C. high and consistent
D. high and different

Completion

______________ **1.** As the circuit breaker carries load current, heat is produced by the ___ losses, which increases the rate at which the lubricants dry out, even if the circuit breakers are exercised regularly.

______________ **2.** Each OEM recommends a specific lubricant for its circuit breakers: one for the current path and one for the ___.

______________ **3.** A(n) ___ lubricant is a lubricant composed of emulsifiers and oil.

______________ **4.** If the contacts do not open far enough, a circuit breaker will suffer a(n) ___ condition in which it cannot interrupt a fault and is destroyed.

______________ **5.** The adjustment point for contact sequence is the same as for contact pressure, which is the stud between the ___ link and the contact assembly.

______________ **6.** ___ closing allows the circuit breaker contacts to be closed slowly by bypassing the spring-charging mechanism.

______________ **7.** Before starting any maintenance procedure on a circuit breaker, the circuit breaker must always be cycled to ensure that the springs are discharged and the contacts are ___.

______________ **8.** The ___ lever of a prop-and-roller mechanism is used to change motion in one direction to another direction, usually at a 90° angle.

______________ **9.** With the circuit breaker in the reset position, the trip latch and the secondary latch are reset to the same position as when the circuit breaker is in the ___ position.

______________ **10.** A medium-voltage circuit breaker in good operating condition should have contact resistance test readings less than ___ μΩ.

Short Answer

1. What are six common reasons for circuit breaker failure?

2. List the three primary and common adjustments on low- and medium-voltage air circuit breaker contacts.

3. What are three common circuit breaker electrical problems?

4. Why should not unapproved lubricants be used on operating mechanisms, even if they seem to work at the time of application?

5. Why should all three contacts make at about the same time when a circuit breaker closes?

Matching

Simple Prop-and-Roller Operating Mechanism Components

_______________ 1. Main roller

_______________ 2. Trip bar

_______________ 3. Arc extinguisher/chute

_______________ 4. Prop

_______________ 5. Linkage

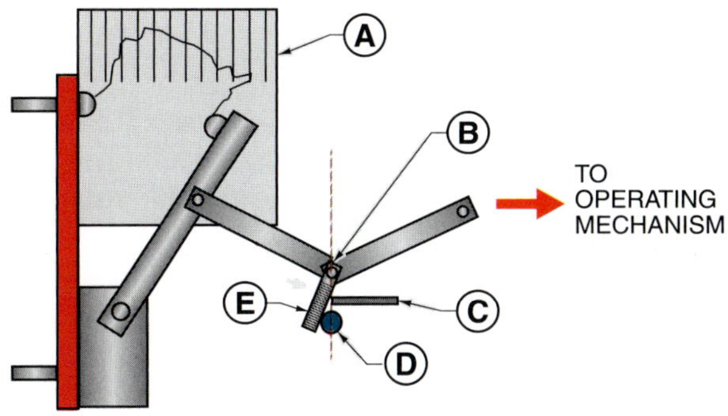

Modern Prop-and-Roller Operating Mechanism Components

_______________ 1. Main (cam) roller

_______________ 2. Bell crank lever

_______________ 3. Insulating coupling

_______________ 4. Secondary latch roller

_______________ 5. Prop

_______________ 6. Secondary latch

_______________ 7. Trip latch

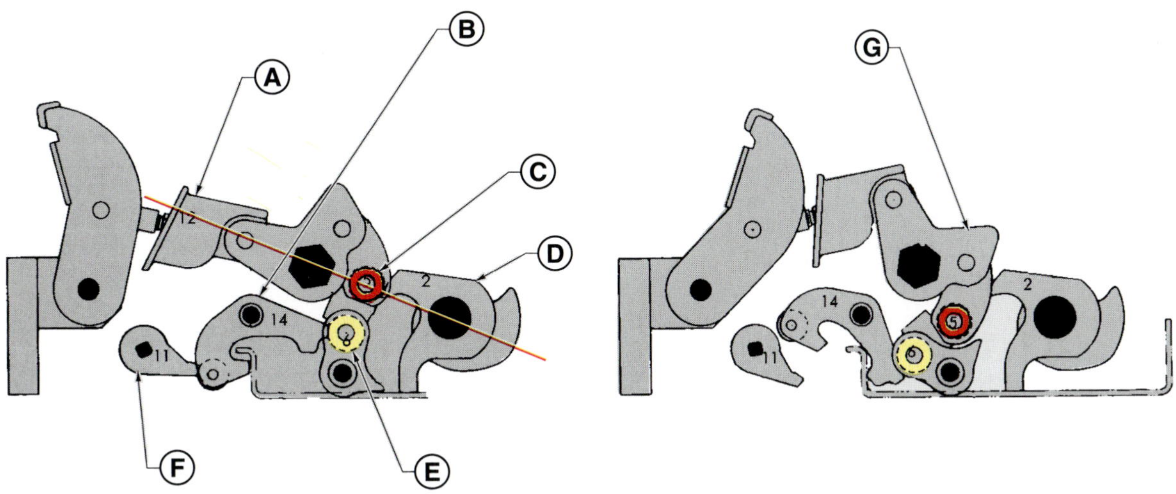

APPENDIX

HIPOT TEST SETS WITH TRUE DC OUTPUT (FROM GE 86132)	
Original Equipment Manufacturer	**Model Number**
Hipotronics	860PL
Hipotronics	880PL
Hipotronics	7BT 60A
Biddle (AVO)	222060
General Electric	282A2610P001

DC HIPOT TESTER RECOMMENDATIONS – G.E.	
Manufacturer	**Model Number**
Hipotronics	860PL
Hipotronics	880PL
Hipotronics	7BT 60A
Biddle (AVO)	222060
General Electric	282A2610P001

ELECTRICAL EQUIPMENT FAILURE CAUSED BY INSULATION FAILURE	
Equipment Type	**Insulation Failure**
Transformers	84%
Circuit breakers	21%
Disconnect switches	15%
Insulated switchgear bus	95%
Bus duct	90%
Cable	89%
Cable joints (splices)	91%
Cable terminations	87%

ANSI/NETA MTS-2011 TABLE 100.14.1 — TEMPERATURE CORRECTION FACTORS			
Temperature		Multiplier	
°C	°F	Apparatus Containing Oil-Immersed Insulation	Apparatus Containing Solid Insulation
–10	14	0.125	0.25
–5	23	0.180	0.32
0	32	0.25	0.40
5	41	0.56	0.50
10	50	0.50	0.63
15	59	0.75	0.81
20	68	1.00	1.00
25	77	1.40	1.25
30	86	1.98	1.58
35	95	2.80	2.00
40	104	3.95	2.50
45	113	5.60	3.15
50	122	7.85	3.98
55	131	11.20	5.00
60	140	15.85	6.30
65	149	22.40	7.90
70	158	31.75	10.00
75	167	44.70	12.60
80	176	63.50	15.80
85	185	89.789	20.00
90	194	127.00	25.20
95	203	180.09	31.60
100	212	254.00	40.00
105	221	359.15	50.40
110	230	509.00	63.20

SWITCHGEAR WITHSTAND VOLTAGES – ANSI/NETA MTS-2011 TABLE 100.2			
Type of Switchgear	Rated Maximum Voltage*	Maximum Test Voltage†	
		AC	DC
Low-voltage power circuit breaker switchgear	0.254/0.508/0.635	1.6	2.3
Metal-clad switchgear	4.76	14	20
	8.25	27	37
	15.0	27	37
	27.0	45	—
	38.0	60	—
Station-type cubicle switchgear	15.5	37	—
	38.0	60	—
	72.5	120	—
Metal-enclosed interrupter switchgear	4.76	14	20
	8.25	19	27
	15.0	27	37
	27.0	45	—
	38.0	60	—

* in kV/rms
† in kV

TEMPERATURE CORRECTION FACTORS (ANSI/NETA MTS – 2011 TABLE 100.14.1)			
Temperature		Multiplier	
°C	°F	Apparatus Containing Oil-Immersed Insulation	Apparatus Containing Solid Insulation
–10	14	0.125	0.25
–5	23	0.180	0.32
0	32	0.25	0.40
5	41	0.36	0.50
10	50	0.50	0.63
15	59	0.75	0.81
20	68	1.00	1.00
25	77	1.40	1.25
30	86	1.98	1.58
35	95	2.80	2.00
40	104	3.95	2.50
45	113	5.60	3.15
50	122	7.85	3.98
55	131	11.20	5.00
60	140	15.85	6.30
65	149	22.40	7.90
70	158	31.75	10.00
75	167	44.70	12.60
80	176	63.50	15.80
85	185	89.789	20.00
90	194	127.00	25.20
95	203	180.00	31.60
100	212	254.00	40.00
105	221	359.15	50.40
110	230	509.00	63.20

CIRCUIT BREAKER GROUND FAULT DEFEAT		
Manufacturer	Trip Unit Model	Ground Fault Defeat
Allis-Chalmers	AG, AG1, AG2, DG, DG1	Short Terminals 2–4
Allis-Chalmers	4WAG, 4WDG	Short Terminals 9–4
BBC (ITE)	Power Shield	Short Terminals 1–2
Carriere/Multilin	FB - 600	Short Terminals 4–5
FPE	SS, SD	Move wire from "G" to "N"
General Electric	Power Sensor	Short Terminal 1–5 0.15 μF capacitor under breaker
General Electric	SST, ECS, MicroVersaTrip	Need special cable
General Electric	MicroVersaTrip RMS-9	Use manufacturer's secondary injection test set "TV-RMS"
General Electric	MicroVersa Trip PM+	Use manufacturer's secondary injection test set "TV-RMS"
Siemens-Allis	Static Trip II	Short Terminals 4–9
Siemens-Allis	Static Trip III	Short Terminals 4–5
SureTrip	ST - 003	Short Terminals O–N
Westinghouse	Amptector I and II	Move wire from G to N
Westinghouse	Digitrip	Move wire from G to N

EQUIPMENT FAILURES DUE TO INSULATION FAILURE	
Component	**Insulation Failure***
Transformers	84
Circuit breakers	21
Disconnect switches	15
Insulated switchgear bus	95
Bus duct	90
Cable	89
Cable joints (splices)	91
Cable terminations	87

* in %

MINIMUM INSULATION RESISTANCE FOR CONTROL POWER TRANSFORMERS TABLE 100.5 – ANSI/NETA MTS-2011			
Transformer Coil Rating Type*	**Minimum DC Test Voltage**	**Recommended Minimum Insulation Resistance†**	
		Liquid Filled	**Dry**
0 – 600	1000	100	500
601 – 5000	2500	1000	5000
Greater than 5000	5000	5000	25,000

*** in MΩ**
† in yr

RUBBER INSULATING GLOVE VOLTAGE CLASSIFICATIONS	
Class	**Max Use Level***
0	1000
1	7500
2	17,000
3	26,000
4	36,000

* in V

MINIMUM INSULATION RESISTANCE FOR CONTROL POWER TRANSFORMERS – ANSI/NETA MTS-2011 TABLE 100.5			
Transformer Coil Rating Type*	**Minimum DC Test Voltage**	**Recommended MInimum Insulation Resistance†**	
		Liquid Filled	**Dry**
0 – 600	1000	100	T500
601 – 5000	2500	1000	5000
> 5000	5000	5000	25,000

* in V
† in MΩ

ABB/ITE OD-5 LTD TIME-CURRENT CHARACTERISTIC CURVE

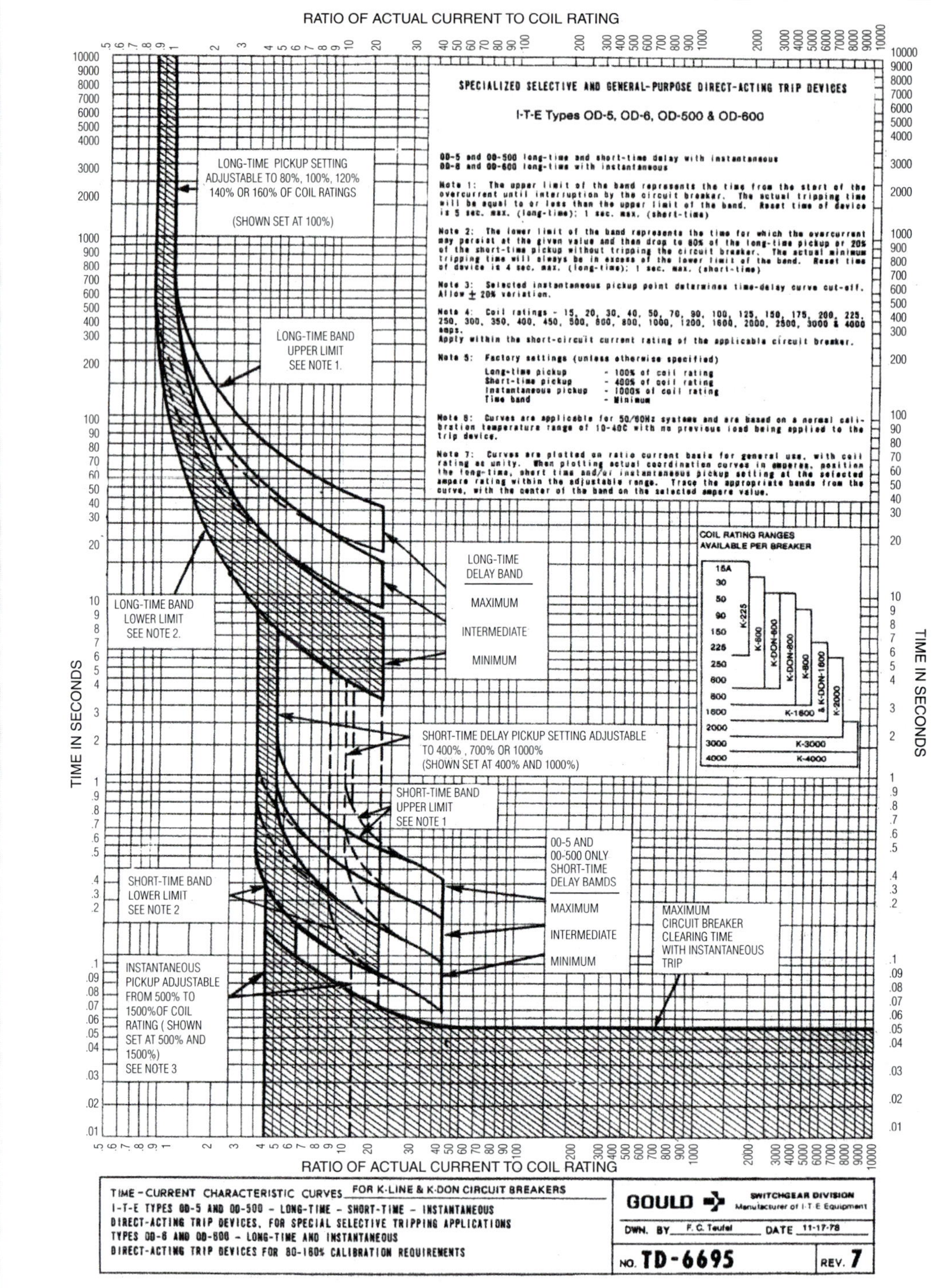

ABB Power T&D

SS-5 TIME-CURRENT CHARACTERISTIC CURVE

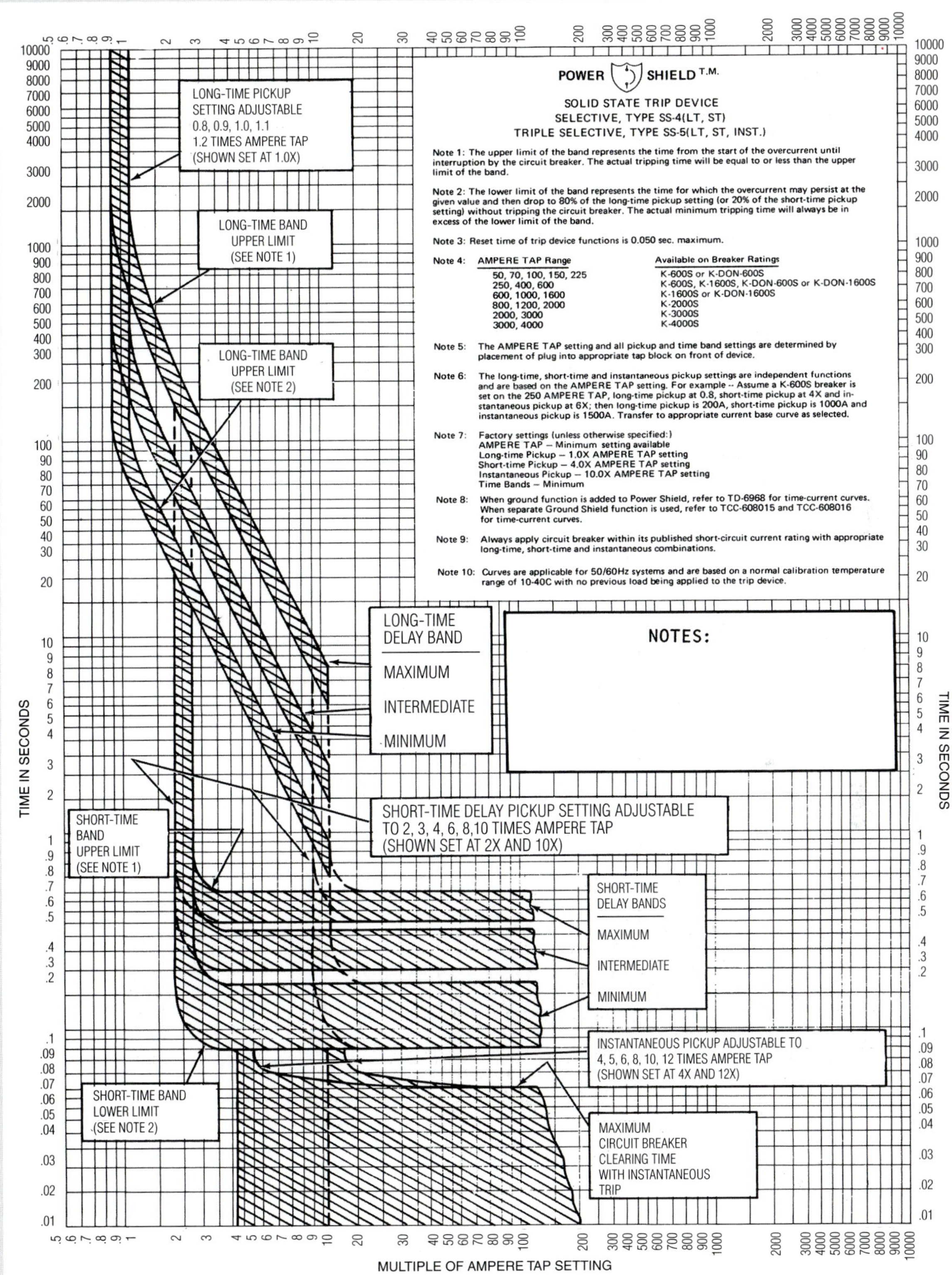

ANSI/NETA MTS-2011 RELATED PROCEDURES

7. INSPECTION AND TEST PROCEDURES

7.1 Switchgear and Switchboard Assemblies

1. Visual and Mechanical Inspection

1. Inspect physical, electrical, and mechanical condition including evidence of moisture or corona.
2. Inspect anchorage, alignment, grounding, and required area clearances.
3. Prior to cleaning the unit, perform as-found tests, if required.
4. Clean the unit.
5. Verify that fuse and/or circuit breaker sizes and types correspond to drawings and coordination study as well as to the circuit breaker's address for microprocessor-communication packages.
6. Verify that current and voltage transformer ratios correspond to drawings.
7. Inspect bolted electrical connections for high resistance using one or more of the following methods:
    1. Use of a low-resistance ohmmeter in accordance with Section 7.1.2.
    2. Verify tightness of accessible bolted electrical connections by calibrated torque-wrench method in accordance with manufacturer's published data or Table 100.12.
    3. Perform a thermographic survey in accordance with Section 9.
8. Confirm correct operation and sequencing of electrical and mechanical interlock systems.
    1. Attempt closure on locked-open devices. Attempt to open locked-closed devices.
    2. Make key exchange with all devices included in the interlock scheme as applicable.
9. Use appropriate lubrication on moving current-carrying parts and on moving and sliding surfaces.
10. Verify correct barrier and shutter installation and operation.
11. Exercise all active components.
12. Inspect mechanical indicating devices for correct operation.
13. Verify that filters are in place and/or vents are clear.
14. Perform visual and mechanical inspection of instrument transformers in accordance with Section 7.10.

* Optional

7. INSPECTION AND TEST PROCEDURES

7.1 Switchgear and Switchboard Assemblies (*continued*)

15. Inspect control power transformers.

    1. Inspect for physical damage, cracked insulation, broken leads, tightness of connections, defective wiring, and overall general condition.

    2. Verify that primary and secondary fuse ratings or circuit breakers match drawings.

    3. Verify correct functioning of drawout disconnecting and grounding contacts and interlocks.

16. Perform as-left tests.

2. Electrical Tests

1. Perform resistance measurements through bolted electrical connections with a low-resistance ohmmeter in accordance with Section 7.1.1, if applicable.

2. Perform insulation–resistance tests for one minute on each bus section, phase-to-phase and phase-to-ground. Apply voltage in accordance with manufacturer's published data. In the absence of manufacturer's published data, use Table 100.1.

*3. Perform a dielectric withstand voltage test on each bus section, each phase-to-ground with phases not under test grounded, in accordance with manufacturer's published data. If manufacturer has no recommendation for this test, it shall be in accordance with Table 100.2. The test voltage shall be applied for one minute. Refer to Section 7.1.3 before performing test.

*4. Perform insulation-resistance tests on control wiring with respect to ground. The applied potential shall be 500 volts dc for 300-volt rated cable and 1000 volts dc for 600-volt rated cable. Test duration shall be one minute. For units with solid-state components or control devices that cannot tolerate the applied voltage, follow manufacturer's recommendation.

5. Perform electrical tests on instrument transformers in accordance with Section 7.10.

6. Perform ground–resistance tests in accordance with Section 7.13.

7. Determine accuracy of all meters and calibrate watthour meters in accordance with Section 7.11.

8. Control Power Transformers

    1. Perform insulation-resistance tests. Perform measurements from winding-to-winding and each winding-to-ground. Test voltages shall be in accordance with manufacturer's published data. In the absence of manufacturer's published data, use Table 100.1.

* Optional

7. INSPECTION AND TEST PROCEDURES

7.1 Switchgear and Switchboard Assemblies (*continued*)

2. Verify correct function of control transfer relays located in switchgear with multiple power sources.

9. Verify operation of switchgear/switchboard heaters and their controller, if applicable.

10. Perform system function tests in accordance with Section 8.

3. Test Values

3.1 Test Values – Visual and Mechanical

1. Compare bolted connection resistance values to values of similar connections. Investigate values which deviate from those of similar bolted connections by more than 50 percent of the lowest value. (7.1.1.7.1)

2. Bolt-torque levels should be in accordance with manufacturer's published data. In the absence of manufacturer's published data, use Table 100.12. (7.1.1.7.2)

3. Results of the thermographic survey shall be in accordance with Section 9. (7.1.1.7.3)

3.2 Test Values – Electrical

1. Compare bolted connection resistance values to values of similar connections. Investigate values which deviate from those of similar bolted connections by more than 50 percent of the lowest value.

2. Insulation-resistance values of bus insulation should be in accordance with manufacturer's published data. In the absence of manufacturer's published data, use Table 100.1. Values of insulation resistance less than this table or manufacturer's recommendations should be investigated. Dielectric withstand voltage tests should not proceed until insulation-resistance levels are raised above minimum values.

3. If no evidence of distress or insulation failure is observed by the end of the total time of voltage application during the test, the test dielectric withstand voltage specimen is considered to have passed the test.

4. Minimum insulation-resistance values of control wiring should be comparable to previously obtained results but not less than two megohms.

5. Results of electrical tests on instrument transformers should be in accordance with Section 7.10.

6. Results of ground resistance tests should be in accordance with Section 7.13.

7. Accuracy of meters should be in accordance with Section 7.11.

* Optional

7. INSPECTION AND TEST PROCEDURES

7.1 Switchgear and Switchboard Assemblies (*continued*)

8. Control Power Transformers

   1. Insulation-resistance values of control power transformers should be in accordance with manufacturer's published data. In the absence of manufacturer's published data, use Table 100.5. Values of insulation resistance less than this table or manufacturer's recommendations should be investigated.

   2. Control transfer relays should perform as designed.

9. Heaters should be operational.

10. Results of system function tests shall be in accordance with Section 8.

* Optional

7. INSPECTION AND TEST PROCEDURES

7.6.1.1 Circuit Breakers, Air, Insulated-Case/Molded-Case

1. Visual and Mechanical Inspection

1. Inspect physical and mechanical condition.
2. Inspect anchorage and alignment.
3. Prior to cleaning the unit, perform as-found tests, if required.
4. Clean the unit.
5. Operate the circuit breaker to insure smooth operation.
6. Inspect bolted electrical connections for high resistance using one or more of the following methods:
    1. Use of a low-resistance ohmmeter in accordance with Section 7.6.1.1.2.
    2. Verify tightness of accessible bolted electrical connections by calibrated torque-wrench method in accordance with manufacturer's published data or Table 100.12.
    3. Perform a thermographic survey in accordance with Section 9.
7. Inspect operating mechanism, contacts, and arc chutes in unsealed units.
8. Perform adjustments for final protective device settings in accordance with coordination study provided by end user.
9. Perform as-left tests.

2. Electrical Tests

1. Perform resistance measurements through bolted connections with a low-resistance ohmmeter, if applicable, in accordance with Section 7.6.1.1.1.
2. Perform insulation-resistance tests for one minute on each pole, phase-to-phase and phase-to-ground with the circuit breaker closed, and across each open pole. Apply voltage in accordance with manufacturer's published data. In the absence of manufacturer's published data, use Table 100.1.
3. Perform a contact/pole-resistance test.

* Optional

7. INSPECTION AND TEST PROCEDURES

7.6.1.1 Circuit Breakers, Air, Insulated-Case/Molded-Case (*continued*)

*4. Perform insulation-resistance tests on all control wiring with respect to ground. The applied potential shall be 500 volts dc for 300-volt rated cable and 1000 volts dc for 600-volt rated cable. Test duration shall be one minute. For units with solid-state components, follow manufacturer's recommendation.

5. Determine long-time pickup and delay by primary current injection.

6. Determine short-time pickup and delay by primary current injection.

7. Determine ground-fault pickup delay by primary current injection.

8. Determine instantaneous pickup current by primary injection.

*9. Test functions of the trip unit by means of secondary injection.

10. Perform minimum pickup voltage test on shunt trip and close coils in accordance with Table 100.20.

11. Verify correct operation of auxiliary features such as trip and pickup indicators, zone interlocking, electrical close and trip operation, trip-free, antipump function, and trip unit battery condition.

12. Reset all trip logs and indicators.

13. Verify operation of charging mechanism.

3. Test Values

3.1 Test Values – Visual and Mechanical

1. Compare bolted connection resistance values to values of similar connections. Investigate values which deviate from those of similar bolted connections by more than 50 percent of the lowest value. (7.6.1.1.6.1)

2. Bolt-torque levels should be in accordance with manufacturer's published data. In the absence of manufacturer's published data, use Table 100.12. (7.6.1.1.6.2)

3. Results of the thermographic survey shall be in accordance with Section 9. (7.6.1.1.6.3)

4. Settings shall comply with coordination study recommendations. (7.6.1.1.8)

* Optional

7. INSPECTION AND TEST PROCEDURES

7.6.1.1 Circuit Breakers, Air, Insulated-Case/Molded-Case (*continued*)

3.2 Test Values – Electrical

1. Compare bolted connection resistance values to values of similar connections. Investigate values which deviate from those of similar bolted connections by more than 50 percent of the lowest value.

2. Insulation-resistance values should be in accordance with manufacturer's published data. In the absence of manufacturer's published data, use Table 100.1. Values of insulation resistance less than this table or manufacturer's recommendations should be investigated.

3. Microhm or dc millivolt drop values should not exceed the high levels of the normal range as indicated in the manufacturer's published data. If manufacturer's data is not available, investigate values that deviate from adjacent poles or similar breakers by more than 50 percent of the lowest value.

4. Insulation-resistance values of control wiring should be comparable to previously obtained results but not less than two megohms.

5. Long-time pickup values should be as specified, and the trip characteristic should not exceed manufacturer's published time-current characteristic tolerance band, including adjustment factors. If manufacturer's curves are not available, trip times should not exceed the value shown in Table 100.7.

6. Short-time pickup values should be as specified, and the trip characteristic should not exceed manufacturer's published time-current tolerance band.

7. Ground fault pickup values should be as specified, and the trip characteristic should not exceed manufacturer's published time-current tolerance band.

8. Instantaneous pickup values of molded-case circuit breakers should fall within manufacturer's published tolerances and/or Table 100.8.

9. Pickup values and trip characteristics should be within manufacturer's published tolerances.

10. Minimum pickup voltage on shunt trip and close coils should be in accordance with manufacturer's published data. In the absence of manufacturer's published data, refer to Table 100.20.

11. Breaker open, close, trip, trip-free, antipump, and auxiliary features should function as designed.

12. Trip logs and indicators are reset.

13. The charging mechanism should operate in accordance with manufacturer's published data.

* Optional

ANSI LOW VOLTAGE

7. INSPECTION AND TEST PROCEDURES

7.6.1.2 Circuit Breakers, Air, Low-Voltage Power

1. Visual and Mechanical Inspection

1. Inspect physical and mechanical condition.
2. Inspect anchorage, alignment, and grounding.
3. Verify that all maintenance devices are available for servicing and operating the breaker.
4. Prior to cleaning the unit, perform as-found tests, if required.
5. Clean the unit.
6. Inspect arc chutes.
7. Inspect moving and stationary contacts for condition, wear, and alignment.
8. Verify that primary and secondary contact wipe and other dimensions vital to satisfactory operation of the breaker are correct.
9. Perform all mechanical operator and contact alignment tests on both the breaker and its operating mechanism in accordance with manufacturer's published data.
10. Inspect bolted electrical connections for high resistance using one or more of the following methods:
    1. Use of a low-resistance ohmmeter in accordance with Section 7.6.1.2.2.
    2. Verify tightness of accessible bolted electrical connections by calibrated torque-wrench method in accordance with manufacturer's published data or Table 100.12.
    3. Perform a thermographic survey in accordance with Section 9.
11. Verify cell fit and element alignment.
12. Verify racking mechanism operation.
13. Use appropriate lubrication on moving current-carrying parts and on moving and sliding surfaces.
14. Perform adjustments for final protective device settings in accordance with coordination study provided by end user.
15. Perform as-left tests.
16. Record as-found and as-left operation counter readings, if applicable.

* Optional

7. INSPECTION AND TEST PROCEDURES

7.6.1.2 Circuit Breakers, Air, Low-Voltage Power (*continued*)

2. Electrical Tests

1. Perform resistance measurements through bolted connections with a low-resistance ohmmeter, if applicable, in accordance with Section 7.6.1.2.1.

2. Perform insulation-resistance tests for one minute on each pole, phase-to-phase and phase-to-ground with the circuit breaker closed, and across each open pole. Apply voltage in accordance with manufacturer's published data. In the absence of manufacturer's published data, use Table 100.1.

3. Perform a contact/pole-resistance test.

*4. Perform insulation-resistance tests on all control wiring with respect to ground. The applied potential shall be 500 volts dc for 300-volt rated cable and 1000 volts dc for 600-volt rated cable. Test duration shall be one minute. For units with solid-state components or control devices that cannot tolerate the applied voltage, follow manufacturer's recommendation.

5. Determine long-time pickup and delay by primary current injection.

6. Determine short-time pickup and delay by primary current injection.

7. Determine ground-fault pickup and delay by primary current injection.

8. Determine instantaneous pickup current by primary current injection.

*9. Test functions of the trip unit by means of secondary injection.

10. Perform minimum pickup voltage test on shunt trip and close coils in accordance with Table 100.20.

11. Verify correct operation of auxiliary features such as trip and pickup indicators, zone interlocking, electrical close and trip operation, trip-free, antipump function, and trip unit battery condition.

12. Reset all trip logs and indicators.

13. Verify operation of charging mechanism.

* Optional

7. INSPECTION AND TEST PROCEDURES

7.6.1.2 Circuit Breakers, Air, Low-Voltage Power (*continued*)

3. Test Values

3.1 Test Values – Visual and Mechanical

1. Compare bolted connection resistance values to values of similar connections. Investigate values which deviate from those of similar bolted connections by more than 50 percent of the lowest value. (7.6.1.2.1.10.1)

2. Bolt-torque levels should be in accordance with manufacturer's published data. In the absence of manufacturer's published data, use Table 100.12. (7.6.1.2.1.10.2)

3. Results of the thermographic survey shall be in accordance with Section 9. (7.6.1.2.1.10.3)

4. Settings shall comply with coordination study recommendations. (7.6.1.2.1.15)

5. Operations counter should advance one digit per close-open cycle. (7.6.1.2.1.16)

3.2 Test Values – Electrical

1. Compare bolted connection resistance values to values of similar connections. Investigate values which deviate from those of similar bolted connections by more than 50 percent of the lowest value.

2. Insulation-resistance values of breakers should be in accordance with manufacturer's published data. In the absence of manufacturer's published data, use Table 100.1. Values of insulation resistance less than this table or manufacturer's recommendations should be investigated.

3. Microhm or dc millivolt drop values should not exceed the high levels of the normal range as indicated in the manufacturer's published data. If manufacturer's data is not available, investigate values that deviate from adjacent poles or similar breakers by more than 50 percent of the lowest value.

4. Insulation-resistance values of control wiring should be comparable to previously obtained results but not less than two megohms.

5. Long-time pickup values should be as specified, and the trip characteristic shall not exceed manufacturer's published time-current characteristic tolerance band.

6. Short-time pickup values should be as specified, and the trip characteristic should not exceed manufacturer's published time-current tolerance band.

7. Ground fault pickup values should be as specified, and the trip characteristic should not exceed manufacturer's published time-current tolerance band.

8. Instantaneous pickup values should be within the tolerances of manufacturer's published data.

* Optional

7. INSPECTION AND TEST PROCEDURES

7.6.1.2 Circuit Breakers, Air, Low-Voltage Power (*continued*)

9. Pickup values and trip characteristic should be as specified and within manufacturer's published tolerances.

10. Minimum pickup voltage on shunt trip and close coils should be in accordance with manufacturer's published data. In the absence of manufacturer's published data, refer to Table 100.20.

11. Auxiliary features should operate in accordance with manufacturer's published data.

12. Trip logs and indicators are reset.

13. The charging mechanism should operate in accordance with manufacturer's published data.

* Optional

7. INSPECTION AND TEST PROCEDURES

7.6.1.3 Circuit Breakers, Air, Medium-Voltage

1. Visual and Mechanical Inspection

1. Inspect physical and mechanical condition.
2. Inspect anchorage, alignment, and grounding.
3. Verify that all maintenance devices are available for servicing and operating the breaker.

*4. Perform operator analysis (first-trip) test.

5. Prior to cleaning the unit, perform as-found tests, if required.
6. Clean the unit.
7. Inspect arc chutes.
8. Inspect moving and stationary contacts for condition, wear, and alignment.
9. If recommended by manufacturer, slow close/open breaker and check for binding, friction, contact alignment, contact sequence, and penetration. Verify that contact sequence is in accordance with manufacturer's published data. In the absence of manufacturer's published data, refer to ANSI/IEEE C37.04.
10. Perform all mechanical operation tests on the operating mechanism in accordance with manufacturer's published data.
11. Inspect bolted electrical connections for high resistance using one or more of the following methods:
    1. Use of a low-resistance ohmmeter in accordance with Section 7.6.1.3.2.
    2. Verify tightness of accessible bolted electrical connections by calibrated torque-wrench method in accordance with manufacturer's published data or Table 100.12.
    3. Perform a thermographic survey in accordance with Section 9.
12. Verify cell fit and element alignment.
13. Verify racking mechanism operation.
14. Inspect puffer operation.
15. Use appropriate lubrication on moving current-carrying parts and on moving and sliding surfaces.

* Optional

7. INSPECTION AND TEST PROCEDURES

7.6.1.3 Circuit Breakers, Air, Medium-Voltage (*continued*)

*16. Perform time-travel analysis.

17. Perform as-left tests.

18. Record as-found and as-left operation-counter readings.

2. Electrical Tests

1. Perform resistance measurements through bolted connections with a low-resistance ohmmeter, if applicable. See Section 7.6.1.3.1.

2. Perform insulation-resistance tests for one minute on each pole, phase-to-phase and phase-to-ground with the circuit breaker closed, and across each open pole. Apply voltage in accordance with manufacturer's published data. In the absence of manufacturer's published data, use Table 100.1.

*3. Perform insulation-resistance tests on all control wiring with respect to ground. The applied potential shall be 500 volts dc for 300-volt rated cable and 1000 volts dc for 600-volt rated cable. Test duration shall be one minute. For units with solid-state components or control devices that cannot tolerate the applied voltage, follow manufacturer's recommendation.

4. Perform a contact/pole-resistance test.

5. With the breaker in a test position, perform the following tests:

   1. Trip and close breaker with the control switch.

   2. Trip breaker by operating each of its protective relays.

   3. Verify mechanism charge, trip-free, and antipump functions.

*6. Perform minimum pickup voltage tests on trip and close coils in accordance with Table 100.20.

*7. Perform power-factor or dissipation-factor tests with breaker in both the open and closed positions.

*8. Perform power-factor or dissipation-factor tests on each bushing equipped with a power-factor/ capacitance tap. In the absence of a power-factor/ capacitance tap, perform hot-collar tests. These tests shall be in accordance with the test equipment manufacturer's published data.

*9. Perform a dielectric withstand voltage test on each phase with the circuit breaker closed and the poles not under test grounded. Test voltage should be in accordance with manufacturer's published data. In the absence of manufacturer's published data, use Table 100.19.

* Optional

NETA

Page 74

ANSI/NETA MTS-2011

7. INSPECTION AND TEST PROCEDURES

7.6.1.3 Circuit Breakers, Air, Medium-Voltage (*continued*)

10. Verify blowout coil circuit continuity.

11. Verify operation of cubicle space heaters, if applicable.

*12. Test instrument transformers in accordance with Section 7.10.

3. Test Values

3.1 Test Values – Visual and Mechanical

1. Compare first-trip operation time and trip-coil current waveform to manufacturer's published data. In the absence of manufacturer's published data, compare first-trip operation time and trip-coil current waveform to previously obtained results. (7.6.1.3.1.4)

2. Compare bolted connection resistance values to values of similar connections. Investigate values which deviate from those of similar bolted connections by more than 50 percent of the lowest value. (7.6.1.3.1.11.1)

3. Bolt-torque levels should be in accordance with manufacturer's published data. In the absence of manufacturer's published data, use Table 100.12. (7.6.1.3.1.11.2)

4. Results of the thermographic survey shall be in accordance with Section 9. (7.6.1.3.1.11.3)

5. Compare travel and velocity values to manufacturer's published data and previous test data. (7.6.1.3.1.16)

6. Operations counter should advance one digit per close-open cycle. (7.6.1.2.1.18)

3.2 Test Values – Electrical

1. Compare bolted connection resistance values to values of similar connections. Investigate values which deviate from those of similar bolted connections by more than 50 percent of the lowest value.

2. Insulation-resistance values of circuit breakers should be in accordance with manufacturer's published data. In the absence of manufacturer's published data, use Table 100.1. Values of insulation resistance less than this table or manufacturer's recommendations should be investigated.

3. Insulation-resistance values of control wiring should be comparable to previously obtained results but not less than two megohms.

4. Microhm or dc millivolt drop values shall not exceed the high levels of the normal range as indicated in the manufacturer's published data. If manufacturer's data is not available,

* Optional

7. INSPECTION AND TEST PROCEDURES

7.6.1.3 Circuit Breakers, Air, Medium-Voltage (*continued*)

investigate values that deviate from adjacent poles or similar breakers by more than 50 percent of the lowest value.

5. Breaker mechanism charge, close, open, trip, trip-free, and antipump features shall function as designed.

6. Minimum pickup for trip and close coils shall be in accordance with manufacturer's published data. In the absence of manufacturer's data, refer to Table 100.20.

7. Power-factor or dissipation-factor values shall be compared with previous test results of similar breakers or manufacturer's published data.

8. Power-factor or dissipation-factor and capacitance values should be within ten percent of nameplate rating for bushings. Hot collar tests are evaluated on a milliampere/milliwatt loss basis, and the results should be compared to values of similar bushings.

9. If no evidence of distress or insulation failure is observed by the end of the total time of voltage application during the dielectric withstand voltage test, the circuit breaker is considered to have passed the test.

10. The blowout coil circuit should exhibit continuity.

11. Cubicle space heaters should be operational.

12. The results of instrument transformer tests shall be in accordance with Section 7.10.

* Optional

ANSI/NETA

7. INSPECTION AND TEST PROCEDURES

7.6.3 Circuit Breakers, Vacuum, Medium-Voltage

1. Visual and Mechanical Inspection

1. Inspect physical and mechanical condition.
2. Inspect anchorage, alignment, and grounding.
3. Verify that all maintenance devices are available for servicing and operating the breaker.

*4. Perform operator analysis (first-trip) test.

5. Prior to cleaning the unit, perform as-found tests, if required.
6. Clean the unit.
7. Inspect vacuum bottle assemblies.
8. Measure critical distances such as contact gap as recommended by the manufacturer.
9. If recommended by the manufacturer, slow close/open the breaker and check for binding, friction, contact alignment, contact sequence, and penetration.
10. Perform all mechanical operation tests on the operating mechanism in accordance with manufacturer's published data.
11. Inspect bolted electrical connections for high resistance using one or more of the following methods:
    1. Use of a low-resistance ohmmeter in accordance with Section 7.6.3.2.2.
    2. Verify tightness of accessible bolted electrical connections by calibrated torque-wrench method in accordance with manufacturer's published data. In the absence of manufacturer's published data, use Table 100.12.
    3. Perform a thermographic survey in accordance with Section 9.
12. Verify cell fit and element alignment.
13. Verify racking mechanism operation.
14. Inspect vacuum bellows operation.
15. Use appropriate lubrication on moving current-carrying parts and on moving and sliding surfaces.

* Optional

7. INSPECTION AND TEST PROCEDURES

7.6.3 Circuit Breakers, Vacuum, Medium-Voltage (*continued*)

*16. Perform time-travel analysis.

17. Perform as-left tests.

18. Record as-found and as-left operation counter readings.

2. Electrical Tests

1. Perform resistance measurements through bolted connections with a low-resistance ohmmeter, if applicable, in accordance with Section 7.6.3.1.10.

2. Perform insulation-resistance tests for one minute on each pole, phase-to-phase and phase-to-ground with circuit breaker closed and across each pole with the breaker open. Apply voltage in accordance with manufacturer's published data. In the absence of manufacturer's published data, use Table 100.1.

*3. Perform insulation-resistance tests on all control wiring with respect to ground. The applied potential shall be 500 volts dc for 300-volt rated cable and 1000 volts dc for 600-volt rated cable. Test duration shall be one minute. For units with solid-state components or control devices that cannot tolerate the applied voltage, follow manufacturer's recommendation.

4. Perform a contact/pole-resistance test.

5. With breaker in a test position, perform the following tests:

    1. Trip and close breaker with the control switch.

    2. Trip breaker by operating each of its protective relays.

    3. Verify mechanism charge, trip-free, and antipump functions.

*6. Perform minimum pickup voltage tests on trip and close coils in accordance with Table 100.20.

*7. Perform power-factor or dissipation-factor tests on each pole with the breaker open and each phase with the breaker closed.

*8. Perform power-factor or dissipation-factor tests on each bushing equipped with a power-factor/ capacitance tap. In the absence of a power-factor/ capacitance tap, perform hot-collar tests. These tests shall be in accordance with the test equipment manufacturer's published data.

9. Perform a vacuum bottle integrity (dielectric withstand voltage) test across each vacuum bottle with the breaker in the open position in strict accordance with manufacturer's published data.

* Optional

7. INSPECTION AND TEST PROCEDURES

7.6.3 Circuit Breakers, Vacuum, Medium-Voltage (*continued*)

*10. Perform a dielectric withstand voltage test on each phase with the circuit breaker closed and the poles not under test grounded. Test voltage should be in accordance with manufacturer's published data. In the absence of manufacturer's published data, use Table 100.19.

11. Verify operation of heaters, if applicable.

12. Test instrument transformers in accordance with Section 7.10.

3. Test Values

3.1 Test Values – Visual and Mechanical

1. Compare first-trip operation time and trip-coil current waveform to manufacturer's published data. In the absence of manufacturer's published data, compare first-trip operation time and trip-coil current waveform to previously obtained results. (7.6.3.1.4)

2. Mechanical operation and contact alignment should be in accordance with manufacturer's published data. (7.6.3.1.9)

3. Compare bolted connection resistance values to values of similar connections. Investigate values which deviate from those of similar bolted connections by more than 50 percent of the lowest value. (7.6.3.1.11.1)

4. Bolt-torque levels should be in accordance with manufacturer's published data. In the absence of manufacturer's published data, use Table 100.12. (7.6.3.1.11.2)

5. Results of the thermographic survey shall be in accordance with Section 9. (7.6.3.1.11.3)

6. Compare travel and velocity values to manufacturer's published data and previous test data. (7.6.3.1.16)

7. Operation counter should advance one digit per close-open cycle. (7.6.3.1.18)

3.2 Test Values – Electrical

1. Compare bolted connection resistance values to values of similar connections. Investigate values which deviate from those of similar bolted connections by more than 50 percent of the lowest value.

2. Insulation-resistance values of circuit breakers should be in accordance with manufacturer's published data. In the absence of manufacturer's published data, use Table 100.1. Values of insulation resistance less than this table or manufacturer's recommendations should be investigated.

* Optional

7. INSPECTION AND TEST PROCEDURES

7.6.3 Circuit Breakers, Vacuum, Medium-Voltage (*continued*)

3. Insulation-resistance values of control wiring should be comparable to previously obtained results but not less than two megohms.

4. Microhm or dc millivolt drop values shall not exceed the high levels of the normal range as indicated in the manufacturer's published data. If manufacturer's data is not available, investigate values that deviate from adjacent poles or similar breakers by more than 50 percent of the lowest value.

5. Breaker mechanism charge, close, open, trip, trip-free, and antipump features shall function as designed.

6. Minimum pickup for trip and close coils shall be in accordance with manufacturer's published data. In the absence of manufacturer's data, refer to Table 100.20.

7. Power-factor or dissipation-factor values shall be compared to manufacturer's published data. In the absence of manufacturer's published data the comparison shall be made to similar breakers.

8. Power-factor or dissipation-factor and capacitance values should be within ten percent of nameplate rating for bushings. Hot collar tests are evaluated on a milliampere/milliwatt loss basis, and the results should be compared to values of similar bushings.

9. If no evidence of distress or insulation failure is observed by the end of the total time of voltage application during the vacuum bottle integrity test, the test specimen is considered to have passed the test.

10. If no evidence of distress or insulation failure is observed by the end of the total time of voltage application during the dielectric withstand voltage test, the test specimen is considered to have passed the test.

11. Heaters should be operational.

12. Results of instrument transformer tests shall be in accordance with Section 7.10.

* Optional

APPENDIX B

Frequency of Maintenance Tests

NETA recognizes that the ideal maintenance program is reliability-based, unique to each plant and to each piece of equipment. In the absence of this information and in response to requests for a maintenance timetable, NETA's Standards Review Council presents the following time-based maintenance schedule and matrix.

One should contact a NETA Accredited Testing Company for a reliability-based evaluation.

The following matrix is to be used in conjunction with Appendix B, Inspections and Tests. Application of the matrix is recognized as a guide only.

Specific condition, criticality, and reliability must be determined to correctly apply the matrix. Application of the matrix, along with the culmination of historical testing data and trending, should provide a quality electrical preventive maintenance program.

MAINTENANCE FREQUENCY MATRIX				
		EQUIPMENT CONDITION		
		POOR	AVERAGE	GOOD
EQUIPMENT RELIABILITY REQUIREMENT	LOW	1.0	2.0	2.5
	MEDIUM	0.50	1.0	1.5
	HIGH	0.25	0.50	0.75

APPENDIX B

Frequency of Maintenance Tests (*continued*)

Inspections and Tests Frequency in Months (Multiply These Values by the Factor in the Maintenance Frequency Matrix)				
Section	**Description**	**Visual**	**Visual & Mechanical**	**Visual & Mechanical & Electrical**
7.1	Switchgear & Switchboard Assemblies	12	12	24
7.2	Transformers			
7.2.1.1	Small Dry-Type Transformers	2	12	36
7.2.1.2	Large Dry-Type Transformers	1	12	24
7.2.2	Liquid-Filled Transformers	1	12	24
	Sampling	–	–	12
7.3	Cables			
7.3.1	Low-Voltage, Low-Energy	–	–	–
7.3.2	Low-Voltage, 600 Volt Maximum	2	12	36
7.3.3	Medium- and High-Voltage	2	12	36
7.4	Metal-Enclosed Busways	2	12	24
	Infrared Only	–	–	12
7.5	Switches			
7.5.1.1	Air, Low-Voltage	2	12	36
7.5.1.2	Air, Medium-Voltage, Metal-Enclosed	–	12	24
7.5.1.3	Air, Medium- and High-Voltage Open	1	12	24
7.5.2	Oil, Medium-Voltage	1	12	24
7.5.3	Vacuum, Medium-Voltage	1	12	24
7.5.4	Medium-Voltage, SF_6	1	12	24
7.5.5	Cutouts	12	24	24
7.6	Circuit Breakers			
7.6.1.1	Air, Insulated-Case/Molded-Case	1	12	36
7.6.1.2	Air, Low-Voltage Power	1	12	36
7.6.1.3	Air, Medium-Voltage	1	12	36
7.6.2	Oil, Medium-Voltage	1	12	36
	Sampling	–	–	12
7.6.2	Oil, High-Voltage	1	12	12
	Sampling	–	–	12
7.6.3	Vacuum, Medium-Voltage	1	12	24
7.6.4	SF_6	1	12	12
7.7	Circuit Switchers	1	12	12
7.8	Network Protectors	12	12	24

Page 244
ANSI/NETA MTS-2011

APPENDIX B

Frequency of Maintenance Tests (*continued*)

Inspections and Tests Frequency in Months (Multiply These Values by the Factor in the Maintenance Frequency Matrix)				
Section	**Description**	**Visual**	**Visual & Mechanical**	**Visual & Mechanical & Electrical**
7.9	Protective Relays			
7.9.1	Electromechanical and Solid State	1	12	12
7.9.2	Microprocessor-Based	1	12	12
7.10	Instrument Transformers	12	12	36
7.11	Metering Devices			
7.11.1	Electromechanical and Solid-State	12	12	36
7.11.2	Microprocessor-Based	12	12	36
7.12	Regulating Apparatus			
7.12.1.1	Step-Voltage Regulators	1	12	24
	Sample Liquid	–	–	12
7.12.1.2	Induction Regulators	12	12	24
7.12.2	Current Regulators	1	12	24
7.12.3	Load Tap-changers	1	12	24
	Sample Liquid	–	–	12
7.13	Grounding Systems	2	12	24
7.14	Ground-Fault Protection Systems	2	12	12
7.15	Rotating Machinery			
7.15.1	AC Induction Motors and Generators	1	12	24
7.15.2	Synchronous Motors and Generators	1	12	24
7.15.3	DC Motors and Generators	1	12	24
7.16	Motor Control			
7.16.1.1	Motor Starters, Low-Voltage	2	12	24
7.16.1.2	Motor Starters, Medium-Voltage	2	12	24
7.16.2.1	Motor Control Centers, Low-Voltage	2	12	24
7.16.2.2	Motor Control Centers, Medium-Voltage	2	12	24
7.17	Adjustable-Speed Drive Systems	1	12	24
7.18	Direct-Current Systems			
7.18.1	Batteries	1	12	12
7.18.2	Battery Chargers	1	12	12
7.18.3	Rectifiers	1	12	24
7.19	Surge Arresters			
7.19.1	Low-Voltage Surge Protection Devices	2	12	24
7.19.2	Medium- and High-Voltage Surge Protection Devices	2	12	24

APPENDIX B

Frequency of Maintenance Tests (*continued*)

Inspections and Tests Frequency in Months (Multiply These Values by the Factor in the Maintenance Frequency Matrix)				
Section	**Description**	**Visual**	**Visual & Mechanical**	**Visual & Mechanical & Electrical**
7.20	Capacitors and Reactors			
7.20.1	Capacitors	1	12	12
7.20.2	Capacitor Control Devices	1	12	12
7.20.3.1	Reactors, (Shunt and Current-Limiting) Dry-Type	2	12	24
7.20.3.2	Reactors, (Shunt and Current-Limiting) Liquid-Filled	1	12	24
	Sampling	–	–	12
7.21	Outdoor Bus Structures	1	12	36
7.22	Emergency Systems			
7.22.1	Engine Generator	1	2	12
	Functional Testing	–	–	2
7.22.2	Uninterruptible Power Systems	1	12	12
	Functional Testing	–	–	2
7.22.3	Automatic Transfer Switches	1	12	12
	Functional Testing	–	–	2
7.23	Telemetry/Pilot Wire SCADA	1	12	12
7.24	Automatic Circuit Reclosers and Line Sectionalizers			
7.24.1	Automatic Circuit Reclosers, Oil/Vacuum	1	12	24
	Sample	–	–	12
7.24.2	Automatic Line Sectionalizers, Oil	1	12	24
	Sample	–	–	12
7.27	EMF Testing	12	12	12

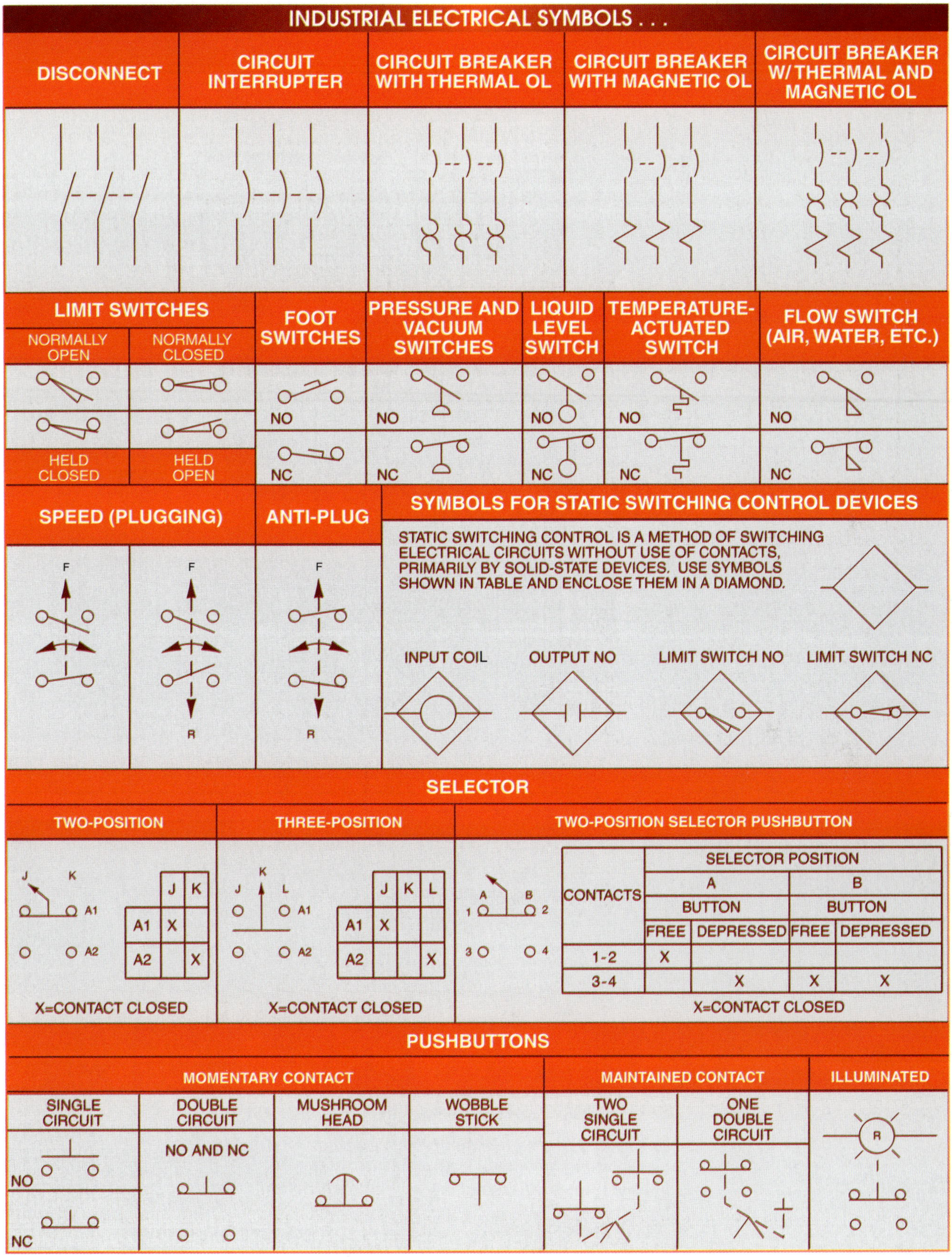
INDUSTRIAL ELECTRICAL SYMBOLS . . .
DISCONNECT
CIRCUIT INTERRUPTER
CIRCUIT BREAKER WITH THERMAL OL
CIRCUIT BREAKER WITH MAGNETIC OL
CIRCUIT BREAKER W/ THERMAL AND MAGNETIC OL
LIMIT SWITCHES
NORMALLY OPEN
NORMALLY CLOSED
HELD CLOSED
HELD OPEN
FOOT SWITCHES
NO
NC
PRESSURE AND VACUUM SWITCHES
NO
NC
LIQUID LEVEL SWITCH
NO
NC
TEMPERATURE-ACTUATED SWITCH
NO
NC
FLOW SWITCH (AIR, WATER, ETC.)
NO
NC
SPEED (PLUGGING)
F
F
R
ANTI-PLUG
F
R
SYMBOLS FOR STATIC SWITCHING CONTROL DEVICES
STATIC SWITCHING CONTROL IS A METHOD OF SWITCHING ELECTRICAL CIRCUITS WITHOUT USE OF CONTACTS, PRIMARILY BY SOLID-STATE DEVICES. USE SYMBOLS SHOWN IN TABLE AND ENCLOSE THEM IN A DIAMOND.
INPUT COIL
OUTPUT NO
LIMIT SWITCH NO
LIMIT SWITCH NC
SELECTOR
TWO-POSITION
J
K
A1
A2
	J	K
A1	X	
A2		X
X=CONTACT CLOSED
THREE-POSITION
J
K
L
A1
A2
	J	K	L
A1	X		
A2			X
X=CONTACT CLOSED
TWO-POSITION SELECTOR PUSHBUTTON
A
B
1
2
3
4
CONTACTS	SELECTOR POSITION			
	A		B	
	BUTTON		BUTTON	
	FREE	DEPRESSED	FREE	DEPRESSED
1-2	X			
3-4		X	X	X
X=CONTACT CLOSED
PUSHBUTTONS
MOMENTARY CONTACT
SINGLE CIRCUIT
NO
NC
DOUBLE CIRCUIT
NO AND NC
MUSHROOM HEAD
WOBBLE STICK
MAINTAINED CONTACT
TWO SINGLE CIRCUIT
ONE DOUBLE CIRCUIT
ILLUMINATED
R

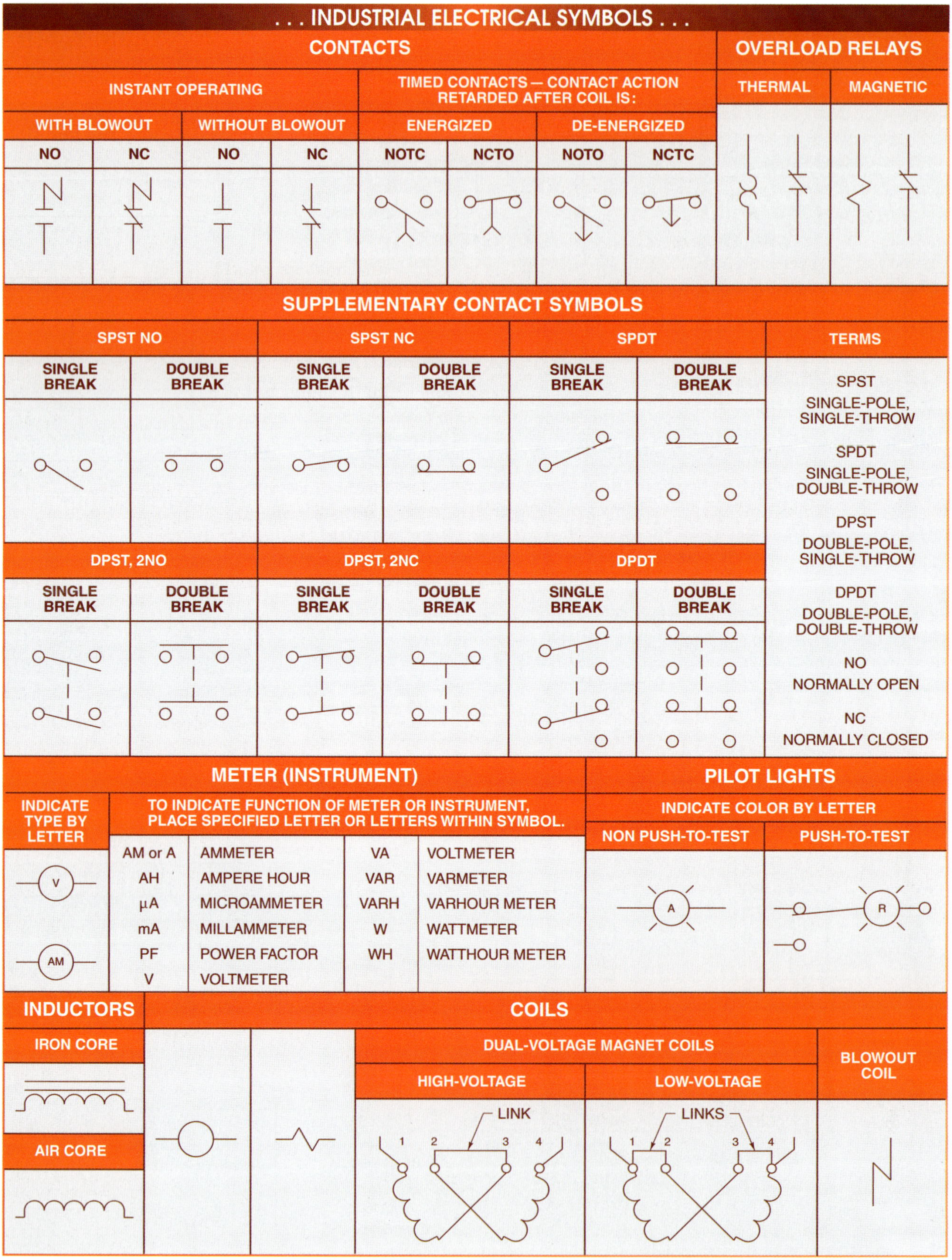
. . . INDUSTRIAL ELECTRICAL SYMBOLS . . .
CONTACTS
INSTANT OPERATING
WITH BLOWOUT
WITHOUT BLOWOUT
NO
NC
NO
NC
TIMED CONTACTS — CONTACT ACTION RETARDED AFTER COIL IS:
ENERGIZED
DE-ENERGIZED
NOTC
NCTO
NOTO
NCTC
OVERLOAD RELAYS
THERMAL
MAGNETIC
SUPPLEMENTARY CONTACT SYMBOLS
SPST NO
SPST NC
SPDT
SINGLE BREAK
DOUBLE BREAK
TERMS
SPST
SINGLE-POLE, SINGLE-THROW
SPDT
SINGLE-POLE, DOUBLE-THROW
DPST
DOUBLE-POLE, SINGLE-THROW
DPDT
DOUBLE-POLE, DOUBLE-THROW
NO
NORMALLY OPEN
NC
NORMALLY CLOSED
DPST, 2NO
DPST, 2NC
DPDT
METER (INSTRUMENT)
INDICATE TYPE BY LETTER
V
AM
TO INDICATE FUNCTION OF METER OR INSTRUMENT, PLACE SPECIFIED LETTER OR LETTERS WITHIN SYMBOL.
AM or A AMMETER
AH AMPERE HOUR
µA MICROAMMETER
mA MILLAMMETER
PF POWER FACTOR
V VOLTMETER
VA VOLTMETER
VAR VARMETER
VARH VARHOUR METER
W WATTMETER
WH WATTHOUR METER
PILOT LIGHTS
INDICATE COLOR BY LETTER
NON PUSH-TO-TEST
PUSH-TO-TEST
A
R
INDUCTORS
IRON CORE
AIR CORE
COILS
DUAL-VOLTAGE MAGNET COILS
HIGH-VOLTAGE
LOW-VOLTAGE
LINK
LINKS
1
2
3
4
BLOWOUT COIL

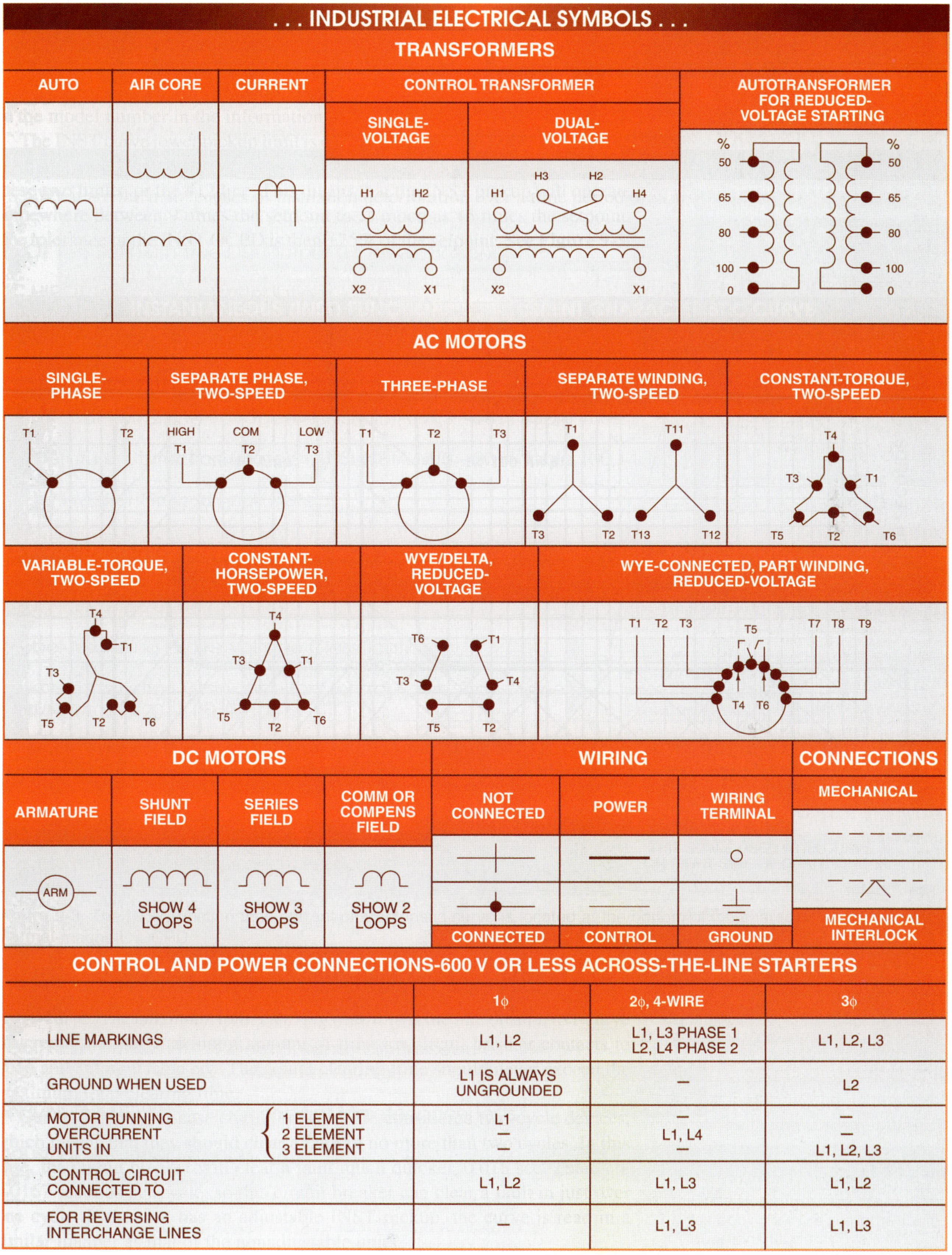

CONTROL AND POWER CONNECTIONS-600 V OR LESS ACROSS-THE-LINE STARTERS

		1ϕ	2ϕ, 4-WIRE	3ϕ
LINE MARKINGS		L1, L2	L1, L3 PHASE 1 L2, L4 PHASE 2	L1, L2, L3
GROUND WHEN USED		L1 IS ALWAYS UNGROUNDED	—	L2
MOTOR RUNNING OVERCURRENT UNITS IN	1 ELEMENT	L1	—	—
	2 ELEMENT	—	L1, L4	—
	3 ELEMENT	—	—	L1, L2, L3
CONTROL CIRCUIT CONNECTED TO		L1, L2	L1, L3	L1, L2
FOR REVERSING INTERCHANGE LINES		—	L1, L3	L1, L3

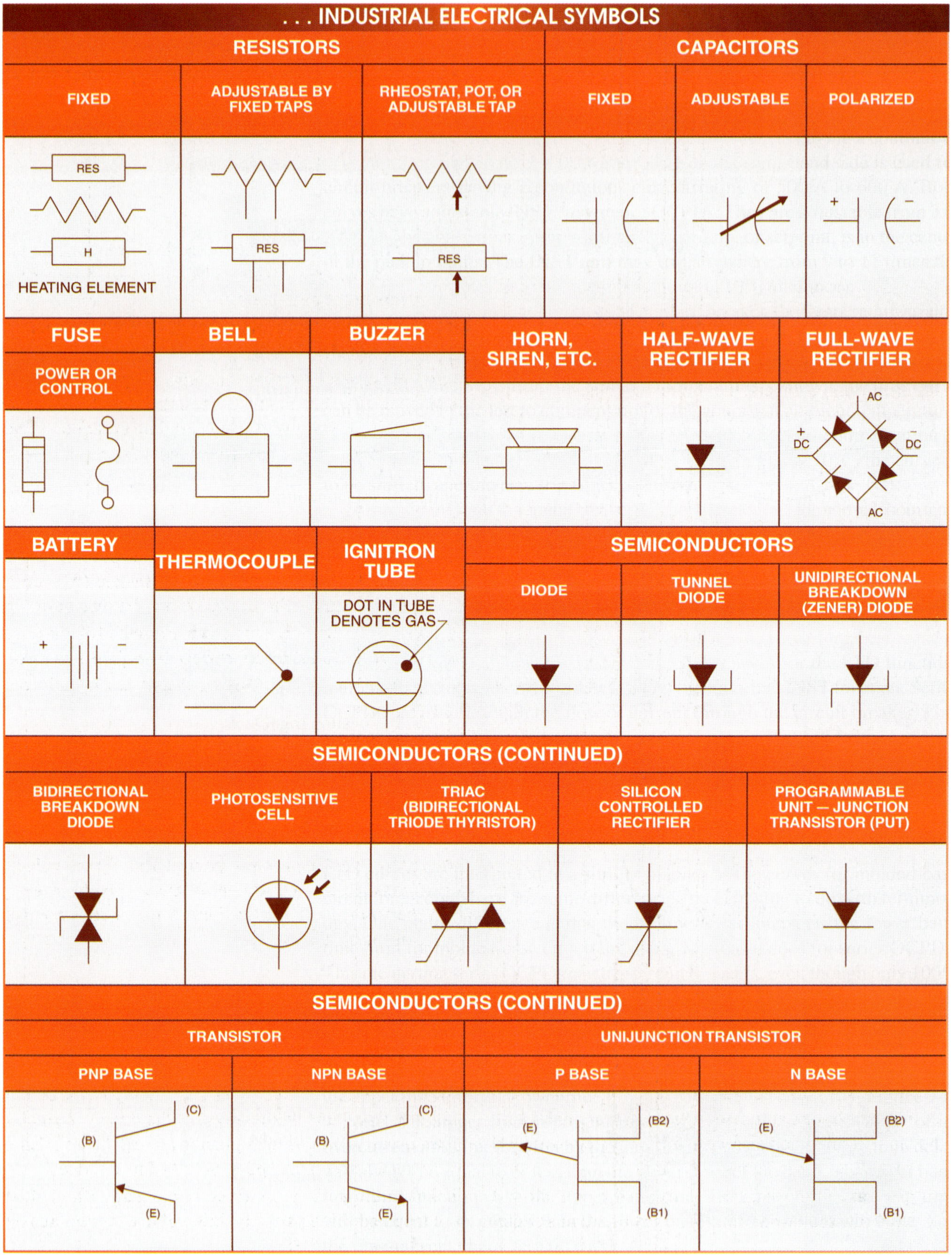
. . . INDUSTRIAL ELECTRICAL SYMBOLS
RESISTORS
FIXED
RES
H
HEATING ELEMENT
ADJUSTABLE BY FIXED TAPS
RES
RHEOSTAT, POT, OR ADJUSTABLE TAP
RES
CAPACITORS
FIXED
ADJUSTABLE
POLARIZED
+
−
FUSE
POWER OR CONTROL
BELL
BUZZER
HORN, SIREN, ETC.
HALF-WAVE RECTIFIER
FULL-WAVE RECTIFIER
AC
+ DC
− DC
AC
BATTERY
+
−
THERMOCOUPLE
IGNITRON TUBE
DOT IN TUBE DENOTES GAS
SEMICONDUCTORS
DIODE
TUNNEL DIODE
UNIDIRECTIONAL BREAKDOWN (ZENER) DIODE
SEMICONDUCTORS (CONTINUED)
BIDIRECTIONAL BREAKDOWN DIODE
PHOTOSENSITIVE CELL
TRIAC (BIDIRECTIONAL TRIODE THYRISTOR)
SILICON CONTROLLED RECTIFIER
PROGRAMMABLE UNIT — JUNCTION TRANSISTOR (PUT)
SEMICONDUCTORS (CONTINUED)
TRANSISTOR
PNP BASE
(C)
(B)
(E)
NPN BASE
(C)
(B)
(E)
UNIJUNCTION TRANSISTOR
P BASE
(E)
(B2)
(B1)
N BASE
(E)
(B2)
(B1)

GLOSSARY

A

alternating current (AC): Current that reverses its direction of flow at regular intervals.

arc: A sustained discharge of electricity across a gap in a circuit or between electrodes, usually accompanied by the electrodes (contacts) being vaporized and/or melted by the extreme heat of the arc.

arc chute: A structure that contains arc dividers. Also known as an arc extinguisher.

arc extinguisher: *See* arc chute.

arcing contact: The contact of a switching device on which an arc is drawn after the main contacts have parted.

B

backboard assembly: The section of a circuit breaker that supports stationary contacts, the primary line, and load disconnects.

bell alarm and lockout: A device installed on LVPCBs to send an alarm and/or to lock out the circuit breaker when it trips.

buffer assembly: A mechanical component that absorbs the shock of a mechanism when a circuit breaker opens and prevents the contact assemblies from overdriving when a circuit breaker closes.

C

capacitance charging current (I_C): Current that flows upon the application of DC voltage to charge the capacitance of the electrical system under test.

circuit breaker: An electrical device that opens and closes a circuit by nonautomatic means and automatically opens a circuit when a predetermined current overload is reached, without damage to itself.

contact gap: The distance between the two closest points of the moving and stationary contacts in a circuit breaker.

contact sequence: The difference in contact spacing as arcing (moving) contacts make contact with stationary contacts in a circuit breaker.

contact wipe: The amount of overtravel a contact assembly has once it initially makes contact.

current sensor: A sensor that reduces load current to a smaller value that is sent to the solid-state logic circuit where, if the value of current and time duration exceed its set points, will send a trip signal to a trip coil.

D

dielectric absorption current (I_{DA}): The polarizing current that is drawn by the insulation system (dielectric) to align the dipole molecules within the insulation.

dipole molecule: A molecule that is affected by the magnetic field created by current flow through its insulation.

direct current (DC): Current that flows in only one direction.

driving pawl: A circuit breaker component that advances the drive gear to charge closing springs.

E

energized electrical work permit (EEWP): A document issued by an employer to allow specific work to be performed on energized electrical equipment with various limitations.

F

fifth wheel: A device with a swiveling, weight-bearing connection used to move heavy loads.

floor tripper: An interlock that protrudes from the bottom of the circuit breaker and causes the circuit breaker to trip open and the closing springs to discharge when a circuit breaker is racked in or out of its cubicle.

frame rating: The continuous current rating of all current-carrying parts of a low-voltage circuit breaker, excluding its OCPD.

fuse: An electrical overcurrent protective device with a fusible portion that is heated and broken by the passage of excessive current.

G

guard circuit: A function on a megohmmeter used to eliminate unwanted return current from test measurements.

H

hipot tester: A test instrument that measures insulation resistance by measuring leakage current.

I

incident energy: The heat from an electrical arc; proportional to time.

Institute of Electrical and Electronics Engineers (IEEE): A professional association that is dedicated to advancing technological innovation and excellence of electrical and electronic equipment.

insulated-case circuit breaker (ICCB): A circuit breaker that is similar in construction to an MCCB but typically uses an electronic or digital OCPD and has much higher interrupting ratings.

interphase barrier: An electrical safety device that prevents contact with energized electrical components and protects against spark scatter.

interrupting rating: *See* short-circuit current rating.

L

latching pawl: A circuit breaker component that locks the drive gear in place when charging the closing springs.

leakage current (I_L): Resistive current that flows through insulation and is the variable that is measured with an insulation resistance test instrument.

lower band time limit: The time that represents the minimum operating time on a time-current characteristic curve.

M

maintenance closing: *See* slow closing.

maximum total clearing time: The time required to interrupt an arc and is considered to be from when the arc starts until it is completely extinguished.

maximum total clearing time line: The line on an INST curve that represents the maximum amount of time for circuit breaker contacts to open and extinguish an arc.

medium-voltage power circuit breaker (MVPCB): A circuit breaker that operates with voltages ranging from 2300 V to 69,000 V.

megavolt ampere (MVA) rating: The rated voltage times the rated interrupting current of a circuit breaker.

megohmmeter: A high-resistance ohmmeter used to measure insulation deterioration on various conductors by measuring high-resistance values during high-voltage test conditions.

micro-ohmmeter: A device used to measure the resistance of an electrical conductor down to the micro-ohm range.

molded-case circuit breaker (MCCB): A circuit breaker that uses a molded case to house and support its current-carrying components as well as to be a part of the insulation system.

N

nameplate: A metal plate attached to a device, such as a circuit breaker, that lists its technical specifications.

O

organic lubricant: A lubricant composed of emulsifiers and oil.

overcurrent: Electrical current in excess of the equipment limit, total amperage load of a circuit, or conductor or equipment rating.

overcurrent protective device (OCPD): A circuit breaker, fuse, or switch with an element that disconnects or discontinues current flow when the amount of current exceeds the design load. Also known as a trip device.

overtravel: The distance a circuit breaker mechanism travels once the contacts touch and the mechanism latches.

P

persistence: The amount of time vaporized metal remains a vapor between the contact faces in a medium-voltage circuit breaker.

personal protective equipment (PPE): Clothing, glasses, gloves, hard hats, respirators, or other safety devices designed to protect workers against safety hazards in the work area and from injury.

potentiometer: A variable-resistance electric device that divides voltage proportionally between two circuits.

predictive maintenance (PdM): The monitoring of wear conditions and equipment operation characteristics for comparison against a predetermined tolerance to predict potential malfunctions or failures.

preventive maintenance (PM): Scheduled work required to keep equipment in peak operating condition.

Q

qualified person: A person who has special knowledge, training, and experience in the installation, programming, maintenance, and troubleshooting of electrical circuit breaker equipment and also has the safety skills and knowledge that allow him/her to perform the task safely.

R

rated temperature rise: The temperature rise of an electrical device or component above ambient, or surrounding air, temperature.

restrike: A resumption of current between the circuit breaker's contacts if the circuit breaker fails to clear an arc.

S

safety interlock device: A device used to ensure that a circuit breaker does not get racked while the circuit breaker is in the closed position.

schematic diagram: A drawing that shows electrical system circuitry with symbols that depict electrical devices and lines representing conductors.

secondary injection test: A test used only on solid-state and digital OCPDs in which test current is injected directly into the logic and timing circuits.

serious violation: A safety violation in which there is a substantial probability that death or serious physical harm could result and the employer knew or should have known of the hazard.

short circuit: An unintentional connection of two ungrounded conductors that have a potential difference between them.

short-circuit current rating: The maximum short-circuit current a circuit breaker can safely clear without external damage. Also known as the interrupting rating.

slow closing: A method of mechanically blocking closing springs from operating when a circuit breaker is manually closed. Also known as maintenance closing.

spark scatter: The molten metal and debris that may be ejected from an arc chute when a circuit breaker interrupts a short circuit at or very near its rating.

spreader bar: A rigging device that is suspended from a crane hook to support lifting lines at various locations.

T

thermal imager: A device that detects heat patterns in the infrared-wavelength spectrum without making direct contact with the workpiece.

thermal-magnetic OCPD: An OCPD that reacts to the heat created by the copper loss (I^2R) when current passes through a conductor.

treeing: Irreversible internal degradation caused by the formation of conductive carbonized pathways.

trip device: *See* overcurrent protective device (OCPD).

U

undervoltage OCPD: A device that senses when the voltage of a circuit drops below a predetermined value.

upper band time limit: The time that represents the maximum operating time on a time-current characteristic curve.

W

willful violation: A safety violation committed with an intentional disregard of, or plain indifference to, the requirements of the Occupational Safety and Health Act and regulations.

wipe: The additional compression of a preloaded spring that is used to apply force to the vacuum interrupter contacts and to provide opening kick-off force.

Z

zero crossover point: The point at which current alternates and begins to change direction.

INDEX

**Page numbers in italic refer to figures.*

D

E

F

M

N

O

P

Q

R

S

T

U

V

W

Z